Methods and Tools in Biosciences and Medicine

Microinjection,
edited by Juan Carlos Lacal, Rosario Perona and James Feramisco, 1999
DNA Profiling and DNA Fingerprinting,
edited by Jörg Epplen and Thomas Lubjuhn, 1999
Animal Toxins – Facts and Protocols,
edited by Hervé Rochat and Marie-France Martin-Eauclaire, 2000
Methods in Non-Aqueous Enzymology,
edited by Munishwar Nath Gupta, 2000
Techniques in Molecular Systematics and Evolution
edited by Rob DeSalle, Gonzalo Giribet and Ward Wheeler, 2002
Methods for Affinity-based Separations of Enzymes and Proteins
edited by Munishwar Nath Gupta, 2002

Techniques in Molecular Systematics and Evolution

Edited by

Rob DeSalle
Gonzalo Giribet
Ward Wheeler

Birkhäuser Verlag
Basel · Boston · Berlin

Editors:
Dr. Rob DeSalle
Dr. Gonzalo Giribet
Dr. Ward C. Wheeler
American Museum of Natural History
Department of Entomology
Central Park West at 79th Street
New York, NY 10024–5192
USA

Library of Congress Cataloging-in-Publication Data
Techniques in molecular systematics and evolution / edited by Rob DeSalle, Gonzalo Giribet,
Ward Wheeler.
    p. ; cm. – (Methods and tools in biosciences and medicine)
   ISBN 3764362561 (hard cover : alk. paper) – ISBN 0-8176-6256-1 (hard cover : alk. paper) –
ISBN 376436257X (soft cover : alk. paper) – ISBN 0-8176-6257-X (soft cover : alk. paper)
   1. Molecular evolution–Methodology. I. DSalle, Rob. II. Giribet, Gonzalo. III.
Wheeler, Ward. IV. Series.
   [DNLM: 1. Evolution, Molecular. 2. Data Interpretation, Statistical. 3.
Genomics–methods. 4. Phylogeny. QH 367.5 T255 2001]
   QH506.T435 2001
   572.8'38–d21   2001035041

Deutsche Bibliothek Cataloging-in-Publication Data
Techniques in molecular systematics and evolution / ed. by Rob DeSalle
.... – Basel ; Boston ; Berlin : Birkhäuser, 2001
   (Methods and tools in biosciences and medicine)
   ISBN 3-7643-6256-1
   ISBN 3-7643-6257-X

ISBN 3–7643–6256–1 Birkhäuser Verlag, Basel – Boston – Berlin
ISBN 3–7643–6257-X Birkhäuser Verlag, Basel – Boston – Berlin

© 2002 Birkhäuser Verlag, PO Box 133, CH-4010 Basel, Switzerland
Member of the BertelsmannSpringer Publishing Group
Printed on acid-free paper produced from chlorine-free pulp. TCF ∞
Printed in Germany
ISBN 3–7643–6256–1
ISBN 3–7643–6257-X

9 8 7 6 5 4 3 2 1   http://www.birkhauser.ch

Contents

List of Contributors

ARENAS-MENA, CESAR, Division of Biology 156–29, California Institute of Technology, Pasadena, CA 91125, USA

BAKER, RICHARD H., The Galton Laboratory, University College London, 4 Stephenson Way, London NW1 2HE, UK; e-mail: richard.baker@ucl.ac.uk

BIRNBAUM, KENNETH D., Department of Biology, New York University, New York, NY 10003; e-mail: kdb4348@nyu.edu

BONACUM, JAMES, Department of Biology, Yale University, New Haven, CT 06511, USA; e-mail: james.bonacum@yale.edu

BONWICH, ELIZABETH, Division of Invertebrate Zoology, American Museum of Natural History, Central Park West at 79th Street, New York, NY 10024–5192, USA; e-mail: bonwich@amnh.org

DESALLE, ROB, Division of Invertebrate Zoology, American Museum of Natural History, Central Park West at 79th Street, New York, NY 10024–5192, USA; e-mail: desalle@amnh.org

DOUKAKIS, PHAEDRA, Department of Biology, Yale University, New Haven, CT 06511, USA; e-mail: doukakis@amnh.org

EGAN, MARY, Division of Invertebrate Zoology, American Museum of Natural History, Central Park West at 79th Street, New York, NY 10024–5192, USA; e-mail: egan@amnh.org

GATES, RUTH D., Department of Organismic Biology, Ecology and Evolution, University of California Los Angeles, 621 Charles E. Young Drive South, Los Angeles, CA 90095–1606, USA; e-mail: rgates@ucla.edu

GATESY, JOHN E., Department of Molecular Biology, University of Wyoming, Box 3944, Laramie, WY 82071, USA; e-mail: hayashi@amnh.org

GIRIBET, GONZALO, Department of Organismic and Evolutionary Biology, Museum of Comparative Zoology, Harvard University, 26 Oxford Street, Cambridge, MA 02138, USA; e-mail: ggiribet@oeb.harvard.edu

GOLOBOFF, PABLO A., INSUE, Consejo Nacional de Investigaciones Científicas y Técnicas, Miguel Lillo 205, 4000 S.M. de Tucumán, Argentina; e-mail: instlillo@infovia.com.ar

GREENWOOD, ALEX D., Department of Mammalogy, American Museum of Natural History, Central Park West at 79th Street, New York, NY 10024–5192, USA; e-mail: alexgr@amnh.org

HADRYS, THORSTEN, Division of Invertebrates, American Museum of Natural History, Central Park West at 79th Street, New York, NY 10024–5192, USA; e-mail: hadrys@amnh.org

HANNER, ROBERT, Division of Invertebrate Zoology, American Museum of Natural History, Central Park West at 79th Street, New York, NY 10024–5192, USA; e-mail: hanner@amnh.org

JACOBS, DAVID K., Department of Organismic Biology, Ecology and Evolution, University of California Los Angeles, 621 Charles E. Young Drive South, Los Angeles, CA 90095–1606, USA; e-mail: djacobs@ucla.edu

KIZIRIAN, DAVID, Natural History Museum of Los Angeles County, 900 Exposition Boulevard, Los Angeles, CA 90007, USA; e-mail: kizirian@nhm.org

NISHIGUCHI, MICHELE K., Department of Biology, New Mexico State University, MSC 3AF, Box 30001, Las Cruces, NM 88003–8001, USA; e-mail: nish@nmsu.edu

O'GRADY, PATRICK M., Division of Invertebrate Zoology, American Museum of Natural History, Central Park West at 79th Street, New York, NY 10024–5192, USA; e-mail: ogrady@amnh.org

PHILLIPS, ALOYSIUS J., Department of Biological Sciences, Columbia University, New York, NY 10027, USA; e-mail: ajp26@columbia.edu

PRENDINI, LORENZO, Percy FitzPatrick Institute, University of Cape Town, Rondebosch 7701, South Africa; e-mail: lprendin@botzoo.uct.ac.za

REMSEN, JAMES, Department of Biology, New York University, New York, NY 10003; e-mail: remsen@amnh.org

ROSENBAUM, HOWARD C., Wildlife Conservation Society, Science Resource Center, 2300 Southern Blvd., Bronx, NY 10460, USA; e-mail: hcr@amnh.org

SIDDALL, MARK E., Division of Invertebrate Zoology, American Museum of Natural History, Central Park West at 79th Street, New York, NY 10024–5192, USA; e-mail: siddall@amnh.org

STARK, JULIAN, Department of Biology, New York University, New York, NY 10003, USA; e-mail: stark@amnh.org

TORRES, ELIZABETH, Department of Biology and Microbiology, California State University, Los Angeles, 5151 State University Drive, Los Angeles, CA 90032, USA; e-mail: etorre11@calstatela.edu

JOHN W. WENZEL, Department of Entomology, Ohio State University, Columbus, OH 43210, USA; e-mail: jww+@osu.edu

WARD C. WHEELER, Division of Invertebrate Zoology, American Museum of Natural History, Central Park West at 79th Street, New York, NY 10024, USA; e-mail: wheeler@amnh.org

MICHAEL F. WHITING, Department of Zoology, Brigham Young University, Provo, UT 84602, USA; e-mail: Michael_Whiting@byu.edu

WRAY, CHARLES G., The Jackson Laboratory, 600 Main Street, Bar Harbor, Maine 04609, USA; e-mail: cgw@informatics.jax.org

WYNER, YAEL, Department of Biology, New York University, New York, NY 10003, USA; e-mail: wyner@amnh.org

General Introduction

Molecular approaches are now soundly entrenched in evolution, systematics and ecology. The amount of information that can be obtained using these techniques has increased by exponential bounds. Analytical problems accompany the increase in amount of data collected for various studies. In addition to sequence data, techniques using molecular visualization of developmental processes have also advanced. The need for more rapid and efficient methods of data acquisition and analysis therefore grows more and more acute in evolutionary biology.

This manual attempts to present some of these techniques that have been developed over the past few years for data acquisition. Data analysis is also an important part of this manual and in this respect we have tried to place the analysis of molecular data in a character-based context as well as present some of the more recent advances in molecular systematic approaches. Consequently we have targeted two audiences with this manual. Evolutionary biologists looking for the methods to acquire data are our first target. We also target molecular biologists who might be looking for information on how data are analyzed in an evolutionary context. To aid the user of this manual we have included URL information for web-located sites wherever possible. We have also tried to incorporate approaches that will push the amount of information systematists will gather in the next few years by using high throughput methods and access to genomics information.

The editors of this manual wish to thank all of the authors of chapters and several individuals who read all or part of the manual. These individuals include Sharon Jansa, Valerie Schawaroch and James A. Waschek. All of the authors reviewed the chapters of their fellow authors and we thank the authors for these tasks too. While we relied heavily on these reviews to improve the quality of the chapters in this manual, any errors in content or otherwise are the editors'. The reader will, no doubt, note that the majority of the authors of chapters in this manual are either current or former American Museum of Natural History (AMNH) Molecular Laboratories researchers or close colleagues of AMNH researchers. In this respect we thank the AMNH in general and Mike Novacek in particular for their continued support of the work of the editors and authors of this volume. We hope that the reader of the manual finds the material useful and organized in an accessible manner.

June, 2001
Rob DeSalle, Gonzalo Giribet and Ward C. Wheeler

Part I
Analytical Methods

Introduction to Part I

We have put together a series of chapters for the first part of this manual that explain clearly the philosophical and mechanical issues relevant to making inferences from molecular data in an evolutionary context. Because we feel strongly that it is most important to understand the basis of phylogenetic analysis before delving into how to collect the data, Wenzel starts this part of the manual with a primer of philosophical issues relevant to analyzing character data. Siddall then describes the basic approaches that are utilized in phylogenetic analysis. We hope these two chapters give a sound foundation for why certain kinds of analyses are used in phylogenetic studies. Two problems that have been made acute by the proliferation of molecular data concern alignment and large data sets. Wheeler addresses the former issue with a discussion of an analysis technique that takes advantage of Fitch optimization approaches to generate "alignment free" phylogenies and in the latter Goloboff summarizes the adjustments phylogenetic analysis methods have made to accommodate extremely large data sets. Issues of robustness of inferences are next addressed by Siddall as he examines the analytical and philosophical problems involved in using measures of support in phylogenetic analysis. One of the outcomes of being able to generate data for multiple gene regions is the recognition that different kinds of phylogenetic signal may emanate from different partitions. To address this problem, O'Grady, Remsen and Gatesy describe the various methods that have been developed over the past five years to examine data partitions. The impact of genomics on systematics and evolutionary studies is the subject of the next two chapters of the manual. Wray and Phillips tackle the role of genome databases and sub-genomic sequencing respectively. To conclude this part of the manual, Baker summarizes the approaches currently available to make inferences about the comparative method and Doukakis, Birnbaum and Rosenbaum summarize the methods used to analyze populations at the molecular level.

Rob DeSalle, Gonzalo Giribet
and Ward C. Wheeler

Phylogenetic Analysis: The Basic Method

John W. Wenzel

Contents

1   Introduction

Systematics has a unique place among natural sciences because it is the most evolutionary of all sciences. It is strictly historical rather than experimental. Of course, many of the restrictions on systematics are found elsewhere. For example, neither systematists nor astronomers can perform manipulative experiments to see if what they observe will happen again in the evolution of life or stars. Sciences that focus on history have special problems and appropriate epistemologies (rules about how we establish the validity of what we know, see [1]). Systematics is unique in requiring that every practicing authority know *all* the work that ever came before, both good and bad. In physics,

Methods and Tools in Biosciences and Medicine
Techniques in molecular systematics and evolution, ed. by Rob DeSalle et al.
© 2002 Birkhäuser Verlag Basel/Switzerland

chemistry, geology, or many other fields, bad work is buried and we can ignore it. That is not true in systematics. Even a terrible systematist is immortal in the sense that someone else must always know what were the names he used, and with what are they synonymous today, and if a third author used some name, then was it according to the concept in question or another idea? To some this seems like nonsense, but it is easily understood by a few examples. Americans have a popular bird called a robin, but this is actually the same as what Europeans call a thrush, and not at all what Europeans call a robin. Australians call some of their trees "pines", but these trees are very unlike pines from Europe. Maybe no one cares about the robin or the pine, so why do we care about this form of excessive traditionalism in systematics? The reason we care is that non-experimental knowledge about the natural world is something that comes slowly, and we want to be sure that good observations connected to names used 100 years ago in a different country are still useful to us today. What, exactly, is a "red panda?" Systematics is the only science that requires consideration of both good and bad, forever.

The necessity to understand everything that came before would seem to impress upon people the necessity of getting it right, because bad work doesn't die with the author, and no one wants to be known as a bad worker for generations hence. Yet the fact that systematics is not an experimental science with conclusive demonstrations seems to permit pretenders plenty of room to make a mess of something that formerly was in order. In fact, the more important an issue is, the more likely people are to show up and say that the last 200 years of science are wrong, because they get an immediate, big-bang pay-off for saying so. Widely discussed papers and perhaps grant awards may follow. Of course, the real systematist counts such a poor study among all the others he knows, and continues ahead slowly. So, what can be done about this problem? The most obvious thing is to make the reconstruction of evolutionary trees as scientific in its method as it can be. That way, if something is wrong, we will discover it and correct it promptly. This concept has ramifications more deep than it seems at first, for it requires us to be cautious about the rules of epistemology. Claims to knowledge, such as "all insects found in India are new species because they are found in India" are not believable, nor are claims that "all third positions of a codon evolve rapidly because they are third positions". These are the sorts of claims that will be ridiculed by systematists born tomorrow because these claims are not demonstrations of anything; they are only our own effort to extend our knowledge outside the realm of its appropriate use. We need to be strict about the validity of our claims to knowledge, and narrow about what we are trying to discover. In this way we ensure that the work we do now can be accepted by future generations.

This book is oriented toward cladistic methods because they best satisfy the challenges of dealing with a fountain of modern evolutionary data. The origin of cladistic methods is that systematic data are considered in some ways as any other scientific data are considered. Models come and go with fashion, politics and technology, but data are only data, and so have greater longevity. So, in cladistic philosophy, we treat all data alike, at least at the starting point. This is a problem too, because no one actually believes that all data are alike, but at least we know that this is the place to start. For my own part, I am interested in behavioral data, but for the purpose of scientific method, I agree to afford my favorite data no special status. Some people might say that this compromises our ability to say what we know about behavior, but in fact it ensures that we only say things that we can demonstrate regarding the behavior of interest in the taxa in question. It is interesting that in this paradigm we learned ten years ago that behavior is much more hierarchical than we thought during the process-oriented 1970s and 1980s [2–6], and more similar to the ideas posed by the classical ethologists [7] or systematists [8]. We would not have thought this according to the late twentieth century *status quo*. What will we learn when we expose molecular data to the same method? Great things, I hope.

This spare treatment of the way cladistic systematics is done can stand on its own as a recipe, a scientific method. Following the recipe are some comments regarding general philosophy and the place of models of causality. In this regard, it is worthwhile to note that the critics of the cladistic method rarely offer philosophical support for their positions other than the mere assertion that what they do is more scientific, sometimes because of case-specific contingencies and complex mechanics. With that challenge in mind, the reader should remember that a method derives importance from generality and beauty from simplicity. This is why cladistic methods were adopted in the first place.

2   Methods

This chapter is written with the assumption that the readers are familiar with the use and significance of such concepts as homoplasy and paraphyly, and the cladistic perspective that groups should be formed on the basis of synapomorphy only. These basic ideas are explained in many other places and are not treated here [9–13]. Neither is there an explanation of modern molecular vocabulary or techniques. This chapter deals with the fundamental mechanics of reconstructing the pattern of evolutionary history.

2.1   The matrix

There are three primary declarations that must be made even before any reconstruction, and these are critical to the outcome of the analysis. They are statements of homology, character coding and weight. After these issues are decided, an analysis produces a preliminary result and some of the original declarations may be revisited. In some studies it is important to re-examine statements of homology and repeat the analysis with new views of the data learned from the initial analysis. Weighting may become a dynamic feature of the search that approximates an optimal value. Regardless of how the analysis proceeds, it is eventually necessary to choose a root for the trees. Multiple trees must be reconciled somehow, and authors may want to measure support for individual nodes in a summary tree. An outline of the method is as offered in Protocol 1.

Protocol 1 Outline of the cladistic method

Prologue: Choose a problem (taxonomic or evolutionary) that seems to relate to a monophyletic group. Sample the taxa as densely as is practical, trying to capture the range of variation.

I. Establish a matrix

A. Decide character homology from gross variation
  1. within characters, decide state homology
B. Code characters
  1. decide on additivity
  2. decide on dependence (polarity not part of this)
C. Choose outgroups to provide root (therefore polarity)
  1. two or more real taxa are preferable
    a. closer is better than more distant

II. Establish weighting scheme

A. Differential or uniform
  1. character, state, or branch
  2. static or dynamic
    a. single function or iterative

III. Calculate shortest Manhattan distance through matrix

A. Take consensus of multiple trees
  1. decide on strict or Adams
  2. decide on optimization
    a. flat choice (all fast, all slow)
    b. logical choice (e. g., derive complexity once, lose it many times)
B. Report tree statistics
  1. length (shortest, consensus)
  2. CI, RI
  3. support:
    a. apomorphy list
    b. bootstrap
    c. parsimony jackknife
    d. Bremer support

IV. Go to step I: Reconsider initial conditions and repeat

Epilogue: Publish consensus trees and those of special interest (certain optimizations, topologies, partitions, roots, weighting, etc.). Reserve judgement where analysis is weak. Recognize monophyletic groups to conserve taxonomy when possible. Discuss unexpected results in terms of characters and scholarship of systematic work that came before.

2.2   Homology

The first step in any analysis is to define the data by reducing them to statements of "same" and "other". This entails statements of homology, defining what aspects of variation constitute different states of the same thing and what aspects illustrate different things. Morphological examples are good for making a general point. For the purpose of coding limb variation among tetrapods, the front leg of a lizard, the wing of a bird, the flipper of a whale, and the wing of a bat are all alike in being "front leg", although they vary in form. The statement offered in the matrix is that a leg is a leg, no matter what it looks like. Calling them all different states of "front leg" is a statement of homology. Of course, if we proceed to examine in detail what these legs are, we will decide that beyond "front leg" they are not really the same thing, and the wing of the bird and the wing of the bat are different things, not really homologous as "wing". There are a number of discussions of cladistic perspectives in defining homology [2, 14–18], and the philosophical issues of how similar is "same" are not to be overlooked, but molecular data (at least sequence data) would seem to be too simple to require much discussion. After all, an A is an A, a C is a C, and so forth. But this supposition that similarity reveals homology is overly optimistic.

The issue of what happens when two As are separately derived will be deferred to the section on homoplasy (p. 24), but here we need to consider how we came to judge that the two As in question are supposed to represent the same position in the first place. In the simplest situation, we have specific primers or other good landmarks to start from and as we walk down the sequence we find all taxa have sequences of the same length, and perhaps many codons match exactly, so it is fairly safe to assume that the two As in question are the same position in the gene. In a less ideal situation, perhaps there are sequences of different length, but it seems nonetheless easy to see which codons were dropped or added in some taxa, so that we can see that one taxon has codons numbered (in the translated product) 12, 13, 14, 15, 16 and a different taxon has only 12, 13, 15, 16. In that case, base 43 (the first of codon 15) in the first taxon corresponds to base 40 in the second, and there are no bases in the second that correspond to positions 40, 41 and 42 of the first (because codon 14 does not exist in the second taxon). It gets worse if we lose landmarks, particularly with noncoding genes, because there are many ways to find correspondences between sequences. We can propose either that dissimilar bases are the same character because they are the same position in the raw data, (A at position 40 in one taxon is homologous with C at position 40 in another) or that we resolve length differences to make bases of similar identity homologous even if they are not in the same position (A at position 40 in one taxon is homologous to A in postion 43 in another, and we postulate that other bases exist only in some taxa, or even that dissimilar bases at different positions are homologs: A at 40 here equals C at 43 there). This becomes a problem in alignment of the sequences under consideration. This problem is very severe both philosophically and analytically, and a detailed treatment of it is available in [19], but for the current purpose it is necessary to state that making statements of homology among bases can be a serious hurdle. It is a common occurrence that a well-meaning researcher should produce an alignment (statements of homology) that yields one phylogeny and a different study gets a different phylogeny with the same data because a different alignment is deemed preferable (for example, see [20] *versus* [21]). Furthermore, whether to use DNA sequence or amino acid sequence is not clear as amino acids would seem to do away with problems of noisy mutations, but it is also possible that the amino acid sequence would differ more than the DNA sequence does between taxa [22]. These are problems of defining the data themselves and any method, cladistic or not, will have to solve these issues as the first step in analysis. Early errors are manifested throughout the analysis, and there is no method that guards against "garbage in, garbage out."

There are several classes of data that are inappropriate to begin with if one aspires to retrace an evolutionary history of given taxa. These would include DNA-DNA hybridization (similarity), RFLPs, RAPDs, or ISSRs because the proposals of homology are so tenuous. Exactly what is alike and why? In each of these, many different processes could have produced a given level of similarity, products of similar length, or similar composition [23]. A well-known example is that there are many different ways to lose a restriction site, but the products appear identical despite their different histories. Because phylogenies are supposed to reconstruct history, it is best to lay aside these sources of data in favor of, say, coding sequence data (or morphology, or behavior) in which identical distributions of closely defined characters are best explained as shared ancestry. Or, looking at it a different way, even if an author shows that an analysis of RFLPs matches the accepted phylogeny generated by something else, that does not demonstrate that other RFLPs are adequate alone to *derive* a phylogeny in a different case. Getting the "right" answer is easy when two very different kinds of data agree, so the only real concern is what to do when data disagree. Whichever data are preferred when different classes are in conflict (the coding sequence would be preferred) is what should be used at all times. Approaching the problem a third way, we can only validate RFLPs when we have sequence, at which time we no longer need RFLPs because we have sequence, so why not just start with sequence? Of course, a study may appear to be more modern or more sophisticated by using new techniques that have recently become fashionable, but science as a whole has rarely benefitted from fashion, and what is fashionable is likely to be wrong if it disagrees with 200 years of careful work. From the perspective of statements of homology in a molecular context, sequence data are the thing to use.

2.3 Character Coding

Additivity: Character coding is the most difficult part of a morphological analysis, but it can be quite easy for molecular data. The states are easily defined (say, A, C, G and T, or perhaps the amino acids translated from codons). In addition to the primary statements of homology, the initial conditions of a tree search include statements about how states within a character relate to each other. The main decision here is whether to dictate that a certain path through the alternative states is preferred or if all possible paths are of equal cost. In the first case, if states are connected such that some states are between others (say A-G-T-C) then the calculations will assume that it is one step to go between adjacent states, and additional steps are tallied as the computer moves through the line as specified. In the example here, it is one step from A to G. If a change between two sister taxa runs from A to T, then that is two steps (one from A to G, and one from G to T) and so forth. This kind of coding is called "additive" to refer to the mathematical properties (Hennig86 and related programs), or "ordered" to refer to the declaration of adjacent states (PAUP and relatives).

If, on the other hand, all paths are considered of equal cost, then it is always only one step between any two states. The state A can go equally easily to G or T, and so this is "nonadditive," or "unordered". There is a general agreement that sequence data should be analyzed as nonadditive although it is common to declare morphological or behavioral characters as additive. There is also a general misunderstanding that to declare characters as nonadditive requires less of a statement about evolution, but this is not accurate because nonadditivity declares that any state can change to any other with equal cost, which is indeed a bold proposal. The preference for nonadditive analysis of molecular data springs rather from the idea that there is no way to determine which states should be intermediate between which other states. Assumptions about the cost of changing states can be adjusted, however, under schemes of weighting (below).

In other forms of data, but rarely in molecular data, it is necessary to decide if there is a dependency among characters, if certain characters should be coded as homologs in hypothetical classes and then differentiated afterward. This would be like deciding that bird wings, whale flippers and turtle legs are all coded alike in the column that defines "four legs" for tetrapods. A separate character defines the differences within "four legs" and between these different states. While there are not many dependencies in sequence data, some dependencies may occur in certain molecular weighting schemes (below).

2.4 Choosing outgroups

Outgroup comparison provides the root for the tree [24]. A tree can always be rerooted among any of the taxa in the analysis (mathematically, the computer doesn't care at all about the significance of the outgroup), but it is ordinary to declare the outgroups prior to tree search for convenience of plotting the results of the search. There is a misconception that it is important to have the sistergroup to the ingroup serve as the outgroup. Although this is a good idea, it is not necessary. It is best to have several outgroups that represent successively more distant lineages when possible. This is better than two taxa from a clade that is the sister to the ingroup, because those two outgroups can be expected to collapse to a single lineage rather than retaining status as "outgroup" and "more distant outgroup". Using successively more distant lineages will permit the two main functions of outgroup comparison to be fulfilled. The first is the demonstration that the ingroup is monophyletic. If the ingroup taxa all form a single clade with several successive outgroups outside them at the base of the tree, then this indicates that the ingroup is monophyletic. If an outgroup falls among the ingroup taxa, or if the ingroup taxa include the base of the tree in an unresolved polytomy, then monophyly is not indicated. The evidence for monophyly is, of course, more convincing if very close relatives of the ingroup taxa are used, but in the absence of such evidence, it can still be demonstrated that the ingroup is monophyletic *relative* to the outgroups that were employed. A single outgroup (or a pair of species from a sister clade, as mentioned above) does not demonstrate monophyly because it could not come out in any more than one position. However, even a single outgroup will serve for the second function of outgroup comparison, which is the dictation of character polarity (of course, several outgroups are better because a single outgroup may have its own derived state of some characters). Comparison with the outgroup determines what character states are plesiomorphic (ancestral) in the ingroup and what are apomorphic (derived). The node that unites the ingroup with the outgroup presents the reconstruction of a hypothetical ancestor that represents all states that are symplesiomorphies of the ingroup. The next node up the tree (the mutual ancestor of the ingroup and only the ingroup when it is monophyletic) presents the first synapomorphies of the ingroup. If only ingroup taxa are considered, no polarity of the resulting tree can be inferred. The resolution at the base of the tree is always a three-way split, and polarity or relationship concerning the bottom three taxa cannot be determined without adding another, more distant outgroup (at which point the three-way split is displaced basally and involves two of the original three lineages).

  In studies where many outgroups are used because the root of the ingroup is poorly known (many taxa are good candidates), workers sometimes worry that the topology of the tree concerning only the outgroups may not make much sense. Although this can be something to consider, it is more important to realize that the study at hand is only a study of the ingroup, and only the relationships among the ingroup taxa are expected to be addressed. A study of the outgroup taxa themselves would necessarily involve different taxa and could be expected to give a different result.

One final concern addresses the length of the branches and placement of the root of the tree in the network of ingroup taxa. If there is a problem with the erroneous placement of taxa due to long branches in the tree [25], it can be expected that often the branch leading from the ingroup to the outgroups will be long, and hence it is possible that a connection from the ingroup to the outgroups may be problematic [26]. Spurious rooting of trees can have serious consequences for demonstrations of monophyly, particularly for taxa falling between the real root and the mistaken root.

2.5 Weighting

In English, "to weigh" is a common word meaning to *measure* weight, whereas "to weight" is more rare, and means to *assign* weight. When an author "weights" his data, he assigns them certain values. All data are weighted (characters left out are weighted zero), but people often say they used "unweighted" data to mean that the characters and changes were weighted equally, i. e. that their data were not weighted differentially. Whether differential weighting is a good thing or not is controversial. Many studies employ some kind of differential weighting, although some arguments hold that any form of differential weighting is unacceptable [27]. In addition to philosophical problems, all weighting schemes have some practical drawbacks, and the descriptions below do not constitute endorsement of the procedures.

There are three fundamental kinds of differential weighting: character weighting, state weighting and branch weighting. The first kind of weighting is that characters can be assigned differential weights according to their presumed importance. This kind of weighting is traditional and still common in systematics in general. It amounts to a decision that the states of a certain character are so informative that changes among the states are made to be very expensive to retain homology of the states as much as possible. Perhaps in some osteological study the reconstruction does not always return amniote tetrapods as a clade, and the faith that Amniota is a clade is so strong that the researcher chooses to weigh the character "amnion" 10 times as much as other characters. This means that a single step across the "present/absent" states of "amnion" equals 10 such steps in any other character. This will do a lot to keep amniotes as a clade because a lot of other characters will have to conspire to undo the special cost of 10 that a paraphyletic Amniota would have at minimum. In a molecular framework, one might decide that changes among the paired bases in stem regions of a ribosome mean something more than changes in open loop regions (because stem regions require compensatory changes, etc.), and so stem regions could be weighted more. Such differential weighting has been shown to return reasonable results when equal weighting does not [28]. A recommendation consistent with this would be to weight third positions less than first or second positions of the codon, because it is presumed that there is more evolutionary noise at the third position ([29]:53); but see [30–32]. Fundamentally, the statement is that some characters are simply more important than others, regardless of the state in which you find them. The obvious shortcoming is that if all third positions are declared to be less informative than others by *a priori* weighting, it may be impossible to discover what data they really obtain because they were never given a chance to show what they mean. Two solutions to this problem (successive approximations and implied weighting) are explained in the section "Implementing weighting" below.

Alternatively, certain types of *changes* can be assigned differential weights regardless of what character exhibits the change. This is state weighting. Although it is virtually unknown in traditional data sets, it is common in molecular systematics. As an example, there is a widespread affection for giving less weight to transitions (changes among the purines A and G, or among the pyrimidines C and T) than transversions (changes between purines and pyrimidines). In this perspective, it does not matter what character (base) changes, but whether the change was a transition or a transversion. If, at a certain point in tree building, there is a choice between base 113 changing from A to G, or base 2014 changing from C to A, the preference is to change 113 because it weighs less *due to the states that are involved*. At this moment, base 2014 weighs more because it marks a transversion. It will weigh less than base 113 elsewhere in the tree if the changes are of opposite class there. The weights have nothing to do with the characters, and everything to do with the states that are involved in the change. Some kinds of weighting can have dependencies as well. Perhaps a worker will treat third positions as different from others, and then within third positions there is a distinction between transitions and transversions, or perhaps changes between G and C weigh differently than between A and T depending on the frequency of the bases. Analogous to character weighting, there is the hazard that *a priori* statements devalue *all* transitions, including those that may be rare and informative.

The differential weighting schemes discussed above (stems versus loops, or transitions versus tranversions) use the observed data through simultaneous comparison of homologues (is base 113 alike in these taxa? what about base 2014?). A third kind of weighting may incorporate the aspects of state weighting, but also accounts for the length of the branch on the reconstructed tree to serve as an estimator of the probability that other changes in question will happen. In general, the length of the branch leading from W to X (accumulated by whatever bases or states) is compared to the length of the branch leading from W to Y (accumulated perhaps by completely different bases and states) to estimate where the cost of placing taxon Z is least. Even if all W, X and Y have the same state and Z has an alternative state, Z's placement will count differently depending on the branch length of W-X or W-Y. There are many proposals, but in general the idea is to expect more evolution where there has already been a lot of evolution, as opposed to where little has been observed. This "maximum likelihood" school aspires to correct for the probability that high rates of change could produce bases of the same identity (A and A) as homoplasies. In this philosophy, the length of the branches in the reconstructed tree is itself taken as a form of empirical data and as a predictor of the likelihood that some kinds of changes will happen or not, with more homoplasy expected on longer branches. These weighting schemes are usually derived from some model of evolution that may rely on an assertion about the evolutionary process in general or perhaps on the data at hand. It appears that such weighting functions are growing in complexity and in contingent, case-specific detail, so there is every reason to believe that models promoted today will be replaced by others in the future. The origin of the modern efforts in maximum likelihood are derived from arguments that (under circumstances of few, dissimilar, terminal taxa and highly variable rates of evolution) raw data may be misinformative about relationships due to peculiar distributions of homoplasy. Strict parsimony may get the "wrong" tree ("wrong" here meaning the tree suggested by the data, which are themselves misleading).

Implementing weighting

Most weighting schemes, such as those discussed above, are implemented *a priori*. There are a few dynamic procedures that rely on changing weights of characters as the tree search is performed. One method that is used widely is *successive approximations weighting*. The concept behind this method is that the tree derived from equally weighted data suggests which characters are more reliable than which others, and this might be used to revise the weights assigned to the characters and therefore possibly change the topology of the tree itself. In this procedure, a most parsimonious tree (or set of most parsimonious trees) is produced, and then either the consistency index (CI) or retention index (RI) of each character serves as a weight for another search (see Tree Statistics, below, for these indices of "fit"). A character that serves as an uncontroverted synapomorphy (no reversals, no parallelism, therefore no homoplasy) will have a CI or RI of 1.0, whereas a character that includes some homoplasy will have lower fit (say, 0.5). If the fit of each character is used as a weight in a second search, then the characters of higher fit (less homoplasy) will contribute more than the characters of lower fit (more homoplasy), possibly changing the tree, or at least helping to choose among trees that are the same length for unweighted data but have different fits [33]. This procedure may be repeated, reweighting and searching several times until the results stabilize on one or a few solutions. The method is recursive, but not circular (*contra* [34]) because it may yield different sets of trees in subsequent iterations. These solutions have several properties deemed to be desirable: starting with equally weighted data permits any conspiracy of strong or weak characters to build initial trees; the successive weights of the characters are determined by fit to the *other* characters in the analysis, and thus do not require any special argument about the path of evolution or individual characters *per se*; the trees that belong to the final set are internally consistent in the sense that the phylogenetic analysis itself implies that some characters are more reliable than others, and taking that into account returns the final phylogenetic analysis.

When reporting the length of a successive weighting tree, it is important to set the data back to equal weights (rather than the index used during the procedure) and count the steps of equally weighted data on the tree. It is not necessary to examine all of the possible solutions in detail to see how the final tree compares to the solutions of the initial, equally weighted run. If the final tree (or trees) is the same length as the trees of the initial set of most parsimonious trees, then the final tree was among the original solutions and the procedure merely chose it over the others by virtue of its high internal consistency. It is also possible that the final tree will be *longer* than the initial trees when data of equal weights are plotted. This means that the final tree was not in the initial set. Because the final tree is from a *differentially weighted* data set, however, it does not compare directly to the trees from an equally weighted data set, and it is still preferred from the point of view of taking a tree of best fit. As an exercise, applying the final weights of the data to the initial trees will show that most of them are even longer than the final tree was, which demonstrates why the final tree was chosen. Common programs such as Nona, Hennig86 and PAUP* all permit successive weighting, although the details vary. For example, Hennig86 permits only ten different categories of weights (0 to 9), which tends to decrease poor characters to zero rapidly, whereas both PAUP* and Nona have finer scales (0 to 100) that tend to keep poor characters in the analysis at low weights. This difference, of course, will influence what trees are returned.

An alternative to successive approximations weighting that also tends to be faster computationally is *implied weighting* [35], sometimes called *Goloboff weighting,* as implemented in Piwe [also called Peewee, but originally an acronym for (P)arsimony under (I)mplied (WE)ights] and PAUP*. The principle is fundamentally the same as above, using the weights implied by the analysis itself to determine which of the alternative solutions should be preferred. The procedure differs in that implied weighting is practiced during the first tree search. As taxa are added to the network they are placed such that alternative positions are evaluated according to whether they require compromising a character that has proven reliable so far (high fit), or compromising a character that has already been shown to include more homoplasy (lower fit). By choosing to add more steps to characters that have already accumulated extra steps (and therefore preserve those that have higher consistency or fit), the procedure builds a tree that is based on the weights implied by the analysis itself. Implied weighting frequently produces the tree that would be the final tree under successive approximations, but much more rapidly because it may do so on the first search. As with successive approximations, the resultant tree may be longer than the most parsimonious tree built by equally weighted data, but it is still a better "fit" according to the principle of the weighting scheme.

However, implied weighting may produce a tree that is very different from successive approximations. One reason for this is that the "fit" of a character to a tree is a concave function of the level of homoplasy. Consider, for simplicity, the

CI of a binary character. As a minimum, there is one change necessary to explain the data. If the character is perfect and has only one change, then the CI is [(minimum number of steps required for a character)/(number of steps observed on the tree)], 1/1, or 100% (see also discussion in Tree Statistics, below). With one homoplasious event (two parallel derivations, or a derivation followed by a loss), the CI is 1/2, or 50%. For two homoplasies, it is 1/3, or 33%; for three it is 25%, for four 20%, and so forth. The series, then, is a decreasing concave function. By modifying the steepness of the concavity (parameter "k" in Piwe) the weighting function can be made more or less severe (high values of k give less severe weighting, low values give more severe weighting). It is not clear how to choose one or another value for k, but the default value in Piwe is more severe than the comparable operation in successive approximations. A practical ramification of this is seen by comparing characters with several extra steps, say two versus four. Successive weighting gives them similar low weights (say 0.33 versus 0.20) whereas implied weighting may tend to pile up extra steps in the weakest characters (moving them from a fit of 0.2 to 0.17 or 0.14) and stress the others, perhaps dramatically (keeping them at a fit of 0.5 or 1.0). When the analysis contains lots of homoplasy among many characters, it can be expected that successive weighting and implied weighting will tend to give different results depending on the value specified for k.

2.6 The tree

Shortest Manhattan Distance
Having coded the characters and determined a weighting scheme (equal or differential, and how it is to be implemented), the next step is to find the optimal solution to the matrix. The preferred method is a parsimony tree, which is the best summary of the data in question [36]. Controversy on this issue is more voluminous than informative. Claims to knowledge (say, that transversions are better than transitions) must rely on other data from other studies, and if the point of the exercise is to learn about the data at hand for the taxa under consideration, then we must permit these data to inform us about themselves or it will be impossible to discover anything different from what we "knew" already. The most parsimonious tree (or trees) is the optimal solution; it is closely akin to the least squares regression line and is the solution that is most consistent with scientific method (see [11]). Other solutions are no more attractive than graphical plots that deviate from the least squares line to go through some favorite point. Although it may sound like preaching, it cannot be repeated too many times that the parsimony tree is to be preferred over all others. If there is no other reason, then it should be preferred by Occam's Razor simply because there are an infinity of trees that are *less* parsimonious, and choosing among them is an issue of *ad hoc* reasoning. Arguments against the

most parsimonious trees are not arguments against parsimony, but rather arguments against the data themselves.

The parsimony solution is a tree that maps the shortest Manhattan distance through the matrix. This is called a Manhattan distance to suggest city blocks in which moving a block East (character X changes states) and then a block North (character Y changes states) results in the same solution as moving a block North (character Y) and a block East (character X). Thus, although some paths are shorter than others, there may be many paths of equal length connecting several points. This means that there may be many ways to reconstruct hypothetical ancestors and, hence, relationships. Some critics see the possibility of multiple solutions as being undesirable, but in fact multiple solutions indicate that the data themselves are ambiguous, and the solution to ambiguous data is not definitive. This is the appropriate answer in terms of scientific method: ambiguous data, ambiguous solution. To derive a definitive solution from ambiguous data is outside the realm of science, and I have elsewhere described such practice as "clairvoyance" [37].

2.7 Multiple trees

This section addresses what to do with multiple trees from the same analysis. If a worker is comparing different trees from different matrices, there can be little said as a general path other than to combine the matrices and do a single analysis of the combined data. This controversial perspective, sometimes termed "total evidence," rests on the easily demonstrable principles that the best answers are derived from including the most evidence [38]. Proponents of the contrasting view, that some data should be preferred over others [39] have yet to make a clear statement as to what method can recognize "good" data, and recent work suggests that some kinds of "bad" data may not be as dangerous as previously thought (see [40], for discussion). The desire for favoring "good" data is often associated with a desire to find the "truth", truth apparently being something that each researcher may recognize without relying on the weight of evidence. Yet proponents of the position that "truth" is the goal have offered surprisingly little philosophical discussion for their views, demoting philosophical counterarguments to footnotes ([41]:426) and generally ignoring critics. Here, we discuss only multiple trees from a single matrix.

Using parsimony as an optimality criterion does not necessarily produce unambiguous results. When there is more than one tree of the same shortest length, this means that there is more than one interpretation of the matrix. Although some procedures are designed to choose among such trees (see successive approximations, above), it is often necessary to accept certain ambiguities as part of the final result. Several steps help limit or summarize the conflict inherent in multiple most parsimonious trees.

The most conservative method (and most valid in the spirit of rejecting a null hypothesis) is to calculate the *strict consensus* tree. This is the tree that presents only the clades found in all of the equally most parsimonious trees. Lack of resolution can be from two causes: either there were no data to address the problem of relationships at that node ("soft" polytomies), or the data support contradictory relationships that cannot be reconciled ("hard" polytomies). The strict consensus tree provides a form of confidence regarding the result, because any clade that exists on the strict consensus tree is supported by 100% of the most parsimonious trees, so in a sense the confidence is 100% regarding those clades in light of the matrix at hand. The consensus tree must be longer than the set of most parsimonious solutions, so the degree of similarity of the various most parsimonious trees can be indicated by the difference in length between the consensus tree and the most parsimonious trees that went into it. Of course, everyone wonders about the relative strength of some clades *versus* others, and this challenge confronts consensus cladograms along with unique solutions. The solution to that issue is addressed in "Tree statistics" (below).

One disadvantage of the strict consensus tree is that if a single taxon "X" is either poorly described in the matrix (lots of missing values) or has a peculiar combination of states for the characters in question, X may be able to fit into the tree at several locations that are very different on an otherwise stable topology. Taxon X will go a long way toward destroying a consensus of clades across the tree. This problem can be alleviated in several ways. The most obvious is to leave X out of the analysis, create the strict consensus tree, and then illustrate on that tree the multiple positions of X. This amounts to saying that the topology is relatively well established but for the placement of X, which is, of course, the real answer. Allowing X to destroy many clades captures less of the real result of the analysis. Of course, an *ad hoc* justification of why X was omitted rather than included, or a justification for why X was omitted instead of a different taxon, will have to pass the scrutiny of the community at large.

Another form of consensus tree is the *Adams consensus*, which can help preserve some resolution from the analysis despite having a problematic taxon [42]. Although the procedure can be somewhat laborious, the idea is to preserve topology as much as possible and to remove the problem taxon to the base of the smallest clade that always includes it. For example, consider an ingroup that can be represented by two equally parsimonious trees A(B(C(DE))) and (AE)(B(CD)). The strict consensus would be completely unresolved ABCDE, but Adams consensus would remove E to the base of the ingroup and represent the solution as EA(B(CD)), preserving the part of the hierarchy that is the same in both schemes, namely A(B(CD)). This form of consensus is useful, but it is not often used in modern analyses. A form of consensus that sometimes appears, but is actually pointless, is the *majority rule* consensus, which is based on the idea that clades that show up in more than half the consensus trees are somehow better supported. This is not true, because all the most parsimonious trees are equal in their status of optimal solutions, and a single solution with certain clades is just as good as the many solutions that have different clades. Additionally, the logic of the argument is flawed from the start, because instability of a different clade generates the many trees upon which a given clade is stable. Stability and support are not the same thing, and majority rule consensus confuses the two. A nice illustration of the various consensus methods is available in Schuh ([13]:148).

Optimization
One of the biggest problems for those who seek to choose among tree topologies is the problem of character optimization. The center of this problem is whether it is better to derive the same character state twice (two forward evolutionary steps), or derive it once and then change back (step forward and step back). In some programs the choice is made as a blanket statement. For example, in PAUP jargon, "acctran" (accelerated transformation) will choose to make characters arise more basally in the tree and then be lost later, while "deltran" (delayed transformation) will choose to make characters arise more distally and separately. This choice operates for all characters at once. Some programs permit optimizations to be customized for each character, at least after the tree search. This permits a researcher to have some characters derived twice, and others derived once and lost, on an *ad hoc* basis. Such a choice among optimization will change the placement of synapomorphies so that certain branches may no longer be supported. Thus, it is common that many trees are optimization dependent, and in some cases these may be excluded from a set of final trees so that the conclusions are based on trees that are supported under all optimizations.

2.8 Tree statistics

Length
The simplest statistic is tree length, which is the total number of steps on the tree when all characters are plotted. This should be reported because it is the most direct index of whether a given tree should be preferred over an alternative (the shortest is preferred). If there is more than one most parsimonious tree (the shortest length), then it is wise to report the number of trees discovered for the shortest length, and then also the length of the consensus tree, which is always greater. If the final most parsimonious tree was derived from differentially weighted data, modest complications are introduced. When using a dynamic procedure that produces differential weighting, as with successive approximations weighting or implied weighting, the length of the tree is usually reported as the number of steps on the matrix of *equally weighted data* rather than the final matrix of differentially weighted data. This may require using a command to reset all weights to 1 (as in Hennig86). If the iterative successive approximations procedure has merely chosen one of

many most parsimonious trees that came from the first analysis of equally weighted data, then the length of the final tree will be the same as the length of the set of original trees (because, of course, the final tree actually was one of the original trees). If a novel tree was chosen (that is, a tree that was not among the set of most parsimonious trees derived from the first analysis of equally weighted data), then the final tree will have a length somewhat greater than the most parsimonious trees when weights are reset to 1. Although this seems like a departure from the optimality criterion of parsimony, it is important to remember that the final tree *is* the preferred parsimonious tree for the differentially weighted matrix. The same logic applies to implied weighting and trees built in Piwe.

CI, RI

Two statistics reveal something about the level of homoplasy on the tree. The consistency index (CI) is simply (minimum number of steps required for a character)/(number of steps observed on the tree) [43]. Thus, a character in two states requires at a minimum one step, and if only one step is observed on the tree, then the CI is 1/1, or 100%, or 1.0. If there are two steps (indicating a parallelism or a reversal), then the CI is 1/2 or 0.5. As the number of homoplasious steps increases, the CI decreases as a concave function. Thus, the average CI for the whole matrix is informative about the level of homoplasy reflected by the tree. The retention index (RI) includes as part of the calculation the departure from the worst possible case [44]. For example, if a matrix of six taxa has three of them with state 0 and three with state 1, then the worst possible case (a completely unresolved bush of six taxa) would have three steps. We can calculate RI using the equation (worst possible-observed)/(worst possible-minimum). If the actual tree has two steps, then we would get (3–2)/(3–1) = 0.66. Note that this value is higher than the CI (which would be 1/2 or 0.5), because it accounts for the fact that the situation could have been worse. Usually, both CI and RI are reported. It is important to note that while high values of these statistics are considered desirable, there is actually nothing wrong with low values: these are only indicators of homoplasy. Homoplasy itself is not necessarily related to the degree to which a matrix can form a tree [45] and generally increases as the number of taxa increase [46], so comparisons between matrices or studies are not very meaningful.

Support

Regardless of how definitive the matrix is regarding a preferred tree, it is now customary to provide some measure of support for each branch in the tree. There are four common ways of doing this: showing apomorphies, generating bootstrap statistics, generating parsimony jackknife values, or calculating Bremer support values (more on support can be found in Chapter 5). The most traditional is to plot or list the number of synapomorphies for each branch. This simple approach relies on the idea that more character support is better than less, or that certain characters are "good" characters. This is commonly used in morphological analyses, and researchers are expected to consider the logical strength of different synapomorphies. For example, a clade held together by a character like "placenta" will be considered better supported than a clade marked by several characters like "hairs straight". Unfortunately, molecular apomorphies are not as easily interpretable (what is the relative value of different state changes in a DNA sequence?) and the sometimes high level of homoplasy implies that some clades may rely on weak apomorphies, thereby frustrating intuitive interpretation of character plots. Each of the three methods explained below is designed to provide a numerical index that indicates branch-by-branch support of the tree.

- Bootstrap values rely upon resampling (with replacement) from the original data matrix [47]. Characters are randomly chosen to compose a matrix the same size as the original matrix, but of course some characters will be chosen more than once and some not at all. The tree that results will be missing some of the clades that were in the most parsimonious tree. The procedure is repeated many times, and the results are compared to the most parsimonious tree to see how various clades fared in the sampling procedure. If a clade is well supported, then sampling from the original data matrix should produce that clade often. A bootstrap value of 90% means that the clade appeared in 90% of the samples, and presumably it is better supported than a clade with a bootstrap of 80%. The benefits of this method are that the bootstrap is easy to calculate and widely used (hence it is familiar). The problems are many: because all characters are sampled equally, uninformative characters can have an effect (by excluding informative characters), which is neither logical nor desirable, hence all uninformative characters need to be removed before the procedure. Also, tree balance plays a part in the frequency of a clade appearing in the bootstrap tree, such that clades in asymmetrical trees are easier to support than clades in balanced trees with the same number of apomorphies (which, on the evidence alone, should not be the case) (see also Chapter 5). Third, although a clade with many taxa actually has higher *empirical* support than a clade with few taxa, the large clade is penalized because there are more ways to break up that clade. Furthermore, because bootstraps are calculated as a zero-sum game, strong support for one place in the tree is interpreted as weak support elsewhere, regardless of empirical values (hence, they do not indicate any absolute level of support, which was basically the original challenge). Many cladists do not like bootstraps, but (given the *caveats* mentioned above), bootstraps may approximate a general sense of the weight of the original matrix for each clade in the optimal tree. Bootstraps are best considered analogous to a weather vane that shows the direction of the wind: they give us an idea of the direction, but the measure is rough. A bootstrap of 90% does not equate with a P value of 0.90. Also, because the calculation of a new tree is required for each sampled matrix, generating good bootstraps (say, based on 1000 replicates) can be quite time-consuming. Finally, although some people seem to think that nodes with low bootstraps should be dissolved (as if there is no confidence in the proposal of relationship when the bootstrap is below, say 75%), this is not a good idea. It is important to remember that the most parsimonious tree (or consensus of most parsimonious trees) is the optimal estimate of relationship, however strong or weak. Dissolving clades that are poorly supported will simply produce a suboptimal tree for no good reason. So, be content that these comparative measures are only useful in a general way, and not to "confirm" or "refute" certain clades.

- The parsimony jackknife is similar to the bootstrap in that it relies on resampling the original matrix, but it differs in several ways [48]. A bootstrap is derived from random sampling (with replacement) from the original matrix. The probability that any individual character is eliminated from the matrix approaches 1/e if there were infinite samples, and the parsimony jackknife procedure simply employs this binomial probability to admit or exclude characters from the sampled matrix on a one-by-one basis (hence, they are not competing with each other, as in the bootstrap). Making a tree with the resulting matrix overcomes some of the problems of the bootstrap. For example, including uninformative characters does not affect whether informative characters will be used in the sampled matrix. Sampling is repeated many times (say, 10,000) to provide yet more accurate approximations of what would obtain with an infinite number of trials. The values returned are read just like a bootstrap; 90% means that 90% of the jackknifed trees contained that clade. Farris's JAC program is astonishingly fast, and it keeps track of where it was, so that if a trial of 1,000 replications seems not enough, the next search starts at 1,001 retaining previous searches. The JAC program does not do very thorough searches, however, so the jackknifed trees are often more poorly resolved than those obtained by the similar utility in PAUP*. Farris's XAC program does include more thorough searches and generally results in greater resolution for the jackknifed tree.

- Bremer support is based on the idea that if the most parsimonious tree is optimal, then any tree that does not contain all the clades of the most parsimonious tree *must* be longer [49, 50]. Thus, an index of clade support could be provided by measuring the distance between the tree of interest and the shortest tree that *doesn't* have a given clade. If the Bremer support of a clade is 0, that means that there is a tree that is the same length as the tree of interest, and does not contain the clade in question. This is why Bremer support should be calculated on a strict consensus tree, because only those clades are found in all of the most parsimonious trees and could be expected to have Bremer support values larger than zero. If the Bremer support for a certain clade is 1, then that means that there is a tree that is only 1 step suboptimal and does not contain the clade of interest. The advantage of Bremer support is that it tells us what we really want to know, which is the price to pay against the optimal solution if we are to accept a solution without the clade of interest. The problems with Bremer support are two-fold. Most severe is that it is very burdensome to calculate all trees 1 step away from most parsimonious tree, plus all those 2 steps away from most parsimonious tree, and those 3 steps away from most parsimonious tree, and so forth, until we are satisfied that we have a good comparison. A simple and crude approximation to Bremer support is to dissolve the node of interest, so that the tree shows all resolution except the one clade in question. This produces a bias that may overestimate Bremer support because there may be very different topologies that are shorter (a common result in the primary search!), but such a tree will not be considered when we only dissolve a node on a single topology we already have. Unfortunately, even thoroughly researched Bremer support values labelled on a tree may not be attainable simultaneously (that is, there is no tree that sacrifices only two clades of interest by adding only the extra steps suggested by the sum of the Bremer values of each clade). Also, it is not really clear what is a strong value and what is a weak value. It appears that Bremer support of 3 or better does correspond to high bootstrap values [51] and is rather stable against random data interrupting the clade [40]. It seems safe to say that Bremer support of 5 is "high", so there is not much need to calculate trees longer than that. This should be true in an absolute sense, so it shouldn't really matter if the tree in question has a total length of 50 or 5000.

3   Philosophical Perspective

The method offered above is useful as a recipe for deriving a phylogenetic tree, and its philosophical justification makes it much stronger than any other. Accordingly, it is appropriate to have some discussion of the philosophical basis and how it compares to competing methods in this regard. The importance of the scientific method leads some cladistic authors to offer thorough discussions of the philosophy of science, or of the scientific method itself. These are important, and anyone who adopts cladistic methods (and especially those who reject cladistic methods) would do well to be sure of the philosophical implications of that decision. An abbreviated treatment is offered here. Good, general papers that deal with cladistic perspectives include Farris [11, 36], Gaffney [9], Platnick [52], Brady [53], Kluge and Wolfe [38], Frost and Kluge [54], Kluge [27, 55], and Schuh ([13]: 44–59).

Certainly one of the goals of the recent revolution in phylogenetic methods is to make phylogenetic reconstruction satisfy the general principles of scientific method. The phylogenies we draw should be based on empirical evidence and should represent testable hypotheses. Because systematic data are acquired by observation rather than manipulative experimentation, some scientists consider systematics to be barely more than straight description, but this view is mistaken. If we were to stop at the data matrix itself (a raw tally of similarity and difference among the taxa), then we would be performing only straight description, but the process of building a tree includes all the same elements of classical hypothesis testing as are found in manipulative experimentation. Hypothesis testing is revealed every time there is an event of homoplasy in the tree (and, of course, homoplasy is only a property of trees; data matrices themselves do not actually exhibit homoplasy any more than they exhibit monophyly). For example, every event of reversal or convergence as illustrated on a tree is a falsification of the postulate that similar states in the original data matrix are homologues. The tree itself is tested by every character in the matrix. The shortest trees are those that are best corroborated, and longer trees for the same data have more homoplasy, hence more proposals of homology have been rejected, and therefore the trees themselves are rejected in favor of the shortest tree. It is important to point out that absolute level of support for the most parsimonious tree (or set of equally most parsimonious trees) does not enter into the decision to prefer it over competing trees. Whether support is strong or weak, the most parsimonious tree is always better corroborated than any longer tree, and so it is to be prefered with respect to any other hypothesis of relationship. To use Farris's [11] analogy, the least squares regression line is preferred over other lines through a cloud of points, and that preference does not rely on whether the relationship is strong or weak. So it is with parsimony trees.

Hypothesis testing operates to reject one hypothesis in favor of another, and as a result homoplasy itself is not enough to reject a given cladogram; there has to be another competing cladogram that has less homoplasy to replace the one we reject. There is sometimes some confusion as to what homoplasy represents in a classical Popperian paradigm, but I think the answer is quite simple. If the most parsimonious cladogram represents an hypothesis analogous to "all swans are white", then homoplasy might be analogous to the statement "here is a black swan." As Farris [11] has pointed out, the two possible explanations for this situation are that the hypothesis is wrong or the statement is wrong. In other words, it is possible that a bird mistaken for a black swan is either not a swan or not black, and this is the situation we interpret as homoplasy. Given this situation, then, to propose that *any* homoplasy refutes a given cladogram would be what Popper referred to as "naive empiricism" ([56]:21) thinking, and something to guard against. The correct procedure is to use homoplasy to choose among alternative cladograms, returning us to the use of parsimony as an optimality criterion. Homoplasy serves to reject hypotheses of relationship that are not the most parsimonious, whereas on the most parsimonious cladograms, homoplasious characters are supposed to represent erroneous statements (similar states are not actually homologues).

Because parsimony trees minimize homoplasy, some authors assert that parsimony can be expected to perform poorly if homoplasy is common. Armbruster ([56]:227) concisely captured this view: "Implicit in the parsimony procedure that forms the basis of most phylogenetic analyses is the assumption that homoplasy is relatively uncommon". This may be offered as the reason to eliminate saturated sites from analysis, or possibly dispense with parsimony completely. Although this opinion has some intuitive appeal, it does not stand up to examination. Returning to the analogy with regression analysis, the least squares line minimizes variance regarding departures from the line, but it is completely mute as to whether the variance itself is high or low [11]. Similarly, the most parsimonious tree has the least homoplasy, but this does not reveal whether homoplasy itself is high or low. Indeed, it is common to find studies where the CI is lower than 50%, which means that similarities in the matrix are due more to homoplasy than to uniquely derived synapomorphy, but that does not prevent us from making a credible tree [46]. The only fair conclusion is that the fear of homoplasy *per se* is unreasonable; therefore, objections to certain data or to parsimony analysis, based on fear of homoplasy, are also unreasonable.

4   Models of Causation

A great conflict in modern systematics surrounds the use of models in phylogenetic reconstruction. Put simply, some people want to include models of evolutionary process in their methods of reconstruction, and others want to exclude models as much as possible. The debate is beyond the scope of this chapter, but it seems relevant to answer some concerns regarding the relationship between parsimony methods and other, model-based, methods of reconstruction.

In general, strict Hennigians claim that their use of parsimony is merely an extension of Occam's Razor (general scientific method), and that parsimony methods are preferred because they are less assumption-laden than competing model-driven methods. The idea is that if we study character data to learn about descent with modification, then we should not make assumptions regarding the evolutionary process, or our conclusions may be no more than a circular discovery of our assumptions ([11]:17, [37], among others). Attacks on this position include 1) that we should use what we know from other research about the process (or causes) of evolution to illuminate the study of interest, and 2) that parsimony is itself a model. These will be taken in turn.

First, there is no necessary relationship between recognizing patterns themselves and understanding the processes (causes) that lead to the patterns. For example, it is believed that the ancient Babylonians could predict with great accuracy the timing of high and low tides, but they did not share our modern understanding of the causes of tides. No doubt modern scientists feel they know more when they explain tides through the processes of rotation of the Earth about its axis, and the moon about the Earth, etc., but statement of the pattern itself at the port in question is not improved by an understanding of process. Pertinent to this book is that describing hierarchical differences between taxa was performed well by Linnaeus without a valid theory of process regarding how taxa get to be what they are. Similarly, there is no necessary benefit to incorporating process theories of how nucleotides should change among relatives if the data can be allowed to speak for themselves, simultaneously informing us of what changes occurred and who are the relatives. As Platnick [58] pointed out, we can recognize a clade called "spiders" without any need to know the cause of that clade. Data do not need process theory, data simply exist.

Data are natural, but theories are made by scientists, so the relationship between a process theory and its data is asymmetrical: data do not need theory, but theory requires a certain amount of data to serve as inspiration for the genesis of the theory, and more to provide confirmation of the theory, and still more to launch the calculations to generate values of interest at present to compare with the natural data at hand. Furthermore, there is a hazard in including a process theory when analyzing unfamiliar data, because if the theory is wrong or inaccurate, then a variety of processes that apply only to the theory and not to the data may produce a false answer for comparison with natural data. This is certainly true for statistical tests of character distributions [59] and would hold also for reconstruction in general. In summary, data can be explained without process theories, and process theories can easily produce wrong interpretations. So, it seems that arguments that "we should use what we know about process" are not convincing by themselves.

Second, detractors claim that the parsimony criterion of optimality is some kind of model, but there is a remarkable inability on the part of these critics to specify the model (see [13]:53 and following, for discussion). In contrast to many loosely delivered opinions, there is no assumption that evolution is parsimonious (above). Parsimony is not a model of evolution any more than parsimony is a model of physics, or of chemistry, or geology, or climatology, or oceanography, etc., even though scientists in these fields use the principle of parsimony to derive their proposals. If it were true that "parsimony" should be a "model" of all these things, then it certainly would have to be a very useful model to serve in so many contexts where the alternative models are so different, and clearly it would have to be more robust than a model that says that 3rd positions evolve faster than 2nd positions (which is not very useful to a climatologist). Rather, parsimony is not a model at all, but these various kinds of data simply speak for themselves through parsimony.

There is a simple, if uncomfortable, relationship between parsimony methods and the leading competitors developed in schools of maximum likelihood. These different methods can be partly synonymized if maximum likelihood is taken as a form of weighting (see above). Instead of weighting characters differentially (say, base #362 more than #363) it may weight the kind of change (transversions weigh more than transitions, regardless of position). This might be true for all changes of that type, and so expectations regarding homoplasy would not be permitted to vary from site to site. In some models, it is easier to place a change on long branches than on short, perhaps regardless of the character or state in question. This assumption has some severe implications for whether the primary job of the systematist is to discover patterns or dictate them. However, this simplified representation of maximum likelihood does not do justice to the sophistication of the competing methods, and the debate between likelihoodists and parsimony advocates is evolving so fast that it is hard to know where to start examining it: consult your current literature for the latest quarrel regarding likelihood methods. What I will hazard here is that *any* method that relies on a supreme model of weighting is not far removed from the Evolutionary Taxonomy in which processes of evolution were handsomely mixed with data to come up with an evolutionary tree that included both pattern and process. The long-traditional Evolutionary Taxonomy was largely dead by the late 1980s, and it is peculiar to see a redux of it today under a more genetical mantle. Surely most likelihoodists reject classical Evolutionary Taxonomy, so it is curious that the incorporation of the model should be central to the school.

As I write this, there is a particularly interesting effort to synonymize parsimony methods and likelihood methods, producing a controversy that is unexpected and hard to predict. Tuffley and Steel [60], in a series of mathematical proofs, claim to demonstrate that maximum likelihood estimators collapse to the parsimony method if it is assumed that the characters do not evolve by a process common to all. On the face of it, this seems understandable because there is no assumption of process in parsimony analysis. Thus, because maximum likelihood estimators specify (perhaps accurately, perhaps not) what the process shall be for the characters at hand, parsimony would appear to serve frequently as the preferred method because it is more robust against the exact causes of any particular character state. Later, maybe hoping to mitigate the damage done to the likelihood school, Steel and Penny [61] stated that the earlier result should not be taken as an endorsement of parsimony, because one can argue that it is more parsimonious to have one mechanism (model) for all characters (which is supposed to represent the maximum likelihood school) than a different mechanism for each character (supposedly, the parsimony school). Of course, this claim to deny the obvious conclusion from Tuffley and Steel is wrong, because parsimony only relies on no *common* process and does not assert that there should be *all different* processes. So, if there are 1,000 informative bases and 999 evolve alike and one follows a different process, the assumption of "no common process" is fulfilled, which means that the parameter space in which parsimony operates well as a likelihood estimator would be quite large. This fundamental conclusion was originally offered by Farris [36] as a reason to favor parsimony.

5 Remarks and Conclusions

Cladistic methods are easily executed. The methods are relatively restrictive with respect to models or process theories regarding the origin of character states. Certain decisions may be made on an *ad hoc* basis, but such decisions are explicit and part of a scientific methodology. Parsimony operates to choose among phylogenetic trees in the context of classical hypothesis testing. Cladistic perspectives have sound philosophical foundations, a position not shared by competing methodologies.

Acknowledgments

R. DeSalle and G. Giribet encouraged me to write this brief summary of methods. Some of this material was presented at the Finnish Academy of Sciences in 1999. P. Goloboff and M. Siddall offered useful advice.

References

1   O'Hara RJ (1988) Homage to Clio, or, toward an historical philosophy for evolutionary biology. *Systematic Zoology* 37: 142–155

2   Wenzel JW (1992) Behavioral homology and phylogeny. Annual Review of Ecology and Systematics 23: 361–381

3   Wenzel JW (1993) Application of the biogenetic law to behavioral ontogeny: a test using nest architecture in paper wasps. *Journal of Evolutionary Biology* 6: 229–247

4   de Quieroz A and Wimberger PH (1993). The usefulness of behaviour for phylogeny estimation: Levels of homoplasy in behavioral and morphological characters. *Evolution* 47: 46–60

5   Coddington JA (1990) Cladistics and spider classification: araneomorph phylogeny and the monophyly of orb-weavers. *Acta Zoologica Fennica* 190: 75–87

6   Proctor HC (1991) The evolution of copulation in water mites: a comparative test for nonreversing characters. *Evolution* 45: 558–567

7   Tinbergen N (1953) *Social behaviour in animals, with special reference to vertebrates*. Methuen, London

8   Ross HH (1964) Evolution of caddisworm cases and nets. *American Zoologist* 4: 209–220

9   Gaffney ES (1979) An introduction to the logic of phylogeny reconstruction. In: J Cracraft, N Eldredge (eds) *Phylogenetic Analysis and Paleontology*. Columbia University Press, New York, 79–111

10 Wiley EO (1981) *Phylogenetics: the theory and practice of phylogenetic systematics*. John Wiley and Sons, Toronto

11 Farris JS (1983) The logical basis of phylogenetic analysis. In: NI Platnick, VA Funk (eds) *Advances in Cladistics*, vol. 2. Columbia University Press, New York, 7–36

12 Brooks DR, McLennan DA (1991) *Phylogeny, Ecology and Behavior*. University of Chicago Press, Chicago

13 Schuh R T (2000) *Biological Systematics: Principles and Applications*. Cornell University Press, Ithaca, New York 236

14 Patterson C (1982) Morphological characters and homology. In: KA Joysey, AE Friday (eds) *Problems of Phylogenetic Reconstruction*. Academic Press, London and New York, 21–74

15 de Pinna MCC (1991) Concepts and tests of homology in the cladistic paradigm. *Cladistics* 7: 367–394

16 Nelson G (1994) Homology and systematics. In: BK Hall (ed) *Homology: The Hierarchical Basis of Comparative Biology*. Academic Press, New York, 101–149

17 Miller JS, Wenzel JW (1995) Ecological characters and phylogeny. *Annual Review of Entomology* 40: 389–415

18 Brower AVZ, Schawaroch V (1996) Three steps of homology assessment. *Cladistics* 12: 265–272

19 Phillips A, Janies D, Wheeler W (2000) Multiple Sequence alignement in Phylogenetic analysis. *Molecular phylogenetics and Evolution* 16: 317–330

20 Cameron S A (1991) A new tribal phylogeny of the Apidae inferred from mitochondrial DNA sequences. In: DR Smith (ed) *Diversity in the genus Apis*. Westview Press, Boulder, Colorado

21   Chavarria G, Carpenter JM (1994) "Total evidence" and the evolution of highly social bees. *Cladistics* 10: 229–258

22   Simmons M P (2000) A fundamental problem with amino-acid-sequence characters for phylogenetic analysis. *Cladistics* 16; in press

23   Hillis DM (1994) Homology in molecular biology. In: BK Hall (ed) *Homology: The Hierarchical Basis of Comparative Biology.* Academic Press, New York, 339–369

24   Nixon KC, Carpenter JM (1993) On outgroups. *Cladistics* 9: 413–426

25   Felsenstein J (1978) Cases in which parsimony or compatibility methods will be positively misleading. *Systematic Zoology* 27: 410–410

26   Wheeler WC (1990) Nucleic acid sequence phylogeny and random outgroups. *Cladistics* 6: 363–367

27   Kluge AG (1997) Testability and the refutation and corroboration for cladistic hypotheses. *Cladistics* 13: 81–96

28   Wheeler WC, Honeycutt RL (1988) Paired sequence difference in ribosomal RNAs: Evolutionary and phylogenetic implication. *Molecular Biology and Evolution* 5: 90–96

29   Sidow A and Wilson AC (1990) Compositional statistics–an improvement of the evolutionary parsimony and its application to deep branches in the tree of life. *Journal of Molecular Evolution* 31: 51–68

30   Hassanin A, LeCointre G, Tillier S (1998) The "evolutionary signal" of homoplasy in protein coding gene sequences and its consequences for *a priori* weighting in phylogeny. *Comptes Rendus de l'Academie des Sciences* 321: 611–620

31   Källersjö M, Albert VA, Farris JS (1999) Homoplasy increases phylogenetic signal. *Cladistics* 15: 91–94

32   Björklund M (1999) Are third positions really that bad? A test using vertebrate cytochrome b. *Cladistics* 15: 191–197

33   Carpenter JM (1988) Choosing among equally parsimonious cladograms. *Cladistics* 4: 291–296

34   Swofford DL, Olsen GJ (1990) Phylogeny reconstruction. In: DM Hillis, C Moritz (eds) *Molecular Systematics.* Sinauer, Sunderland, Massachusetts, 411–501

35   Goloboff PA (1993) Estimating character weights during tree search. *Cladistics* 9: 83–91

36   Farris JS (1979) The information content of the phylogenetic system. *Systematic Zoology* 28: 483–519

37   Wenzel JW (1997) When is a phylogenetic test good enough? *Mémoires du Muséum National d'Histoire Naturelle* 173: 31–45

38   Kluge AG, Wolf AJ (1993) What's in a word? *Cladistics* 9: 183–199

39   Bull JJ, Huelsenbeck JP, Cunningham CW, Swofford DL et al. (1993) Partitioning and combining data in phylogenetic analysis. *Systematic Biology* 42: 384–397

40   Wenzel JW, Siddall ME (1999) Noise. *Cladistics* 15: 51–64

41   Swofford DL, Olsen GJ, Waddell PJ, Hillis DM (1996) In: DM Hillis, C Moritz, BK Mable (eds) Molecular Systematics. Sinauer, Sunderland, Massachusetts: 407–514

42   Adams EN (1972) Consensus techniques and the comparison of taxonomic trees. *Systematic Zoology* 21: 390–397

43   Kluge AG, Farris JS (1969) Quantitative phyletics and the evolution of anurans. *Systematic Zoology* 18: 1–32

44   Farris JS (1989) The retention index and rescaled consistency index. *Cladistics* 5: 417–419

45   Goloboff PA (1991) Random data, homoplasy and information. *Cladistics* 7: 395–406

46   Sanderson MJ, Donoghue MJ (1989) Patterns of variation in levels of homoplasy. *Evolution* 43: 1781–1795

47   Felsenstein J (1985) Confidence limits on phylogenies: an approach using the bootstrap. *Evolution* 39: 783–791

48   Farris JS, Albert VA, Källersjö M, Lipscomb D et al. (1996) Parsimony jackknifing outperforms neighbor-joining. *Cladistics* 12: 99–124

49   Bremer K (1988) The limits of amino acid sequence data in angiosperm phylogenetic reconstruction. *Evolution* 42: 795–803

50 Bremer K (1994) Branch support and tree stability. *Cladistics* 10: 295–304

51 Davis JI (1995) A phylogenetic structure for the monocotyledons, as inferred from chloroplast DNA restriction site variation, and a comparison of measures of clade support. *Systematic Botany* 20: 503–527

52 Platnick NI (1979) Philosophy and the transformation of cladistics. *Systematic Zoology* 28: 537–546

53 Brady RH (1985) On the independence of systematics. *Cladistics* 1: 113–126

54 Frost DR, Kluge AG (1994) A consideration of epistemology in systematic biology, with special reference to species. *Cladistics* 10: 259–294

55 Kluge AG (1999) The science of phylogenetic systematics: Explanation, prediction, and test. *Cladistics* 15: 429–436

56 Popper KR (1992) *Conjectures and Refutations* Routledge, New York

57 Armbruster WS (1996) Exaptation, adaptation, and homoplasy: evolution of ecological traits in *Dalechampia* vines. In: MJ Sanderson, L Hufford (eds) *Homoplasy: The Recurrence of Similarity in Evolution*. Academic Press, San Diego, 227–243

58 Platnick NI (1982) Defining characters and evolutionary groups. *Systematic Zoology* 31: 282–284

59 Wenzel JW, Carpenter JM (1994) Comparing methods: adaptive traits and tests of adaptation. In: P Eggleton and R Vane-Wright (eds) *Phylogenetics and Ecology*. The Linnean Society of London, London, 79–101

60 Tuffley C, Steel M (1997) Links between maximum likelihood and maximum parsimony under a simple model of site substitution. *Bulletin of Mathematical Biology* 59: 581–607

61 Steel M, Penny D (2000) Parsimony, likelihood, and the role of models in molecular phylogenetics. *Molecular Biology and Evolution* 17: 839–850

Parsimony Analysis

Mark E. Siddall

Contents

1 Introduction

Parsimony, in general, is nothing more or less than the universal scientific principle that no extraneous causes or forces should be invoked in explanation of a phenomenon than are minimally required to account for the data. Parsimony is not simplicity; this is a conflation that has befuddled some phylogenetic thinking. By way of example, although the General Theory of Relativity is hardly "simple" in its composition, it is more parsimonious than Newtonian mechanics because it explains gravitation, Mercury's perihelion, the contraction of space and time under motion, and a variety of other phenomena under one theory. It is parsimonious in its explanation, though complex in its formulation. In other words, the principle of parsimony is a general epistemological one that is used by all of us in all aspects of our scientific research and our search for explanations of phenomena. The simple act of including two variables in a correlative analysis, for example one of height and weight, entails an underlying appeal to parsimony – the search for one common causal explanation of two phenomena rather than invoking two separate causes. Epidemiology too was born from the parsimony principle. Rather than invoking separate independent causal sins and evils for a multitude of deaths, John Snow sought a single cause for the 1854 London outbreak of cholera. Finding that most of the cases occurred within 250 yards of the well at Cambridge and Broad Streets, he had the pump handle removed and the epidemic subsided. The argument was parsimonious.

Methods and Tools in Biosciences and Medicine
Techniques in molecular systematics and evolution, ed. by Rob DeSalle et al.
© 2002 Birkhäuser Verlag Basel/Switzerland

In the field of systematics, parsimony analysis is the most successful method of phylogeny reconstruction devised. Just as it is more reasonable to conclude that two students with identical essays copied the same source, so too is it reasonable to conclude that chimps and humans share so many features because they are copied from a common ancestor. The argument is sensible because it is parsimonious [1–3]. But, as we shall see, there is more to the application and even the philosophical implications of parsimony in a phylogenetic framework than this. It is the phylogenetic method that permits the input of the broadest array of discrete information. It allows simultaneous analysis of behavioral, morphological and molecular data, and of extant taxa with extinct taxa. Until someone can explain how the frequency and kind of change in eye color in mammals should affect our beliefs about whether the ancestor of mammals had mammary glands, stochastic model-based methods of maximum likelihood will not permit the use of morphology or behavior, nor will it permit the use of fossil taxa. Parsimony is ultimately agnostic and, I would argue, even humble in its refusal to make claims about evolutionary processes that can neither be known for certain or in any detail. Proponents of parsimony procedures pillory the proffering of a plethora of particular processes as little more than pernicious, perverse and presumptuous.

Many readers may wonder at my advocacy for parsimony, and they may also have read or heard of the problems associated with parsimony, in particular concerns for its statistical consistency and rates of change [4–13]. Minimizing the squared deviations about a line to find the best fit to a series of correlated points does not presuppose that those deviations are small. In fact they can be quite large. Similarly, minimizing the number of steps in a most parsimonious tree does not presuppose that those steps should be few, nor does it imply an assumption of parsimonious evolution. For example, it is quite possible for the shortest, most parsimonious tree to require more than two or three times the number of steps as there are informative characters in a matrix. Moreover, even if all but one of the synapomorphies uniting a clade are actually homoplasies, it is quite possible for the method to have behaved correctly [14,15]. The specifics of some of these arguments will be addressed in greater detail below.

2   The Legacy of Willi Hennig

In the middle part of the last century there was a revolution in the way that systematists were approaching their task. The field of numerical taxonomy was born of a desire to make systematics and the assessment of relationships more repeatable and less subjectively determined by the whims of authoritarian views. Many fields of science, and systematics is no exception, have been hampered by such authoritarianism, in which bold statements are reserved for the well-known or the well-heeled. Wegener [16], in part because he was a meteorologist outsider, but also because he was propounding a radical view, was shunned for his suggestions that continents moved, a notion that contradicted prevailing wisdom. So too with systematics; about the same time, the numerical taxonomists were hardly welcomed by the likes of Mayr and Simpson. Their heresy was to suggest that similarities among species should be scored and analyzed mathematically. Their crime was to put the science of systematics in the hands of the novitiate. Prior to this revolution, and largely ignored elsewhere, a German entomologist, Willi Hennig [17], had come to similar conclusions about assessing the historical relationships of taxa and the grouping of organisms in a phylogenetic system. Eventually the numerical taxonomists paid attention to this work and, after considerable squabbling, Hennig's [17–19] methods superseded previous phenetic numerical methods. An interesting history of this can be found in Farris and Kluge [20].

Hennig's approach differed substantively from that of the phenetic methods in terms of how similarities are to be viewed. The phenetic approach held that any and all similarity was pertinent to assessing relationships. This only seemed reasonable in that humans are more similar on the whole to chimps than to alligators. Hennig's method differed in one crucial respect: the only similarities that are useful in determining relationships are shared derived similarities. There was a directionality to the argument of similarity, one that was more consistent with the principle of descent with modification. Thus, for example, having the derived state of a backbone is a reason to place mammals, birds, fish, frogs and such in a group called Vertebrata, but the lack of a backbone, being the ancestral condition, is not a reason to group jellyfish, nematodes and protists in a group. Invertebrata is not a natural group because you can't point to an ancestor of those organisms that is not also an ancestor of vertebrates. The most recent ancestor of vertebrates, on the other hand, was not an ancestor of any invertebrate. Hennig's intent was to create a system that was natural, one that reflected the real patterns of descent with modification that lies at the core of evolutionary biology: a phylogenetic system.

Shared derived similarities are synapomorphies, ancestral similarities are symplesiomorphies. Grouping of organisms into natural taxa is based on synapomorphies. Consider the matrix of familiar taxa and characters in Table 1.

The entries in the table can be converted to numerical placeholders as seen in Table 2.

Merely as a convention, the plesiomorphic states are assigned a zero. Thus, the lack of an amnion, being the original state, is assigned a zero wherever it occurs in a taxon. So, too, the presence of scales, being ancestral among vertebrates, is assigned a zero. Now, if we simply ignored the directionality of similarities, and if we looked at overall similarity counting both the zeros and the ones, you'll notice that the budgy is considerably different from all of the other species. The bird's similarity values range from 23% to 69% when compared to other species in the matrix, whereas every other taxon has a similarity that is at least 85% to some other taxon (Tab. 3). If we grouped on the basis of overall absolute similarity (i.e., phenetically), the budgy, being least similar, would be forced to stand on its own (Fig. 1) and the tree would not seem

Table 1 Morphological data

Taxon	amnion	legs	scales	blood	internal nostrils	atrial septum	two temporal fenestrations	hemipenes	gizzard	pedicillate teeth	feathers	wings	vertebrae
perch	no	no	yes	cold	no	no	no	no	no	no	no	no	yes
coelacanth	no	no	yes	cold	yes	yes	no	no	no	no	no	no	yes
salamander	no	yes	no	cold	yes	yes	no	no	no	yes	no	no	yes
frog	no	yes	no	cold	yes	yes	no	no	no	yes	no	no	yes
turtle	yes	yes	yes	cold	yes	yes	no	no	no	no	no	no	yes
human	yes	yes	no	warm	yes	yes	no	no	no	no	no	no	yes
gecko	yes	yes	yes	cold	yes	yes	yes	yes	no	no	no	no	yes
snake	yes	yes	yes	cold	yes	yes	yes	yes	no	no	no	no	yes
alligator	yes	yes	yes	cold	yes	yes	yes	no	yes	no	no	no	yes
budgy	yes	yes	no	warm	yes	yes	yes	no	yes	no	yes	yes	yes

Table 2 Numerical placeholders

	1	2	3	4	5	6	7	8	9	10	11	12	13
perch	0	0	0	0	0	0	0	0	0	0	0	0	0
coelacanth	0	0	0	0	1	1	0	0	0	0	0	0	0
salamander	0	1	1	0	1	1	0	0	0	1	0	0	0
frog	0	1	1	0	1	1	0	0	0	1	0	0	0
turtle	1	1	0	0	1	1	0	0	0	0	0	0	0
human	1	1	1	1	1	1	0	0	0	0	0	0	0
gecko	1	1	0	0	1	1	1	1	0	0	0	0	0
snake	1	1	0	0	1	1	1	1	0	0	0	0	0
alligator	1	1	0	0	1	1	1	0	1	0	0	0	0
budgy	1	1	1	1	1	1	1	0	1	0	1	1	0

to be reflective of what we believe to be true about history, unless you think all vertebrates arose from a feathered ancestor.

Hennig's method of course wouldn't do this. Hennigian Argumentation begins with the enumeration of all of the groups of taxa that each character's apomorphic state(s) suggests is a group. The presence of an amnion suggests a group (Fig. 2a), but the absence of an amnion does not (Fig. 2b), just as each derived state for each character suggests a group (Fig. 2c). Each of the derived states then is a putative synapomorphy. Putative because, as we will see, it may turn out that we cannot reconcile all of the groups in Figure 2 with each other. Prior to there being computational methods that would solve this reconciliation of the various putative homology statements seen in Tables 1 and 2, workers had to proceed by hand. The easiest way to think of this is to start with an

Table 3   Similarity matrix

	perch	coelacanth	salamander	frog	turtle	human	gecko	snake	alligator	budgy
perch		85	62	62	69	54	54	62	54	23
coelacanth			77	77	85	69	69	77	69	38
salamander				100	77	77	54	46	62	46
frog					77	77	54	46	62	46
turtle						85	85	76	85	54
human							69	62	69	69
gecko								92	85	54
snake									76	46
alligator										69
budgy										

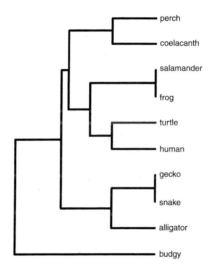

Figure 1   Phenogram of relationships based on absolute similarity.

unresolved tree and then, character by character, start resolving the various groups that each homology statement defines. To do this we need to reconcile as much as we can of the various groups in Figure 2. For example, groups 5 and 6 are identical; group 1 is a subset of those and is compatible with them, and so on. If you ever want to do this by hand, it generally pays to work from the largest groups to the smallest, and Figure 3 demonstrates this process of Hennigian Argumentation. Notice that the addition of characters 11 and 12 have no effect.

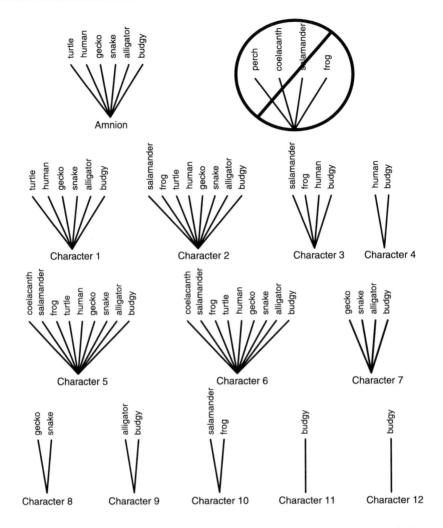

Figure 2 Presence of an apomorphic state (amnion) delimits a putative monophyletic group. Its absence (b) does not. Each character suggests a group based on the apomorphic state.

Being unique to the budgy, (autapomorphies) they are compatible with any topology. But notice too in the last two steps, that characters 3 and 4 have to be placed in more than one location on the eventual tree: scales have to be lost twice and warm bloodedness has to arise twice. In other words, in spite of seeking common causal explanations for all traits scored in the matrix, we find that it is not actually possible to do so for all traits simultaneously on any given tree. These changes in characters that must occur more than once are homoplasies. Homoplasies may take the form of independent origins for the same apomorphic state (as in characters 3 and 4) or may entail reversal back to a more plesiomorphic state. Of course, had the argumentation proceeded by adding these two characters earlier, we would have arrived at a different tree,

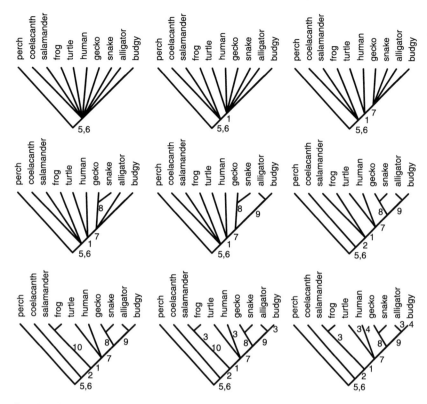

Figure 3   Hennigian argumentation proceeds by incorporating the groups delimited in Figure 2, one at a time until the tree is fully resolved.

one that suggests other characters have to be placed homoplasiously. How should one choose between the competing hypotheses? Inasmuch as we are already explaining such things as multiple taxa having an amnion being the result of descent from a common amniote ancestor (i. e., parsimoniously), again we choose on the basis of parsimony. That tree which invokes the fewest evolutionary transformations, like Newton's principle of admitting "no more causes of natural things than such as are true and sufficient to explain their appearances", is the hypothesis that is to be preferred. Edwards and Cavalli-Sforza [21] too agreed with Hennig that the "most plausible estimate of the evolutionary tree is that which invokes the minimum net amount of evolution".

3   Methods

Many data sets of morphological characters require far too much homoplasy, even on the most parsimonious tree, to permit constructing cladograms by eye effectively. Contemporary phylogenetics, with hundreds or thousands of nucleotides and dozens or even hundreds of taxa, make the Hennigian Argumentation approach all but impossible. In the early 1970s parsimony algorithms were developed that are equivalent to the Hennigian approach [22–25] and that later were incorporated into software packages [26–28]. Two fundamental approaches to counting numbers of steps on trees exist: additive [23] and nonadditive [25]. Briefly, additive characters are those in which the number of steps to go between assigned states is the difference in integer values. Thus, a change from a 0 to a 1 would be one step, a change from a 0 to a 2 would be 2 steps. Non-additive characters require only one step between each state, irrespective of the numerical state assignment: a change from 0 to 1 is one step, 1 to 3 is one step, a to g is one step, c to a is one step and so on. Additive and non-additive procedures also are known as ordered and unordered characters.

The nonadditive method of assessing a tree's length is very straightforward. These algorithms are unlike the Hennigian approach in that they don't build a tree from an unresolved starting point. They assess the cost of each character on a given tree. It may be any tree. Consider the aligned DNA sequences in the following matrix:

```
perch           CAAAAAAAAAA
frog            ACCCACCCACA
salamander      AAACACAGYCC
turtle          CCCTCACTCAC
man             CCCTCCA-CCC
bird            TCCTCCA-CCC
```

and, for now, just the first position, the first character. Fitch Optimization follows here for one possible tree. We can calculate the number of steps in what's called a "downpass" by first grabbing two adjacent terminals on the tree, moving to the node and asking whether or not there is an intersection set for the characters (Fig. 4). Beginning with frog and salamander, and moving to the node that unites them, there is an intersection of **A** and **A**. It is, of course, **A**, so we apply **A** to the node uniting the two taxa. Since there was an intersection, we do nothing to the length, which so far is 0 (zero). Repeating this process for man and bird, matters change. There is no intersection set for **C** and **T**, so we apply the union set to the node (**C,T**) and, because there is no intersection set, we add a length of 1 to the total length for this character. The length of the tree so far is one (1). Continuing down the tree from that node, the intersection of **C**, **T** and **C** is (of course) **C**, so no length is added and **C** is the state-set applied to that node. The length of the tree so far is still one. At the next-to-last node, the intersection set of **A** and **C** is empty (no intersection) so we add a length of 1 to the tree (now total length = 2 for this character), and apply the union set **A**, **C** to that node. Finally, moving out to the last taxon, the intersection set of **A**, **C** and **C** is **C**, so there's no added length to the tree. The total length for this character on this tree is then 2. Repeating this for all characters and summing the lengths

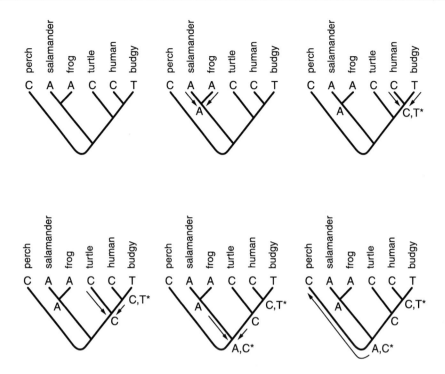

Figure 4   Fitch optimization (asterisks denote steps counted).

required for each of them allows determination of the length of this topology given all of the data.

Being able to determine the cost of a set of characters on a tree is useful enough, but it is not always the best method for arriving at the most parsimonious tree. With a mere 15 taxa there are almost 8 trillion possible trees. Evaluating each and every one of them would be extremely time-consuming. What is needed then is a way to get close to the most parsimonious tree early, and a way to get from there to better and better trees. Getting close to an efficient tree needs an efficient procedure for adding taxa to a growing tree like the Wagner Algorithm. Kluge and Farris [22, 23] showed that in the absence of any homoplasy, the Wagner Algorithm would indeed arrive at the most parsimonious tree. With homoplasy it will at least find an efficient tree. Once that tree is constructed, the length of the tree can be assessed as above and then used as a starting point for heuristic searches for ever better topologies.

Consider again the taxa, characters and states in Tables 1 and 2. We begin by creating a Manhattan distance matrix describing the number of character state differences between each pair of taxa (Tab. 4). We now can pick any two taxa to join. Normally this will be the two closest taxa; however, options in Paup permit connecting the furthest or just adding in the order of appearance. In this case we can choose either salamander-frog or snake-gecko as the closest pair (0 steps, see Fig. 5). Next we need to calculate the patristic distance between each

Table 4 Manhattan distances

	perch	coelacanth	salamander	frog	turtle	man	gecko	snake	alligator	budgy
perch		2	5	5	4	6	6	6	6	3
coelacanth			3	3	2	4	4	4	4	8
salamander				0	3	3	6	6	5	7
frog					3	3	6	6	5	7
turtle						2	2	2	2	6
man							4	4	4	4
gecko								0	2	6
snake									2	6
alligator										4
budgy										

of the yet-to-be-joined taxa to the existing interval connecting salamander and frog denoted INT (salamander:frog). This is simply the absolute value of the sum of the patristic distances of the taxon to be joined to each of the other taxa, less the patristic distance of the interval itself, all divided by two. For example, if we were to add perch to this internode, its distance (D) may be denoted

D[perch to INT(frog:salamander)] = |(perch:salamander + perch:frog – frog:salamander)|/2

or...

$$= |(5 + 5 - 0)|/2$$
$$= 5$$

Repeating this for all available taxa yields the revised reference matrix in Table 5. Coelacanth, turtle and human each have the same and shortest distances (3 steps) to the existing internode such that any of these could be added next. Ideally, you'd want to pursue all possible paths (e. g., mhennig in Hennig86), but here we'll just go with the first one that appears in the matrix (e. g., hennig in Hennig86) and continue adding taxa, admitting that a different order of taxa may lead to a different starting tree. Coelacanth is added to the tree and a hypothetical ancestor (X) is placed at the node where it is connected. Now we need to repeat the process and find the shortest patristic distance among all of the possible additions of other taxa to each of the three intervals

salamander:X
X:frog
X:coelacanth

A

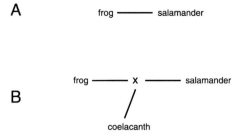

Figure 5 Stepwise addition using the Wagner Algorithm.

B

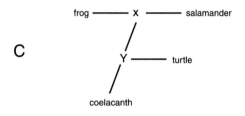

C

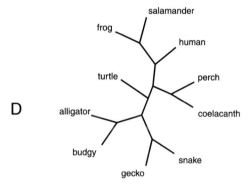

D

Table 5 Revised reference matrix

	perch	coelacanth	turtle	man	gecko	snake	alligator	budgy	Int[SalFrog]
perch		2	4	6	6	6	6	3	5
coelacanth			2	4	4	4	4	8	3
turtle				2	2	2	2	6	3
man					4	4	4	4	3
gecko						0	2	6	6
snake							2	6	6
alligator								4	5
budgy									

but to do this we need to get the state-set for X and incorporate it in the matrix. This is merely the median state for the three closest nodes, in this case,

coelacanth	0 0 0 0 1 1 0 0 0 0 0 0 0
salamander	0 1 1 0 1 1 0 0 0 1 0 0 0
frog	0 1 1 0 1 1 0 0 0 1 0 0 0
X	0 1 1 0 1 1 0 0 0 1 0 0 0

Notice that in this case, the state set for X is the same as for salamander and for frog. We must now calculate the patristic distances of each taxon not yet joined to each of the 3 intervals.

For perch,
    $D(Int\ salamander:X) = |5+5-0|/2 = 5$
    $D(Int\ frog:X) = |5+5-0|/2 = 5$
    $D(Int\ coelacanth:X) = |2+5-3|/2 = 4$
For turtle,
    $D(Int\ salamander:X) = |3+3-0|/2 = 3$
    $D(Int\ frog:X) = |3+3-0|/2 = 3$
    $D(Int\ coelacanth:X) = |2+3-3|/2 = 1$
For man,
    $D(Int\ salamander:X) = |3+3-0|/2 = 3$
    $D(Int\ frog:X) = |3+3-0|/2 = 3$
    $D(Int\ coelacanth:X) = |4+3-3/2 = 2$

and so on. In this case, D [turtle: Int(coelacanth:X)] = 1 is among a few possible shortest next-additions (1 step), so we add turtle to the branch between X and coelacanth, adding a new hypothetical ancestor between them (Y). Without further belaboring this point, we continue this process stepwise until all of the taxa are added and there is a fully resolved tree with all of the taxa (Fig. 5).

Unfortunately, when faced with considerable homoplasy and many taxa, the Wagner algorithm, though it arrives at a relatively short tree, may not arrive at the shortest tree. Moreover, it will arrive at only one tree when there may be multiple equally optimal trees of different shape. In principle it is possible to enumerate all possible trees for a set of taxa and enumerate their lengths, but for more than 10 or 15 taxa this is computationally unfeasible. Trillions of trees would take far too long to evaluate. Working from the principle that trees of similar length have similar shape, there are tree searching algorithms which can take a starting tree and work from its structure to find shorter and shorter trees if they exist.

4   Searching

It is possible with small datasets to evaluate all possible tree topologies. This is done, for example, by adding taxa to the growing tree in all possible locations. However, for more than 10 taxa, exhaustive searching would require evaluating billions of trees.

Specifically, where the number of taxa $t = 4$, there are 3 unrooted trees. The number of possible trees rapidly increases with increasing t. In general, the number of bifurcating rooted trees for t species is given by

$$(2t - 5)!/[2^{t-3}(t - 3)!]$$

for $t > 1$ [29]. This indicates that when $t = 10$, the number is more than two million.

A short-cut around evaluating all possible trees is to enumerate these trees implicitly. If we take as a maximum length, the length obtained from a starting tree, such as the one obtained from the Wagner algorithm, we may be certain that, should the addition of some taxon in a particular place require a longer tree, placement for that taxon and all other larger trees like it need not be evaluated, since adding more taxa with that taxon in that place cannot make the tree shorter. Still, though, this can take a very long time and there still might be "billions and billions" of trees to evaluate. Usually you will be constrained to search with one of the following heuristic methods.

4.1   Nearest-neighbor interchange

Consider the seven-taxon tree with taxa A through G (Fig. 6). It is drawn unrooted. Each internal branch on a tree (e. g., the horizontal one) has 4 attached branches that are each other's nearest neighbors. Consider the branch leading to "C". It is originally connected to the tree such that its nearest neighbor branch is the one leading to (A B). There are two possible nearest neighbor interchanges (arrows) for C around the central branch. The example given shows one such interchange, C with (F G). For any internal branch there are 4 possible nearest neighbor interchanges (NNI). For this tree, then, there are 16 possible nonredundant rearrangements using this method. If one or more of those 16 trees is shorter than the starting tree, we discard the starting tree, keep the shortest and continue the procedure until there is no length improvement.

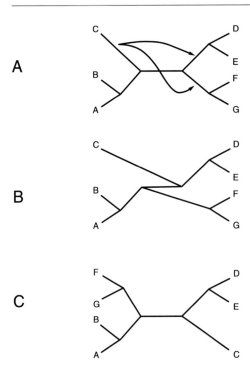

Figure 6   Nearest Neighbor Interchange (NNI) branch swapping.

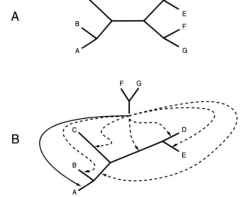

Figure 7   Subtree Pruning Regrafting (SPR) branch swapping.

4.2   Subtree pruning regrafting

Any branch on the tree can be "cut" off, or pruned to create a subtree (Fig. 7). Pruning off **(F G)** for example, leaves a dangling root on the pruned portion that can then be reattached (arrows) to any other branch on the tree. There are as many rearrangements possible as there are branches in the other portion of the tree, one of which is the same as the original tree. For this particular tree, there are approximately 70 rearrangements possible using this method (SPR) of branch swapping. In the example given (solid line with arrow), the subtree is regrafted to the rest of the original tree on the branch leading to **A**. Again, if any of the trees examined is shorter than the starting tree, it is retained and the procedure is repeated until no improvement in tree length can be found.

4.3   Branch-breaking (a.k.a. tree bisection reconnection)

As with SPR, this method can break the tree on any branch. However, then the two subtrees are each considered rootless (Fig. 8), following which, any two branches in the subtrees can be connected. In the example, the tree is split into the two subtrees **((A B) C)** and **((D E) (F G))**, then the branch leading to **A** is connected to the branch leading to **E**. Obviously, there are many, many more ways to rearrange this tree using branch-breaking (or TBR) than there were with NNI or SPR. More extensive methods of rearrangement will take more computational time. However, more extensive methods also will examine broader ranges of related tree shapes and should be expected to do a better job of finding shorter trees. With the speed of today's processors there seems to be no reason to bother with anything other than TBR branch-breaking. Repetitive procedures like bootstrapping and jackknifing (see chapter 5) might use simpler swapping routines (or no swapping at all) depending on the amount of speed and tolerance for error in estimates that can be tolerated. Well-supported clades should appear with marginal rearrangements in either of these resampling regimes.

4.4   The problem of local minima

Because these are heuristic searches, and not exact, it is possible that the order in which the taxa appear in your matrix can bias the direction of the outcome. Recall that there was some order-dependence associated with the step-wise construction in the Wagner algorithm. Very early we could have added either coelacanth, turtle or human to the growing tree. We chose coelacanth only because it was the first to appear in the matrix. Remember too that branch-swapping procedures examine families of trees that are similar in shape (being

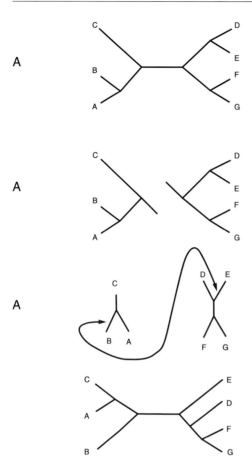

Figure 8 Branch-breaking (a.k.a. TBR branch swapping).

different only by the rearrangements tried). It is possible that the overall most parsimonious tree (the global optimum) has a markedly different shape than the ones achieved by stepwise addition. These are so different, in fact, that none of the rearrangements tried gets close enough to allow finding the globally shortest tree. This amounts to arriving at a locally optimal solution which may not be as short as the globally optimal solution.

Suppose we could arrange trees in such a way that trees of similar shape were adjacent to each other in one dimension (the horizontal in Figure 9). The reality is more complicated and would require multidimensional space, but the analogy should serve. The dot depicts the position in this tree space achieved by a single order-dependent stepwise addition (Figure 9A). From this point, swapping in one direction finds longer trees, swapping in the other quickly finds shorter ones. NNI swapping should, for example, get to the bottom of the peninsula immediately to the right of the dot, but may not get over the next hump. Branch-breaking, on the other hand, should get over that minimal bay and down to the next peninsula to the right. However, for even branch-breaking to escape from that position would require holding on to trees that are

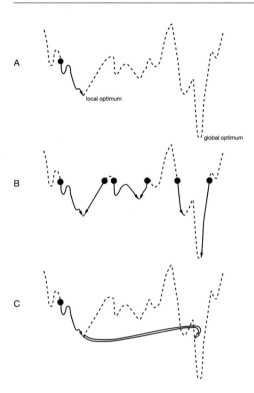

Figure 9 Conceptual tree space with local minima and a global optimum.

considerably longer and it is unlikely to do so. Both methods are likely to get stuck out on a local peninsula that is not the global optimum for the tree space.

Examining this tree space, there are two ways of circumventing this problem (Fig. 9B,C). If we had started with a different tree of the same length but a different shape we might start closer to the global optimum. There are 7 trees in this tree space that have the same length as the one we originally started with (Figure 9B). Two of them are in the peninsula of the global optimum. A way to acquire these multiple starting points is to re-do the analysis with the taxa added in randomly determined orders. This allows you to start in different places in tree-space, and thus increases your chances of (but does not guarantee) arriving at the globally optimal solution. The second way to avoid getting stuck in local minima, suggested by Kevin Nixon, is called the parsimony ratchet [30]. Instead of creating multiple starting points for branch swapping, a search procedure is allowed to hold on to suboptimal trees in order to get over to other peninsulas. By slightly changing the weights of randomly selected characters, researching under these weights, holding those trees for further swapping, resetting the characters back to the original (e. g., equal) weights and swapping again, the ratchet accomplishes this feat. In a sense, the random reweighting warps tree space and the effect of the ratchet is to wormhole the search progress from one peninsula to another without having to circumnavigate the whole coastline (Figure 9C). Newer methods are available with TNT which prove to be faster and even more efficient at finding peninsulas of short trees [31].

Which method should be used and how many replicates should be conducted depends on your patience and the peculiarities of the data. The ratchet is not implemented in PAUP. Both Nona and PAUP have options for random taxon addition search sequences. Arguably conducting 10 of these with TBR branch swapping is a good start. If all 10 searches arrive at the same optimal solution there's likely to be no point in continuing to look. However, if the most optimal solution found is found only once or twice, repeating with 30 or 50 random additions might be wise, or 10 replicates of the ratchet as an alternative.

5   Parsimony Analysis Using Nona (Version 1.9)

Your data need to be in one of two formats. The standard format with character states represented as integers 0 through 9 looks like this:

```
xread
'this is a title'
10 5
one   0000000000
two   1110111010
three 0010103111
four  1121012101
five  1121103111
;
proc/;
```

The "xread" statement tells Nona that this is a data matrix procedure, this is followed by an optional title, the number of characters, the number of taxa and then the matrix. The end of the matrix-statement is a semicolon, the procedure-off "proc/;" statement ends the procedure and restores control to you after the file is processed.

A second format accepts nucleotide sequence data with IUPAC ambiguity codes:

```
dread
gap?
match .
one   aaaaaaaaaa
two   cccacccaca
three aacacagycc
four  cctcactcac
five  cctcca-ccc
;
proc/;
```

The dread directive indicates that these are nucleotide data, the gap statement may take one of two forms: gap? or gap; indicating that gaps should be treated as missing or as a fifth state, respectively. The match statement denotes the symbol used as a substitute for states matching those in the first taxon at the top. Such as,

```
dread
gap?
match .
one aaaaaaaaa
two ccc.ccc.c.
three ..c.c.gycc
four cctc.ctc.c
five cctcc.-ccc
;
proc/;
```

Interleaved formats are not supported by Nona. Execution in Nona can be menu-driven by typing 'menu' at the prompt

> menu

whereupon one is presented with a user-friendly pull-down menu:

```
__NONA_____
|File/Data   f |
|Trees       t |
|Searches    s |
|Options     o |
|  _____  |
```

You may also use Nona's command-line interface (esp. for versions 2.0 and higher) and the equivalent commands will be provided here. Only suitable options are available at any given time using the menu such that the user can't go down blind alleys. For example, with no data set in memory, the Trees and Searches options are disabled. Getting around the menu is easy too. You can either scroll through it with the up or down arrow keys and hit return when the appropriate option is highlighted (such as File/Data), or hit the appropriate key (e. g., f), or you can double click on an item with your mouse. Double clicking on File/Data reveals the next menu:

```
_File/Data _____
|Trees        |Open input file    f |
|Searches     |Open outfile       o |
|Options      |Close outfile  alt-o |
| _____   |Tree save file     s |
             |Execute file       e |
             |                     |
             | _____ |
             |Pack data          p |
             |Save/change data   d |
             |Command line       1 |
             |                     |
             | _____ |
             |Exit               x |
             | _____ |
```

Hitting <esc> will send the user back up to the previous menu. Scrolling to
"Open input file" or hitting an "f" will prompt the user for an input file.
Alternatively, at the command-line prompt ">" you can type "procedure"
followed by the filename.

The "Command" line option above will allow a standard "ccode" for making
characters additive, nonadditive, active, inactive or weighting, as will the
Options... Character... menu:

```
_NONA_____
|File/Data f |
|Trees     t |
|Searches  s |
|Options     _____
| _____|Characters          _____
           |Max. trees          |MAKE ..._____ |
           |Max. partial trees  | active              a  |
           |Tree collapsing     | inactive        alt-a  |
           |Neighborhoods       | additive            d  |
           |Tree comparing      | nonadditive     alt-d  |
           |Pauses on           | _____   |
           |Define constraints  |Weight               w  |
           |Show/save constr's  |Show min/max steps   m  |
           |Clear screen        |Show status          s  |
           |Estimates           | _____   |
           |Stopwatch on  alt-w |
           |User defined      u |
           |No menu           n |
           | _____ |
```

For molecular data, all characters must be nonadditive. This can be easily accomplished by including a ccode – .; statement after the data matrix and before the proc/; statement in your data file. Otherwise you can type 0. after choosing the nonadditive option in the menu above. Note that all items are numbered from 0 in Nona. The dread file-type automatically makes all characters non-additive. You may also type "ccode – ." at the ">" prompt.

Among the most compelling options in the new Nona are the array and simplicity of search options:

```
_NONA_____
|File/Data   f |
|Trees       t |
|Searches      _____
|Options       |Initial_____|
|  _____  |Wagner tree(s)        w |
              |Random adds_SPR        s |
              |Random adds_TBR  alt_s |
              |Swap trees_____|
              |SPR_swapping          b |
              |TBR_swapping      alt_b |
              |Exact solution        e |
              |Mult. cut swapping    m |
              |Find N trees          f |
              |Others_____|
              |Ratchet               h |
              |Quick consensus       q |
              |Quick support     alt_q |
              |Random seed           r |
              |Suboptimal            u |
              |Union construct       k |
              |_____|
```

Unlike other contemporary software, random taxon addition search sequences with (TBR) branch=breaking can be accomplished with a simple key stroke ("alt=s" above) or a mouse click (the default is 10 random additions). From the command line, typing "mult*10;" will accomplish the same procedure. The asterisk indicates TBR swapping (instead of SPR) and the 10 dictates the number of random additions.

As a search proceeds, the user is occasionally told how matters are progressing:

```
Replication 1, 44 steps [seed=1]
Replication 2, 44 steps [seed=19318]
Replication 3, 44 steps [seed=14667]
Replication 4, 44 steps [seed=995]
Replication 5, 44 steps [seed=23277]
Replication 6, 44 steps [seed=9088]
Replication 7, 44 steps [seed=8921]
Replication 8, 44 steps [seed=14464]
Replication 9, 44 steps [seed=15528]
Replication 10, 44 steps [seed=20185]
1 trees of steps=44 retained (*)
Swapping of trees found by 'mult' not necessary
```

and you're also told whether or not additional swapping is fruitless (last line above).

Viewing your tree is easy too, with the Trees... option:

```
_NONA_____
|File/Data   f |
|Trees          _____
|Searches  |View trees           t |
|Options   |Make current         r |
|          |Consensus            c |
|  _____|Diagnosis            d |
           |Evaluation           e |
           |Save (trees)         s |
           |Save (parents')      p |
           |Tree file            f |
           |Edit tree         alt-e |
           |                        |
           | _____|
```

(or typing "vplot" from the command line) which will provide a tree in extended ASCII characters (numbers on the tree are merely node numbers):

```
Tree 0
  __ ₀ ROOT
  |  _₂₇__₁₄ Flam_ng
     |  _₂₆__₂₂__₁₃ Chil_sere
        |     |  _₂₁__₁₁ Chil_cald
        |     |     |  _₁₂ Chil_puer
        |  _₂₅__₃ L_gajardoi
           |  _₂₄__₆ L_epipiptu
              |  _₂₀__₇ L_domeyko
              |     |  _₁₉__₈ L_quilicu
              |     |     |  _₉ L_tofo
              |  _₁₈__₁₇__₅ L_caldera
                 |     |  _₁₀ L_frayjor
                 |  _₁₆__₄ L_longipes
                    |  _₁₅__₁ D_ornatus
                       |  _₂ D_bonarien
```

If there are multiple trees in memory, a consensus can be calculated: that is, a depiction of only those clades that are found in all of the most parsimonious trees. The command line option for this is "nelsen;". Users should carefully read the documentation in Nona.doc and especially in Piwe.doc both for more details and for warnings and idiosyncrasies.

References

1   Sober E (1983) Parsimony methods in systematics – philosophical issues. *Annu. Rev. Ecol. Syst.* 14: 335–357

2   Sober E (1988) *Reconstructing the past: Parsimony, evolution, and inference*. A Bradford Book, MIT Press, Cambridge, MA

3   Sober E (1994) *From a biological point of view: Essays in evolutionary philosophy*. Cambridge Univ. Press, Cambridge

4   Felsenstein J (1978) Cases in which parsimony or compatibility methods will be positively misleading. *Systematic Zoology*, 27: 401–410

5   Goldman N (1990) Maximum likelihood inference of phylogenetic trees, with special reference to a Poisson process model of DNA substitution and to parsimony analyses. *Systematic Zoology*, 39: 345–361

6   Yang Z (1993) Maximum likelihood estimation of phylogeny from DNA squences when substitution rates differ over sites. *Molecular Biology and Evolution*, 10: 1396–1401

7   Yang Z (1996) Phylogenetic analysis using parsimony and likelihood methods. *Journal of Molecular Evolution*, 42: 294–307

8   Zharkikh A, Li WH (1993) Inconsistency of the maximum parsimony method: the case of five taxa with a molecular clock. *Systematic Biology* 42: 113–125

9   Hillis DM and Huelsenbeck JP and Swofford DL (1994) Consistency: Hobgoblin of phylogenetics? *Nature*, 369: 363–364

10 Takezaki N, Nei M (1994) Inconsistency of the maximum parsimony method when the rate of nucleotide substitution

is constant. *Journal of Molecular Evolution*, 39: 210–218

11 Tateno Y, Takezaki N, Nei M (1994) Relative efficiencies of the maximum- likelihood, neighbor-joining, and maximum-parsimony methods when substitution rate varies with site. *Molecular Biology and Evolution*, 11: 261–277

12 Swofford DL, Olsen GJ, Waddell PJ, and Hillis DM (1996) Phylogenetic inference. Pp. 407–514, In: DM Hillis, C Moritz and BK Mable (eds.), *Molecular systematics*. Sinauer Associates, Sunderland, Massachusetts

13 Huelsenbeck JP (1997) In the Felsenstein zone a fly trap? *Syst. Biol.* 46:69–74

14 Farris JS (1983) The logical basis of phylogenetic analysis. In: NI Platnick, VA Funk (eds.), *Advances in cladistics* II. Columbia Univ. Press, New York, 7–36

15 Siddall ME (1998) Success of parsimony in the four-taxon case: Long-branch repulsion by likelihood in the Farris Zone, *Cladistics* 14: 209–220

16 Wegener A (1924) The Origin of Continents and Oceans. Methuen & Co. London, p. 212

17 Hennig W (1950) *Grundzüge einer Theorie der phylogenetischen Systematik*. Deutscher Zentralverlag, Berlin 370p

18 Hennig W (1965) Phylogenetic systematics. *Annu. Rev. Entomol.* 10:97–116

19 Hennig W (1966) *Phylogenetic systematics*. Univ. Illinois Press, Urbana

20 Farris JS, Kluge AG (1997) Parsimony and history. Syst. Biol. 46: 215–218

21 Edwards AWF, Cavalli-Sforza LL (1963) The reconstruction of evolution. *Heredity*, 18: 553 [abstract]

22 Kluge AG, Farris JS (1969) Quantitative phyletics and the evolution of anurans. *Syst. Zool.* 18: 1–32

23 Farris JS (1970) Methods for computing Wagner Trees. *Syst. Zool.* 19: 83–92

24 Farris JS, Kluge AG, Eckardt MJ (1970a) A numerical approach to phylogenetic systematics. *Syst. Zool.* 19: 172–189

25 Fitch WM (1971) Toward defining the course of evolution: Minimum change for a specific tree topology. *Syst. Zool.* 20: 406–416

26 Farris JS (1985) *Hennig 86*. Port Jefferson Station, New York

27 Swofford DL (1991) PAUP Phylogenetic Analysis Using Parsimony, version 3.1.1. Champaign: Illinois Natural History Survey

28 Goloboff PA (1996) NONA- Bastard son of PiWe with No Name

29 Cavalli-Sforza LL, Edwards AWF (1967) Phylogenetic analysis: Models and estimation procedures. *Evolution*, 21:550–570

30 Nixon K (1999) The parsimony ratchet, a new method for rapid parsimony analysis. *Cladistics* 15: 407–414

31 Goloboff PA (1999) Analyzing large data sets in reasonable times: Solutions for composite optima. *Cladistics* 15: 415–428.

Optimization Alignment: Down, Up, Error, and Improvements

Ward C. Wheeler

Contents

1 Introduction

Optimization Alignment (OA) is a method for taking unaligned sequences and creating parsimonious cladograms without the use of multiple alignment. The method consists of two parts. First, a "down-pass" that moves "down" the tree from the terminal taxa (tips) to the root or base of the cladogram and, second, an "up-pass" which moves back up from the base to the tips. The down-pass creates preliminary (i. e., provisional) hypothetical ancestral sequences at the cladogram nodes and generates the cladogram length as a weighted sum of the character transformations (nucleotide substitutions and insertion-deletion events) required by the observed (terminal) sequences. The up-pass takes the information from the down-pass and creates the "final" estimates of the hypothetical ancestral sequences. From these the most parsimonious synapo-morphy scheme can be derived to show which character transformation events characterize the various lineages on the tree. The combination of these two procedures allows phylogenetic searches to take place on unaligned sequence data, resulting in improvements in execution time and quality of results. This process differs from multiple alignment procedures (such as that of Sankoff and Cedergren [1]) in that OA attempts to determine the most parsimonious cost of a

Methods and Tools in Biosciences and Medicine
Techniques in molecular systematics and evolution, ed. by Rob DeSalle et al.
© 2002 Birkhäuser Verlag Basel/Switzerland

cladogram directly, whereas multiple alignment procedures generate column-vector character sets, which are then analyzed phylogenetically in a separate operation.

This OA method was proposed to allow the direct optimization of unequal length sequences on a cladogram [2]. The determination of the length, or cost, of the cladogram is accomplished given the observed sequences and a cost matrix that specifies the costs of all the transformations among nucleotides and insertion-deletion (indel) events. In this sense, the method is a generalization of Sankoff and Rousseau [3], or matrix optimization of character states, but allows for the insertion and deletion of characters. Sankoff and Rousseau [3] expanded the realm of optimization allowing for unequal transformation costs among character states, but still relied on a preexisting alignment to know which states to compare. OA enlarges the world of transformation events that can be optimized on a cladogram, including the creation and destruction of characters. In doing this, OA obviates the need to perform multiple sequence alignment, creating unique, topology-specific homology regimes for each scenario of historical relationship. The method yields better testing of phylogenetic hypotheses since provisional homologies are optimized for each cladogram individually, not *a priori* and universally as with the static homologies of multiple alignment. Furthermore, by treating all sequence variation within the context of topology-specific synapomorphy, hypotheses of molecular variation can be seamlessly integrated with other character variation to yield simultaneous or total evidence analysis. The OA integration of molecular and other character information frequently generates more parsimonious cladograms than multiple sequence alignment [4]; these results often show greater congruence among data sets [5].

Wheeler [2] defined and illustrated OA for a simple case of short sequences, determining the length of a cladogram in terms of nucleotide transformations and indels. Here, I review the method in more detail. First the "down-pass" or initial tree length determination is described in detail, and then the "up-pass" or internal-node sequence reconstruction procedure. Since these procedures yield approximations of the minimal length cladogram (the exact solution is thought to be NP-complete), errors and approximations are introduced and their behavior and techniques for accommodating them are described.

2   Going Down to Get Tree Length

The initial step of any phylogenetic optimization procedure is the "down-pass." The procedure begins at the tips of the tree with the terminal taxa, and moves down through the hypothetical ancestral nodes to the base or root of the tree. This initial or preliminary traverse through the cladogram yields both preliminary character state assignments to the internal (i. e., ancestral) nodes and the cladogram length or cost.

Whether the characters are morphological additive or non-additive characters, unordered or matrix molecular variants [3, 6, 7], the basic operation
begins by choosing an internal node whose two descendants are terminal taxa
(e. g., T1 and T2; Fig. 1). The down-pass character state assignments (preliminary of node A4) are determined from the two descendants and minimize
the amount of change between the ancestor and its two descendants (T1, T2
and A4). This process is repeated until all the internal nodes have been visited
(A4-A1) using the relevant ancestral preliminary states as descendants for more
basal ancestors (e. g., A4 and T2 for A3). The overall cost of the cladogram is the
sum of the costs incurred in determining each ancestral sequence.

Optimization procedures differ in the determination of the ancestral state
reconstructions and how their costs are determined. For simple binary or non-
additive multistate characters, union and intersection operations are performed
on the descendant character states. If the two descendants states agree (identical)
or have common states (non-empty intersection), no cost is incurred and the nodal
reconstructed state is the identical or overlapping state in the two descendants. If
the descendant states disagree (empty intersection), the ancestral state receives
the union of the descendant states and the cost of that reconstruction is increased
by one (e. g., T1 {A} and T2 {C} to yield A4 {M}; Fig. 2).

The general case of multiple states, linked by arbitrary (but metric) transformation cost matrices, can be determined by trying (at least implicitly) all
possible states at all internal nodes and choosing the combination which yields
the most parsimonious result (Fig. 3). As with the simple method mentioned
above, the cost of the determination of ancestral nodes at each node is summed
to get the length of the entire cladogram for that character. For the Sankoff
procedure to function, however, all possible internal states must be known and
defined *a priori*, which allows them all to be considered. In most situations (e. g.,
nucleotide data), this is straightforward, with only five (A, C, G, T and gap) states
possible. Methods have been proposed where the number of possible states can
be arbitrarily large (see [8]), but these character states can still be optimized via
the same process as for the five states of nucleotide data. OA generalizes this

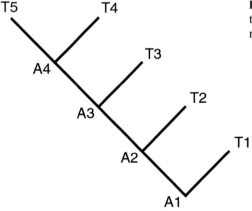

Figure 1 Example cladogram with five
terminal taxa (T1-T5) and four internal
nodes (A1-A4).

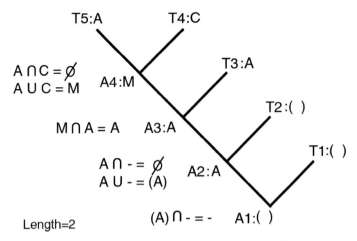

Figure 2 Fitch [7] down-pass for non-additive, or unordered DNA characters with 5 states. IUPAC codes are used to represent nucleotide ambiguity. Parentheses denote the absence of nucleotides at that position (i. e., "gaps").

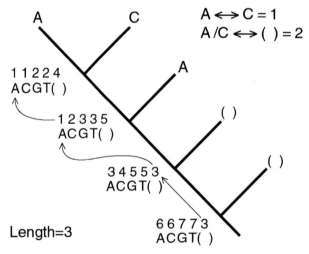

Figure 3 Sankoff and Rousseau [3] optimization for general character transformation matrices. IUPAC codes and parentheses as in figure 2.

procedure by allowing the creation and destruction of characters, but employing a heuristic character optimization algorithm (Heuristic Sankoff Cost (HSC) or Sankoff procedure) to make the calculations tractable.

The Sankoff procedure, though exact, is time-consuming. If there are "s" states, "n" taxa and "m" characters, the cost of determining the length of a cladogram via the exact procedure would be proportional to $2s^2m(n-1)$ (but short-cuts exist; see [9, 10]). A non-additive character has no dependence on the number of states (i. e., would depend only on m and n), much less its square, and can be implemented with much more efficient bitwise operations. An approximate solution can be found, however, by making a simplifying assumption and performing a simple weighted version of non-additive optimization. The simplifying assumption is that we only have to worry about the immediate descendant states. If this is the case, then we can precalculate the outcome of all possible descendant state pairs. For five states (A, C, G, T, and gap) there are 31 possible combinations of states a descendant can exhibit (five single states and 26 combinations). Hence, there are only 961 possible events that can occur at an ancestral node. Furthermore, there are 31 cases where the two descendants are identical; the remainders are not order-dependent so there are only 465 calculations that are required. Each of these would result in a preliminary ancestral state assignment and a cost (the minimum cost transformation implied by the descendant states). These are calculated and stored in a 31×31 table with the descendant states as indices before any optimization or search takes place (Fig. 4). During optimization, the results of every ancestral optimization would just be looked up in the table (Fig. 5). The method is approximate, it ignores intermediate, locally sub-optimal solutions that might be globally more parsimonious later, but can reduce execution time considerably. Solutions can be checked by performing a complete Sankoff down-pass on those rare occasions during a search where a candidate tree is thought to be equally or more parsimonious than the current best [10]. This heuristic procedure was first used in MALIGN [11, 12] and is used in PHAST [10] and POY [13].

The OA procedure relies on a combination of HSC and the Needleman and Wusch (NW) pairwise alignment procedure [14]. The three types of optimization discussed above (Farris, Fitch, and Sankoff), assume that the descendant characters to be compared or optimized are known. In other words, previous optimization schemes assume that it is known which "A" on one sequence corresponds to which "C" in another. This is generally not the case with nucleotide sequence data sets, since the sequences of terminal taxa can vary in length. In other words, these optimization procedures require pre-aligned sequences. However, if a means can be found to create parsimonious preliminary ancestral sequences from two descendant sequences and determine the cost of that creation, the coupling of homology assessment and testing can be made seamless rendering multiple alignment unnecessary.

Determination of preliminary ancestral sequences is approached in the same way that Sankoff optimization looks at all the possible state assignments and chooses the most parsimonious, such that OA looks at all the potential correspondences between the nucleotides in the descendant sequences to determine which scheme of correspondences and transformations yields the most parsimonious preliminary ancestral sequence. This is done in a manner akin to the pair-wise alignment procedure of NW. In this case, the NW procedure is modified from one that maximizes sequence similarity to one that minimizes the cost of the ancestral sequence as approximated by the HSC.

Figure 4 Look-up table including hypothetical ancestral states and cost used in the Heuristic Sankoff Cost (HSC) procedure. IUPAC codes and parentheses as in Figure 2. An IUPAC code with parentheses denotes ambiguity with respect to that IUPAC nucleotide, or ambiguity, and the absence of base or "gap." These costs and states are based on a cost of 3 for indels, 2 for transversions and 1 for transitions.

	A	C	G	T	No Base	M	R	W	(A)	S	Y	(C)	K	(G)	(T)	V	H	(M)	D	(R)	(W)	B	(S)	(Y)	(K)	N	(V)	(H)	(B)	(D)	(N)
A	0	2	1	2	3	0	0	0	0	1	2	2	1	1	2	0	0	0	0	0	0	1	1	2	1	0	0	0	1	0	0
C	M	0	2	1	3	0	2	1	2	0	0	0	1	2	1	0	0	0	1	2	1	0	0	0	1	0	0	0	0	1	0
G	R	S	0	2	3	1	0	1	1	0	2	2	0	0	2	0	1	1	0	0	1	0	0	2	0	0	0	1	0	0	0
T	W	Y	K	0	3	1	2	0	2	1	0	1	0	2	0	1	0	1	0	2	0	0	1	0	0	0	1	0	0	0	0
No Base	(A)	(C)	(G)	(T)	0	3	3	3	0	3	3	0	3	0	0	3	3	0	3	0	0	3	0	0	0	3	0	0	0	0	0
M	A	C	R	Y	(M)	0	0	0	0	0	0	0	1	1	1	0	0	0	0	0	0	0	0	0	1	0	0	0	0	0	0
R	A	V	G	D	(R)	A	0	0	0	0	2	2	0	0	2	0	0	0	0	0	0	0	0	2	0	0	0	0	0	0	0
W	A	Y	R	T	(W)	A	A	0	0	1	0	1	0	1	0	0	0	0	0	0	0	0	1	0	0	0	0	0	0	0	0
(A)	A	M	R	W	No Base	A	A	A	0	1	2	0	1	0	0	0	0	0	0	0	0	1	0	0	0	0	0	0	0	0	0
S	R	C	G	Y	(S)	C	G	N	R	0	0	0	0	0	1	0	0	0	0	0	1	0	0	0	0	0	0	0	0	0	0
Y	H	C	H	T	(Y)	C	N	T	H	C	0	0	0	2	0	0	0	0	0	2	0	0	0	0	0	0	0	0	0	0	0
(C)	M	C	H	Y	No Base	C	V	Y	No Base	C	C	0	1	0	0	0	0	0	1	0	0	0	0	0	0	0	0	0	0	0	0
K	R	Y	G	T	(K)	N	G	T	R	G	T	T	0	0	0	0	0	1	0	0	0	0	0	0	0	0	0	0	0	0	0
(G)	R	S	G	K	No Base	R	G	R	No Base	G	B	No Base	G	0	0	0	1	0	0	0	0	0	0	0	0	0	0	0	0	0	0
(T)	W	Y	W	T	No Base	Y	D	T	No Base	T	No Base	T	No Base	T	0	1	0	0	0	0	0	0	0	0	0	0	0	0	0	0	0
V	A	C	G	Y	(V)	M	R	A	A	S	C	C	G	G	Y	0	0	0	0	0	0	0	0	0	0	0	0	0	0	0	0
H	A	C	R	T	(H)	M	A	W	A	C	Y	C	T	R	T	M	0	0	0	0	0	0	0	0	0	0	0	0	0	0	0
(M)	A	C	R	Y	No Base	M	A	A	(A)	C	C	(C)	N	No Base	No Base	M	M	0	0	0	0	0	0	0	0	0	0	0	0	0	0	0
D	A	Y	G	T	(D)	A	R	W	R	G	T	Y	K	G	T	R	W	A	0	0	0	0	0	0	0	0	0	0	0	0	0	0
(R)	A	V	G	D	No Base	A	R	A	(A)	G	N	No Base	G	(G)	No Base	R	A	(A)	R	0	0	0	0	0	0	0	0	0	0	0	0	0
(W)	A	Y	R	T	No Base	A	A	W	(A)	Y	T	No Base	T	No Base	(T)	A	W	(A)	W	(A)	0	0	0	0	0	0	0	0	0	0	0
B	R	C	G	T	(B)	C	G	T	R	S	Y	C	K	G	T	S	Y	C	K	G	T	0	0	0	0	0	0	0	0	0	0
(S)	R	C	G	Y	No Base	C	G	N	No Base	S	C	C	G	(G)	No Base	S	C	(C)	G	(G)	No Base	S	0	0	0	0	0	0	0	0	0
(Y)	H	C	B	T	No Base	C	N	T	No Base	C	Y	C	T	No Base	(T)	S	Y	(C)	T	No Base	(T)	Y	(C)	0	0	0	0	0	0	0	0
(K)	R	Y	G	T	No Base	N	G	T	No Base	G	T	No Base	K	(G)	(T)	G	T	No Base	K	(G)	(T)	B	(G)	(T)	0	0	0	0	0	0	0
N	A	C	G	T	(N)	M	R	W	A	S	Y	C	K	G	T	V	H	M	D	R	W	B	S	Y	K	0	0	0	0	0	0
(V)	A	C	G	Y	No Base	M	R	A	A	S	C	(C)	G	(G)	No Base	V	M	(M)	R	(R)	(A)	S	(S)	(C)	(G)	V	0	0	0	0	0
(H)	A	C	R	T	No Base	M	A	W	A	C	Y	(C)	T	No Base	(T)	M	H	(M)	W	(A)	(W)	Y	(C)	(Y)	(T)	H	(M)	0	0	0	0
(B)	R	C	G	T	No Base	C	G	T	No Base	S	Y	(C)	K	(G)	(T)	S	Y	(C)	K	(G)	(T)	B	(S)	(Y)	(K)	B	(S)	(Y)	0	0	0
(D)	A	Y	G	T	No Base	A	R	W	(A)	G	T	No Base	K	(G)	(T)	R	W	(A)	D	(R)	(W)	K	(G)	(T)	(K)	D	(R)	(W)	(K)	0	0
(N) or X	A	C	G	T	No Base	N	R	W	(A)	S	Y	(C)	K	(G)	(T)	V	H	(M)	D	(R)	(W)	B	(S)	(Y)	(K)	N	(V)	(H)	(B)	(D)	0

Note- (Base Code) denotes the IUPAC code for a base plus gap ambiguity

    Consider four sequences T1:"AA", T2:"A", T3:"GG" and T4:"G". T1 is defined *a priori* as the outgroup and the indel cost set to 2 and the base change cost to 1 (transition cost = transversion cost = 1). At least initially, assume a candidate tree (T1 (T2 (T3 T4))) (Fig. 6). As mentioned above, the process starts with a node, both of whose descendants are terminals. Here, that is the node with descendants T3 and T4 (A3). In order to determine the lowest cost preliminary hypothetical ancestral sequence, a NW-type procedure is performed with a cost matrix based on the analysis parameters mentioned above (Fig. 7). The NW

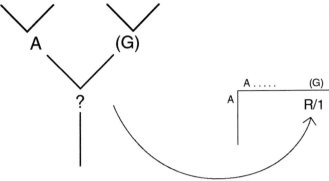

Figure 5 The use of the look-up table (Fig. 4) in the Heuristic Sankoff Cost procedure.

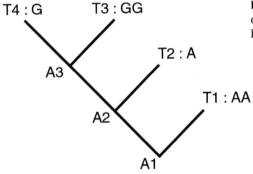

Figure 6 A simple example of four sequences of one to two bases, each related by a cladogram.

	A	C	G	T	( )
A	0	1	1	1	2
C	1	0	1	1	2
G	1	1	0	1	2
T	1	1	1	0	2
( )	2	2	2	2	0

Figure 7 Cost matrix of transformations used to diagnose the cladogram and sequences of Figure 6.

procedure minimizes the cost (in this case) of the nodal sequence by implicitly determining the cost of all possible preliminary reconstructions through a dynamic programming procedure. A matrix is set up and updated via "wavefront" optimization [14–16 and references therein]. This case requires a six cell node (n+1 by m+1, where n and m are the lengths of the descendant sequences) to consider the five possible homology schemes (Fig. 8). The result of this procedure is that there are two possible reconstructions of cost 2 (a single indel) using the HSC. The HSC is used not only to determine the cost of the reconstruction but the state as well. In this simple case, the preliminary ancestral sequence would be ambiguous as to the length (one or two bases) but one of those bases would be a "G" and the ambiguous length would be either a "G" or nothing. The "G" or nothing, represented as (G), is the result of the

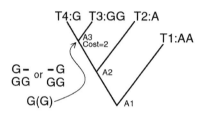

T4: - G
T3: GG   ⇒ G(G)
                                    Cost 2

T4: G -
T3: GG   ⇒ (G)G

T4: G - -
T3: - GG   ⇒ (G)(G)(G)

T4: - G -
T3: GGG   ⇒ (G)(G)(G)   Cost 6

T4: - - G
T3: G G -   ⇒ (G)(G)(G)

Figure 8 The determination of the preliminary sequence for node A3 of the cladogram in Figure 6. There are five possible paths through the matrix and five possible preliminary ancestral sequences.

lowest cost union (via HSC) between the corresponding descendant states of "G" and nothing (a gap if this were an alignment). An exact solution would require that we follow both of these possibilities (and all their multiplicative derivatives), but in the current implementation and description of the method a single preliminary hypothetical ancestral sequence is chosen.

After this node, the process is repeated for all the other unoptimized nodes. The node (A2) has descendants A3 and T2. The process is repeated as above with the descendant sequences (G)G and A. In this case, preliminary node reconstruction would yield "R" ambiguous with respect to A and G, but unambiguous as to sequence length (1) (Fig. 9). The NW and HSC yield a cost of 1 for this node (3 so far after A3 and A2 have been visited) and the ambiguity of length in A2 is resolved since the length of T1 was also 1. The final node (root node) is determined by comparing its descendants A1 ("R") and T1 ("AA"). Following the same NW-HSC process yields an ambiguous preliminary root node assignment of "A(A)" and "(A)A" (there were two paths to get there) with a local cost of 2 and a total cladogram cost of 5 steps (Fig. 10).

This completes the down-pass. Here the total cost (based on an indel cost of two and base change cost of one) was five with two indels and a single base change. In a search, other cladograms would be optimized and another solution of equal cost would have been found. For example, the cladogram (T1 (T3 (T2 T4))) also requires 5 steps, but with three base changes and a single indel.

An obvious conclusion of this dependence on the indel and base change costs is that the preferred (i. e., most parsimonious) cladogram may vary with the parameter values. For this example if the indel cost is increased by one to three, only the cladogram linking T2 and T4 is chosen (length 6), minimizing the now more costly indels. Alternatively, if the gaps cost is reduced by one to one, the cladogram linking T3 and T4 is favored (length 3), minimizing base changes.

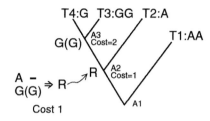

		T2				
	–		A	T2: A – T3: G(G)	R	Cost 1
–	0	→	2	T2: – A T3: G(G)	(G) R	Cost 3
A3 G	↓ 2	↘	1	T2: A – – T3: – G(G)	(A)(G)	Cost 4
(G)	↓ 2	↓	1	T2: – A – T3: G – (G)	(G)(A)	Cost 4
				T2: – – A T3: G(G) –	(G)(A)	Cost 4

Figure 9 The determination of the preliminary sequence for node A2 of the cladogram in Figure 6. There are five possible paths through the matrix and five possible preliminary ancestral sequences.

T4:G T3:GG T2:A

G(G) A3 Cost=2   T1:AA

A –
G(G) ⇒ R   R A2 Cost=1

Cost 1   A1

		A2				
	–		R	A2: R – A1: A A	⇒ A(A)	
–	0		2	A2: – R A1: A A	⇒ (A)A	Cost 2
T1 A	↓ 2	↘	0	A2: R – – A1: – A A	⇒ (R)(A)(A)	
A	↓ 4	↘↓	2	A2: – R – A1: A – A	⇒ (A)(R)(A)	Cost 6
				A2: – – R A1: A A –	⇒ (A)(A)(R)	

Figure 10 The determination of the preliminary sequence for node A1 of the cladogram in Figure 6. There are five possible paths through the matrix and five possible preliminary ancestral sequences.

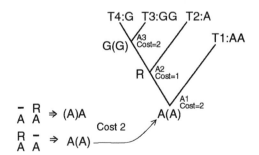

T4:G T3:GG T2:A

G(G) A3 Cost=2   T1:AA

R A2 Cost=1

– R ⇒ (A)A
A A

R – ⇒ A(A)
A A

Cost 2

A1 Cost=2

A(A)

3   Going Up to Get Ancestral States

In order to reconstruct a parsimonious set of character states at the internal (hypothetical) nodes, a second or up-pass is required. This process moves from the root of the cladogram "up" to the tips, incorporating the information from a node's ancestor as well as its descendants. As originally described [2] and implemented in POY [13], the process is extremely simple, basically trying all possible states (based on the down-pass homologies) in turn and keeping those with the lowest cost.

More specifically, the starting point is the root node. There is no true up-pass for this node, since it has no ancestor. The final states for the root node are simply assigned from the preliminary states or the final states of the outgroup taxon. The descendants of this node are then visited. These nodes are the first ones with both descendants and ancestors. The preliminary homologies among the preliminary down-pass states and the two descendant sequences are known (saved) from the down-pass step and the correspondences with the final states of its ancestor are determined by the same NW-HSC process used in the down-pass, but between the preliminary states of the node and the final states of its ancestor. For each position then, the two descendant and ancestral states are known. Each state is tried in turn as the final state, and the most parsimonious set taken as the final state set for that node (Fig. 11). The process then moves on to the next node and so on until final state sets are determined for each non-terminal node. The final state sets for the terminals and the root node are of course identical to the preliminary states.

Given that the final states are based on the same "greedy" short-sighted simplifications of the down-pass (locally lowest cost reconstructions), the final state sets are approximations as well. In addition to erroneously estimating

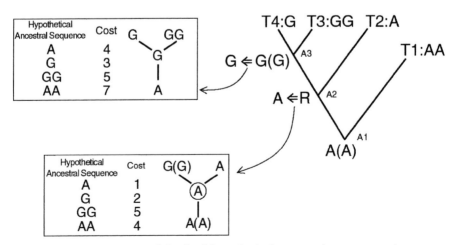

Figure 11   The determination of the final hypothetical ancestral sequences *via* an up-pass.

hypothetical ancestral sequences, this can cause problems with many phylogenetic search shortcuts, which rely on these hypothetical ancestral sequences as a surrogate for the information within their descendant clade.

4 Short-cuts and Errors

This form of sequence optimization makes two sorts of errors. The first concerns the down-pass tree lengths. Since the method determines local node costs based only on the descendant sequence information, the estimated cost must be equal to, or more likely greater than the minimal cost (Fig. 12). This effect can be compounded by the fact that the optimization regime is simultaneously determining homology relationships among the nucleotides as well from this same restricted (only descendant) sequence set. These two factors together ensure that the down-pass tree length is an upper bound on the minimal cost.

The second source of error comes from the establishment of the set of final (up-pass) states for the hypothetical ancestral sequences. Since the preliminary (down-pass) sequences are constructed in a myopic manner, these errors are carried along into the up-pass phase. This can result in both the inclusion of nucleotide states that should not be there, as well as missing those nucleotide states that should be there. When these reconstructions are used as part of short-cut methods during phylogenetic searches, this can cause the short-cuts to over- or underestimate candidate tree lengths. In general, these errors are

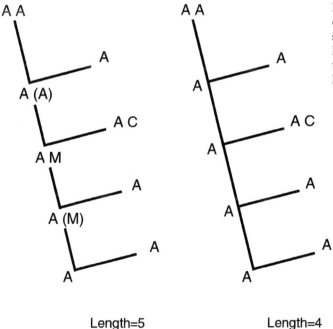

Figure 12 The down-pass calculation of length (left) showing how the value can be overestimated compared to the actual minimum length (right).

not large (less than 1%) and can be accommodated by checking tree length calculations with full down-pass optimization on candidate trees. In POY, the options "slop" and "checkslop" allow the verification of trees within a specified difference in tree length from the current best. Although this slows things down, it increases confidence that the search is proceeding on verified shortest trees.

Another expression of this effect is the apparent dependence of cladogram length on root position. This would seem to be counterintuitive, given that the parameters (indel cost, transition-transversion ratio, etc.) are symmetrical. As mentioned before [2], this is also due to the heuristic tree length procedures. In the example of Figure 12, if rooted at sequence "AA" instead of sequence "A", the minimal tree length of 4 is produced (Fig. 13).

5   Improvements

Many improvements could be made to these procedures. Three general classes would involve multiple solutions, sub-optimal solutions and character-specific virtual roots.

5.1   Multiple solutions

During the down-pass, the consideration of multiple equally parsimonious preliminary sequences would be an obvious step. Since each node would be likely to generate its own multiple solutions which should be multiplied down the cladogram, extremely large potential sets of preliminary sequences could be generated. Although it might not be practical to maintain multiple solutions throughout the entire down-pass, these solutions (or a set of them) could be maintained and considered at the optimization of the next (parent) node (or some other range down the cladogram). The NW-HSC process would be

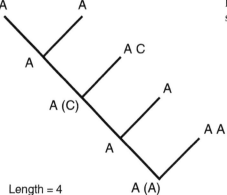

Figure 13   Rerooted diagnosis of Figure 13 showing the achievement of minimum length.

performed on each combination of the candidate preliminary sequences for the two descendants of a node. This would generate a new set of equally costly preliminary sequences for this node and the process would be repeated until the entire cladogram was optimized.

The up-pass determination of the final hypothetical ancestral sequences could also be improved by maintaining multiple equal cost solutions. Here would be the product of the multiple preliminary solutions and those of the final assignment of the parent node, which would be more computationally intensive, but surely tractable. The same decisions would be required on the size of the set of solutions to be maintained (this could be large, or a smaller random sample could be stored) and the reach over which to hold and test these multiple solutions.

5.2   Sub-optimal solutions

One of the principal reasons for the myopia of the down-pass (and up-pass) methods is that they ignore locally sub-optimal solutions which might turn out to be globally optimal. In the example of Figure 12, the assignment of pre-liminary sequence "A" is globally optimal, but ignored as too expensive initially. If some set of sub-optimal solutions were considered, they might prove useful in subsequent optimization stages. Unfortunately, there are many sub-optimal solutions and many of them are sub-optimal for good reasons. It is unclear whether this notion would prove practicable.

5.3   Virtual roots

As the examples of Figures 12 and 13 show, rooting can affect the behavior of the down- and up-pass algorithms. Certain characters might well behave better (i. e., generate shorter cladogram lengths) given certain roots. Unfortunately, it is unlikely that these roots will be identical for all characters. Perhaps a system of "virtual" roots could be erected with each character having its own "best" root, and then optimized on that basis. The overall cladogram would then be more explicitly unrooted and presented in a rooted fashion only at the end of the analysis.

6   Remarks and Conclusions

This explicit discussion of the procedures involved in the diagnosis of sequence data on cladograms shows both the strengths and weaknesses of this approach. Not requiring *a priori* sequence alignment and generating cladogram-specific homology schemes would seem to be strengths. The heuristic nature of the cladogram length and ancestral sequence reconstruction would seem to be weaknesses. These can be improved, however, as described above. Although the problem is unlikely to be solved exactly, improvements along the lines suggested here could well bring incremental benefits and, combined with ideas of others, generate more satisfactory methods and more reliable results.

Acknowledgments

I would like to acknowledge and thank the following for help with this manuscript: James Carpenter, Rob DeSalle, Rasmus Hövmoller, Gonzalo Giribet, Pablo Goloboff, Matthew Gruell, Daniel Janies, Susanne Schulmeister and Steven Thurston.

References

1 Sankoff DD, Cedergren RJ (1983) Simultaneous comparison of three or more sequences related by a tree. In D Sankoff, JB Kruskal (eds): *Time Warps, String Edits, and Macromolecules: the Theory and Practise of Sequence Comparison.* Addison-Wesley, Reading, MA, 253–264

2 Wheeler WC (1996) Optimization Alignment: the end of multiple sequence alignment in phylogenetics? *Cladistics* 12: 1–9

3 Sankoff DD, Rousseau P (1975) Locating the vertices of a Steiner tree in arbitrary space. *Math. Prog.* 9: 240–246

4 Wheeler WC (2000) Heuristic Reconstruction of Hypothetical-Ancestral DNA Sequences: Sequence Alignment versus Direct Optimization. In: R Scotland, RT Pennington (eds) *Homology and Systematics.* Systematics Society, London, 217 106–113

5 Giribet G (2001) Exploring the behavoir of POY; a program for direct optimization of molecular data. *Cladistics* 17: 560–570

6 Farris JS (1970) Methods for computing Wagner trees. *Syst. Zool.* 19: 83–92

7 Fitch WM (1971) Toward defining the course of evolution: Minimum change for a specific tree topology. *Syst. Zool.* 20: 406–416

8 Wheeler WC (1999) Fixed Character States and the Optimization of Molecular Sequence Data. *Cladistics* 15: 379–385

9 Goloboff PA (1994) Character optimization and calculation of tree lengths. *Cladistics* 9: 433–436

10 Goloboff PA (1998) Tree searches under Sankoff parsimony. *Cladistics* 14: 229–237

11 Wheeler WC, Gladstein DS (1994) MALIGN: A multiple sequence alignment program. *J. Hered.* 85: 417–418

12 Wheeler WC, Gladstein DS (1991–1998), *Malign. Program and documentation.* New York, NY. Documentation by Daniel Janies and Ward Wheeler

13 Gladstein DS, Wheeler WC (1997) *"POY: The Optimization of Alignment Characters." Program and Documentation.* New York, NY. Available at "ftp.amnh.org" / pub/molecular

14 Needleman SB, Wunsch CD (1970) A general method applicable to the search

for similarities in the amino acid sequence of two proteins. *J. Mol. Biol.* 48: 443–453

15 Kruskal JB and Sankoff D (1983) An anthology of algorithms and concepts for sequence comparison. In: D Sankoff and JB Kruskal (eds) *Time Warps, String Edits, and Macromolecules: the Theory and Practise of Sequence Comparison.* Addison-Wesley, Reading, MA, 265–310

16 Phillips A, Janies D and Wheeler WC (2000) Multiple sequence alignment in phylogenetic analysis. *Mol. Phyl. Evol.* in press

Techniques for Analyzing Large Data Sets

Pablo A. Goloboff

Contents

1 Introduction

Parsimony problems of medium or large numbers of taxa can be analyzed only by means of trial-and-error or "heuristic" methods. Traditional strategies for finding most parsimonious trees have long been in use, implemented in the programs Hennig86 [1], PAUP [2], and NONA [3]. Although successful for small and medium-sized data sets, these techniques normally fail for analyzing very large data sets, i. e., data sets with 200 or more taxa. This is because rather than simply requiring *more* of the same kind of work used to analyze smaller data sets, very large data sets require the use of *qualitatively different* techniques. The techniques described here have so far been used only for prealigned sequences, but they could be adapted for other methods of analysis, like the direct optimization method of Wheeler [4].

Methods and Tools in Biosciences and Medicine
Techniques in molecular systematics and evolution, ed. by Rob DeSalle et al.
© 2002 Birkhäuser Verlag Basel/Switzerland

2 Traditional Techniques

The two basic heuristic computational techniques for finding most parsimonious trees are wagner trees and branch-swapping. A *wagner tree* is a tree created by sequentially adding the taxa at the most parsimonious available branch. At each point during the addition of taxa, only part of the data are actually used. A taxon may be placed best in some part of the tree when only some taxa are present, but it may be placed best somewhere else when all the taxa are considered. Therefore, which taxa have been added determines the outcome of a wagner tree, so that different addition sequences will lead – for large data sets – to different results. *Branch-swapping* is a widely used technique for improving the trees produced by the wagner method. Branch-swapping takes a tree and evaluates the parsimony of each of a series of branch-rearrangements (discarding, adding, or replacing the new tree if it is, respectively, worse, equal, or better than previously found trees). The number of rearrangements to complete swapping depends strongly on the number of taxa. The most widely used branch-swapping algorithm is "tree bisection reconnection" or TBR ([5], called "branch-breaking" in Hennig86; [1]). In TBR, the tree is clipped in two, and the two subtrees are rejoined in each possible way. The number of rearrangements to complete TBR increases with the cube of the number of taxa, and thus the time needed to complete TBR on a tree of twice the taxa is much more than twice the time. Thus, if a tree of 10 taxa requires **x** rearrangements for complete swapping, a tree of 20 taxa will require **8x**, 40 taxa will require **50x** and 80 taxa will require **400x**. Because of special short-cuts, which allow deducing tree length for rearrangements without unnecessary calculations (see [6, 7] for basic descriptions, and [8], for a description of techniques for multi-character optimization), the rearrangements for larger trees can in many cases be evaluated more quickly than the rearrangements for smaller trees. Therefore, the time for swapping increases in those cases with less than the cube of the number of taxa (although it is still more than the square). In implementations which do not (or cannot) use some of these short-cuts, the time to complete TBR may well increase with the cube of the number of taxa (the use of some of the techniques described here, like sectorial searches and tree-fusing, would be even more beneficial under those circumstances).

For even relatively small data sets (i. e., 30 or 40 taxa), TBR may be unable, given some starting trees, to find the most parsimonious trees. In computer science, this is known as the problem of local optima (known in systematics as the problem of "islands" of trees; [9]) This is easily visualized by thinking of the parsimony of the trees as a "landscape" with peaks and valleys. The goal of the analysis is to get to the highest possible peak; this is done by taking a series of "steps" in several possible directions, going back if the step took us to a lower elevation, continuing from the new point if the step took us higher. Note that if the "steps" with which the swapping algorithm "walks" in this landscape are too short, it may easily get trapped in an isolated peak of non-maximal height. To reach higher peaks, the algorithm would have to descend and then go up again – but the algorithm does not do so, by virtue of its own design.

The two traditional strategies around the problem of local optima for the TBR algorithm are the use of multiple starting points for TBR and the retention of suboptimal trees during swapping. The first is more efficient and is thus the only one that will be considered here. The multiple starting points for TBR are best obtained by doing wagner trees using different addition sequences to create multiple wagner trees. Typically, the addition sequence can be randomized to obtain many different wagner trees to be later input to TBR – this has been termed a "random addition sequence" or RAS. The expectation is that some of the initial trees will eventually be near or on the slopes of the highest peaks. For data sets of 50 to 150 taxa, this method generally works well, although it may require the use of large numbers of RAS+TBR.

The strategy of RAS+TBR, however, is very inefficient for data sets of much larger size. It might appear in principle that larger data sets might simply require a larger number of replications, but the number of RAS+TBR needed to actually find optimal trees for data sets with 500 or more taxa seems to increase exponentially.

3   Composite Optima: Why do Traditional Techniques Fail?

Traditional techniques fail because very large trees can exhibit what Goloboff [10] termed *composite optima*. The TBR algorithm can get stuck in local optima for many data sets with 30–50 taxa. But a tree with (say) 500 taxa has many regions or sectors that can be seen as *sub-problems* of 50 taxa. Each of these sub-problems might have its own "local" and "global" optima. Whether a given sector is in a globally optimal configuration will be, to some extent, independent of whether other sectors in the tree are in their optimal configurations. For a tree to be optimal, all sectors in the tree have to be in a globally optimal configuration, but the chances of achieving this result in a given RAS+TBR may be extremely low. If five sectors of the tree are in an optimal configuration, just starting a new RAS+TBR will possibly place *other* sectors of the tree in optimal configurations, but it is unlikely also to place the *same* five sectors that were optimal again in optimal configurations. Consider the following analogy: you have six dice, and the goal is to achieve the highest sum of values by throwing them. You can either take the six dice and throw all of them at once, in which case the probability of getting the highest value is $(1/6)^6$, or 2 in 100,000. Or, you can use a *divisive* strategy: throw all the dice together only once, and then take each of the six dice and, in turn, throw it 50 times, keeping the highest value in each case. In the first case, you may well not find the highest possible value in 100,000 throws. With the divisive strategy of the second case, you would be almost guaranteed to find the highest possible value with a total of 301 throws.

In the real world, parsimony problems do not have sectors clearly identified as the dice, and the resolution among different sectors is often not really independent. This simply makes the problem more difficult. It is then easy to understand why finding a shortest tree using RAS+TBR may become so difficult for large real data sets. Consider a tree of 500 taxa; such a tree could have 10 different sectors which can have its own local optima; if a given RAS+TBR has a chance of 0.5 to find a globally optimal configuration for a given sector, then the chances of a given RAS+TBR to find a most parsimonious tree are 0.5^{10}, or less than 1 in 1,000. Thus, not only the number of rearrangements necessary to complete TBR swapping on trees with more taxa increases exponentially, but so does the number of replications of RAS+TBR that have to be done in order to find optimal trees.

4   Techniques for Analyzing Large Data Sets

The best way to analyze data sets with composite optima will be by means analogous to the divisive strategy described above for the dice. Re-starting a new replication every time a replication of RAS+TBR gets stuck will simply not do the job in a reasonable time. There are four basic methods that have been proposed to cope with the problem of local optima. The first one to be developed is the parsimony ratchet ([11], originally presented at a symposium in 1998; see [12]). Subsequently developed methods are sectorial-searches, tree-fusing and tree-drifting [10]. The expected difference in performance between the traditional and these new techniques is about as much as one would expect for the two strategies for throwing the dice.

4.1 Ratchet

The ratchet is based on slightly perturbing the data once the TBR gets stuck, repeating a TBR search for the perturbed data using the same tree as starting point, then using that tree for searching again under the original data. The perturbation is normally done by either increasing the weights of a proportion (10 to 15%) of the characters, or by eliminating some characters, as in jack-knifing (but with lower probabilities of deletion). The TBR searches for both the perturbed and the original data must be made saving only one (or very few) trees. The effectiveness of the ratchet is not significantly increased by saving more trees, but run times are (see [11] for details).

The ratchet works because the perturbation phase makes partial changes to the tree, but without changing its entire structure. The changes are made, at each round, to only *part* of the tree, improving, it is hoped, the tree a few parts at a time. In the end, the changes made by the ratchet are determined by character conflict: a given TBR rearrangement can improve the tree for the perturbed data only if some characters actually favor the alternative groupings. Since it is character conflict in the first place that determines the existence of local optima, the ratchet addresses the problem of local optima at its very heart. The ratchet is very effective for finding shortest trees. In the case of the 500-taxon data set of Chase et al. [13], the ratchet can find a shortest tree in about 2 hours (on a 266 MHz pentium II machine). Using only multiple RAS+TBR, it takes from 48 to 72 hours to find minimum length for that data set.

4.2 Sectorial searches

The sectorial searches choose a sector of the tree with a size such that it can be properly handled by the TBR algorithm, create a reduced data set for that part of the tree, and analyze that sector by doing some number of RAS+TBR (without saving multiple trees). Then the best tree for the sector is replaced onto the entire tree. The process is repeated several times, choosing different sectors. The sectors can be chosen at random, or based on a consensus previously calculated by some means. Details are given in Goloboff [10]. Sectorial searches find short trees much more effectively than TBR alone; in the case of Chase et al.'s data set, finding trees under 16225 steps using TBR alone would require using over 10 times more replications than when using sectorial searches, and this would take about 7 times longer.

Sectorial searches alone rarely find an optimal tree for large data sets. Used alone, they are less effective than the ratchet, normally going down to some non-minimal length (much lower than TBR alone), and then they get stuck. Sectorial searches, however, analyze many reduced data sets, which take almost no time at all. They thus have the advantage that they get down to a non-minimal length faster than the ratchet. They are then useful as initial stages of the search, in combination with other methods.

4.3 Tree-fusing

Tree-fusing takes two trees and evaluates all possible exchanges of sub-trees with identical taxon-composition. The sub-tree exchanges that improve the tree are then actually made. See Goloboff [10] for details. Tree-fusing is best done by successively fusing pairs of trees and thus needs several trees as input to produce results; getting those trees will require several replications of RAS+TBR, possibly followed by some other method (like a sectorial search, ratchet, or tree-drifting). Once several close-to-optimal trees have been obtained, tree-fusing produces dramatic improvements in almost no time. It is easy to see why: each of the sectors will be in an optimal configuration in at least some of the trees, and tree-fusing simply merges together those optimal sectors to achieve a globally optimal tree. In this sense, tree-fusing makes it possible to make good use of trees which are not globally optimal, as long as they have at least some sectors in optimal configuration.

4.4   Tree-drifting

Tree-drifting is based on an idea quite similar to that of the ratchet. It is based on doing rounds of TBR, alternatively accepting only optimal, and suboptimal as well as optimal trees. The suboptimal trees are accepted, during the drift phase, with a probability that depends on how suboptimal the trees are. One of the key components of the method is the function for determining the probability of acceptance, which is based on both the absolute step difference and a measure of character conflict (the relative fit difference, which is the ratio of steps gained and saved in all characters, between the two trees; see [14]). Trees as good as or better than the one being swapped are always accepted. Once a given number of rearrangements has been accepted, a round of TBR accepting only optimal trees is made, and the process is repeated (as in the ratchet) a certain number of times. Tree-drifting is about as effective as the ratchet at finding shortest trees, although in current implementations tree-drifting seems to find minimum length about two to three times faster than the ratchet itself. This difference is probably a consequence of the fact that the ratchet analyzes the perturbed data set until completion of TBR, while the equivalent phase in tree-drifting only does a fixed number of replacements. Since there is no point in having the ratchet find the actually optimal trees for the perturbed data, the ratchet could be easily modified such that the perturbed phase finishes as soon as a certain number of rearrangements has been accepted. Most likely this would make the ratchet about as fast as tree-drifting.

4.5   Combined methods

The methods described above can be combined. Thus, the best results have been obtained when RAS+TBR is first followed by sectorial searches, then some drift or ratchet, and the results are fused. Repeating this procedure will sometimes find minimum length much more quickly than other times. If the procedure uses (say) ten initial replications, on occasion the first four or five replications will find a shortest tree, the rest of the time effectively being wasted –at least as far as hitting minimum length is concerned. On other occasions, the ten replications will not be enough to find minimum length, but then there is no point in starting from scratch with another ten replications: maybe just adding a few more, and tree-fusing those new replications with the previous ten ones, will do the job. The most efficient results, unsurprisingly, are then obtained when the methods described above are combined, and the parameters for the entire search are supervised and changed at run time. At each point, the number of initial replications is changed according to how many replications had to be used in previous hits to minimum length; if fewer replications were needed, the number is decreased, and vice versa. Goloboff [10] suggested that it would also be beneficial to change the number of sectorial searches as well, and the number of drift cycles, to be done *within* each replication (although this has not been actually implemented so far).

The process just described in the end also makes it likely that the best results obtained correspond to the actual minimum length. Each hit to minimum length will use as many initial replications as necessary to reproduce the previously found best length; if the length used so far as bound is in fact not optimal, shorter trees will eventually be found. With every certain number of hits to minimum length, the results from all previous replications can be submitted to tree-fusing. If the trees from several independent hits to some length do not produce shorter trees when subject to fusing, it is likely that that length represents indeed the minimum possible (and thus tree-fusing provides an additional criterion, beyond mere convergence, to determine whether the actual minimum length has been found in a particular case). Alternatively, the search parameters can be made very aggressive (i. e., many replications, with lots of drifting and fusing, etc.) at first, to make sure that one has the actual minimum length, and subsequently they can be switched to the more effort-saving strategy when it comes to determining the consensus tree for the data set being analyzed.

4.6 Minimum length: multiple trees or multiple hits?

The approach to parsimony analysis for many years has been that of trying to actually find each and every possible most parsimonious tree for the data. Getting all possible most parsimonious tree for large data sets can be a difficult task (since there can be millions of them). What is more important, for the purpose of taxonomic studies, is that there is absolutely no point in doing so. Since the trees found are to be used to create a (strict) consensus tree, it would be much less wasteful to simply gather the minimum number of trees necessary to produce the same consensus that would be produced by all possible most parsimonious trees. In this sense, it is more fruitful to find additional trees of minimum length by producing new, independent hits to minimum length, than it is to find trees from the same hit by doing TBR saving multiple trees. Doing TBR saving multiple trees will produce, by necessity, trees which are in the same local optimum or island, differing by few rearrangements, while the trees from new hits to minimum length could, potentially, be more different –possibly belonging to different islands. The consensus from a few trees from independent hits to minimum length is likely to be the same as the consensus from every possible most parsimonious tree, especially when the trees are collapsed more stringently. The trees can be collapsed by applying the TBR algorithm, not to find multiple trees, but rather to collapse all the nodes between source and destination node when a rearrangement produces a tree of the same length as the tree being swapped. This allows production of the same results as would be produced by saving large numbers of trees, but more quickly and using less RAM. This is one of the main ideas in Farris et al.'s [15] paper, further explored in Goloboff and Farris [14]. Thus, current implementation of the methods described here exploits this idea. As minimum length is successively hit, the consensus for the results obtained so far can be calculated. The consensus will become less and less resolved with additional hits to minimum length, up to a point, where it will become stable. Once additional hits to minimum length do not further de-resolve the consensus, the search can be stopped, and it is likely that the consensus corresponds to the same consensus that would be obtained if each and every most parsimonious tree was used to produce a consensus. If the user wants more confidence that the actual consensus has been obtained, once the consensus became stable, it is possible to restart calculating a consensus from the new (subsequent) hits to minimum length, until it becomes stable again; the grand consensus of both consensuses is less likely to contain spurious groups (i. e., actually unsupported groups, present in some most parsimonious trees, but not in all of them). For Chase et al.'s data set, when the consensus is calculated every three hits to minimum length, until stability is achieved twice, the analysis takes (on a 266 MHz Pentium II) an average time of only 4 hours (minimum length being hit 20 to 40 times). The exact consensus is obtained 80% of the time, but the 20% of the cases where the consensus is not exact exhibit only one or two spurious nodes. The consensus could be made more reliable by re-calculating it until stability is reached more times, and by re-calculating it less frequently (e. g., every five hits to minimum length, instead of three). This is in stark contrast with a search like Rice et al.'s [16] analysis, based on ca. 9000 trees (found in 3.5 months of analysis) from a single replication, which produced 46 spurious nodes.

5 TNT: Implementation of the New Methods

The techniques described here have been implemented in "Tree analysis using New Technology" (TNT), a new program by P. Goloboff, J. Farris, and K. Nixon [17]. The program is still a prototype, but demonstration versions are available from www.cladistics.com. The program has a full Windows interface (although command-driven versions for other operating systems are anticipated). The input format is as for Hennig86 and NONA (see Siddall, this volume). The program allows the user to change the parameters of the search, either by hand, or by letting the program try to identify the best parameters for a given size of data set and degree of exhaustiveness. In general, a few recommendations can be made. Data sets with fewer than 100 taxa will be difficult to analyze only when extremely incongruent. In those cases, the methods of tree-fusing and sectorial searches perform more poorly (these methods assume that some isolated sectors in the tree can indeed be identified, but this is unlikely to be the case for such data sets). Therefore, smaller data sets are best analyzed by means of extensive ratchet and/or tree-drifting, reducing tree-fusing and sectorial searches to a minimum. Larger data sets can be analyzed with essentially only sectorial searches plus tree-fusing if they are rather clean. However, as data sets become more difficult, it is necessary to increase not only the number of initial replications, but also the exhaustiveness of each replication. This is best done by selecting (at some point in each of the initial replications) sectors of larger size and analyzing them with tree-drift instead of simply RAS+TBR (this is the "DSS" option in the sectorial-search dialogue of TNT). Larger sectors are more likely to identify areas of conflict, and it is less likely that better solutions will be missed, because they would require that some taxon be moved outside the sector being analyzed. After certain number of sector selections are analyzed with tree-drifting, several cycles of global tree-drifting further improve the trees, before submitting them to tree-fusing. The tree-drifting can be done faster if some nodes are constrained during the search (the constraint is created from a consensus of the previous tree and the tree resulting from the perturbed round of TBR; see [10]). This might conceivably decrease the effectiveness of the drift, but it can be countered by doing an unconstrained cycle of drift with some periodicity, and since it means more cycles of drift per unit time, in the end it means an increase in effectiveness. The "hard cycles" option in the "Drift" dialogue box of TNT sets the number of hard drift cycles to do before an unconstrained cycle is done. If large numbers of drift cycles are to be done, it is advisable to set the hard cycles so that a large portion of the drift cycles are constrained (e. g., eight or nine out of ten). For difficult data sets, making the searches more exhaustive will take more time per replication, but in the end will mean that minimum length can be found much more quickly.

The number of hits to re-check for consensus stability and the number of times the consensus should reach stability are changed from the main dialogue box of the "New Technology Search." As discussed above, this determines the reliability of the consensus tree obtained, with larger numbers meaning more reliable results. If the user so prefers, he may simply decide to hit minimum length a certain number of times and then let the program stop.

6   Remarks and Conclusions

New methods for analysis of large data sets perform at speeds that were unimaginable only a few years ago. Parsimony problems of a few hundred taxa had been considered "intractable" by many authors, but they can now be easily analyzed. No doubt the enormous progress made in the last few years in this area has been facilitated by the fact that people have recently started publishing and openly discussing new algorithms and ideas. Although at present it is difficult to predict whether the currently used methods will be further improved, the possibility certainly exists: the field of computational cladistics is still an area of active discussion and ferment.

Acknowledgments

The author wishes to thank Martín Ramírez and Gonzalo Giribet for comments and help during the preparation of the manuscript. Part of the research was carried out with the deeply appreciated support from PICT 98 01–04347 (Agencia Nacional de Promoción Científica y Tecnológica), and from PEI 0324/ 97 (CONICET).

References

1   Farris JS (1988) *HENNIG 86 , v. 1.5*, program and documentation. Port Jefferson, NY

2   Swofford DL (1993) *PAUP: Phylogenetic analysis using parsimony, v. 3.1.1*, program and documentation, Illinois

3   Goloboff PA (1994b) *Nona, v. 1.5.1*, program and documentation. Available at ftp.unt.edu.ar/pub/parsimony

4   Wheeler WC (1996) Optimization alignment: the end of multiple sequence alignment in phylogenetics? *Cladistics* 12: 1–9

5   Swofford D, Olsen G (1990) Phylogeny reconstruction. In: D Hillis and C Moritz (eds.): *Molecular Systematics*. 411–501

6   Goloboff PA (1994a) Character optimization and calculation of tree lengths. *Cladistics* 9: 433–436

7   Goloboff PA (1996) Methods for faster parsimony analysis. *Cladistics* 12: 199–220

8   Moilanen A (1999) Searching for most parsimonious trees with simulated evolutionary optimization. *Cladistics* 15: 39–50

9   Maddison D (1991) The discovery and importance of multiple islands of most parsimonious trees. *Syst. Zool.*, 40: 315–328

10 Goloboff PA (1999) Analyzing large data sets in reasonable times: solutions for composite optima. *Cladistics* 15: 415–428

11 Nixon KC (1999) The Parsimony Ratchet, a new method for rapid parsimony analysis. *Cladistics* 15: 407–414

12 Horovitz I (1999) A report on "One Day Symposium on Numerical Cladistics". *Cladistics* 15: 177–182

13 Chase MW, Soltis DE, Olmstead RG, Morgan D et al. (1993) Phylogenetics of seed plants: An analysis of nucleic sequences from the plastid gene *rbc*L. *Ann. Mo. Bot. Gard.* 80: 528–580

14 Goloboff PA, Farris JS (2001) Methods for quick consensus estimation. *Cladistics* 17: 526–534

15 Farris JS, Albert VA, Källersjö M, Lipscomb, D et al. (1996) Parsimony jackknifing outperforms neighbor-joining. *Cladistics* 12: 99–124

16 Rice KA, Donoghue MJ, Olmstead RG (1997) Analyzing large data sets: *rbc*L 500 revisited. *Syst. Biol.* 46: 554–563

17 Goloboff PA, Farris JS, Nixon KC (1999) *T.N.T.: Tree analysis using New Technology*. Available at www.cladistics.com.

5 Measures of Support

Mark E. Siddall

Contents

1 Introduction

It has become common in phylogenetic papers to report some sort of nodal value regarding the relative strength or 'support' for a particular portion of a cladogram. The most common among these has been the bootstrap, though of late Bremer's [1] direct measure has proven more popular and of unambiguous interpretation. A common feature of these measures is their relation to character information. The bootstrap, for example, purports to examine the relative "confidence" associated with portions of a cladogram in relation to the issue of character sampling. It is of some interest that the degree of corroboration of a phylogenetic hypothesis may actually relate more to the number of taxa included than to the number of characters. This perspective is made most clear in consideration of Popper's [2] view on the issue of corroboration. Specifically, testability, or the degree of corroboration $C_{max} = [1 - (1 \div$ number of competing hypotheses$)]$. Popper [2] referred to this as the level of testability and the formula ensures that a tautology has testability of zero. That is, if there is but one hypothesis, and no competitors, the degree of testability (and by consequence the maximum possible corroboration) is zero. The significance of this perspective to phylogenetic questions is clear when one enumerates the number of possible hypotheses that might compete for data. This number of hypotheses is quite independent of characters and is wholly contingent upon the number of taxa included.

Methods and Tools in Biosciences and Medicine
Techniques in molecular systematics and evolution, ed. by Rob DeSalle et al.
© 2002 Birkhäuser Verlag Basel/Switzerland

2   The Bootstrap

Felsenstein [3] introduced the bootstrap explicitly as a means to assess "confidence limits" on phylogenetic inference. This method enjoys great popularity, in large part due to its being made available in PAUP [3]. The phylogenetic bootstrap (Fig. 1), as formulated by Felsenstein [4], proceeds by resampling characters from the original data matrix, creating a pseudoreplicate matrix and then recalculating the most parsimonious tree, for example, on these pseudoreplicate data. In other words, a pseudoreplicate data set will be comprised of as many characters as there are in the original matrix, but the actual composition of characters will depend on which ones were sampled more than once or not at all. The frequency with which a clade is found across all of the bootstrap

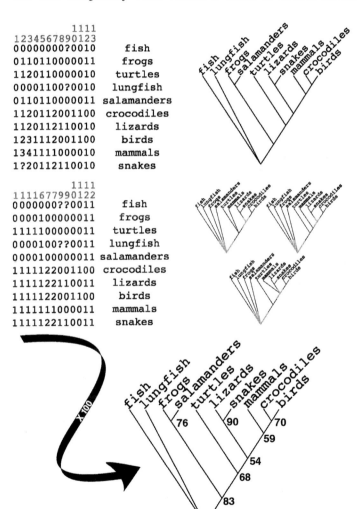

```
         1111
1234567890123
0000000?0010    fish
0110110000011   frogs
1120110000010   turtles
00001100?0010   lungfish
0110110000011   salamanders
1120112001100   crocodiles
1120112110010   lizards
1231112001100   birds
1341111000010   mammals
1?20112110010   snakes

         1111
1111677990122
0000000??0011   fish
0000100000011   frogs
1111100000011   turtles
0000100??0011   lungfish
0000100000011   salamanders
1111122001100   crocodiles
1111122110011   lizards
1111122001100   birds
1111111000011   mammals
1111122110011   snakes
```

Figure 1 The bootstrap resampling routine takes a matrix of c characters which resolves a most parsimonious tree (top) and, for each pseudoreplicate, randomly selects characters for inclusion and recalculation of optimal trees (middle). The frequency that clades are found in many (e. g., 100) pseudoreplicates is taken by some to be a measure of confidence in the veracity of that clade (bottom).

replicates is then to be taken as the confidence that one may have in that clade's veracity. Algorithmically, it is not necessary to actually create replicate datasets in this way. Instead, weighting vectors can be employed to accomplish the same effect. That is, one need only set all weights of all characters to 0 and then continually increment the weight of any character as it happens to be picked in the resampling protocol. This has the same effect as including and excluding characters according to the random sampling routine, and applying a random weight. In a morphological matrix, pseudoreplicates will have four-headed taxa with no vertebrae with which to attach their three sets of forelimbs. Although one can reasonably postulate, in the statistical sense considered for fork length of fishes, the notion that one might have sampled fish lengths differently (having for example only 2 fish of length 12 instead of the four that were observed), the implications of using the bootstrap in relation to phylogenetic data are considerably different. Concerns for the utility and meaningfulness of bootstrap values in phylogenetics have been offered [5–7]. In its original formulation Felsenstein [4] argued that clades that do not appear in at least 95% of bootstrap pseudoreplicates should be considered rejected. Sanderson [5] argued that the 95% rejection level was too conservative and would lead to rejection of well-supported clades. Consequently, there have been various attempts to adjust the 95% rule to a more meaningful value. Rodrigo [8], suggested that the bootstrap replicates themselves should be bootstrapped, Brown [9] suggested permutation of character states to calibrate the bootstrap, Hillis and Bull [10] evaluated the bootstrap proportions and determined that the 95% limit actually corresponded to an observed bootstrap proportion of 0.70.

From the start, the bootstrap may have been ill-conceived as a probabilistic confidence measure for the reasons that Felsenstein [4] outlined when describing the method. That is, unless one can justify the notion of a sampling universe, satisfy the conditions of independent and identical distributions of the information being resampled, and meet the requirement for a random sampling protocol of systematic data, a statistical interpretation of bootstrap values is not possible. This is certainly untenable for morphological data, but those who ascribe to a stochastic view of DNA sequence data may find the arguments more convincing. Personally, I do not [11]. Sanderson [12] considered various criticisms of the bootstrap, suggesting that some common ground could be found and suggesting that these criticisms were neither fatal to nor particular to the use of the bootstrap in phylogenetics.

Sanderson ([12]: 300) conflates accuracy and precision when asserting that the "statistical approach is aimed at estimating the *true* tree, or, more accurately, the tree that would be reconstructed in the absence of sampling and other *stochastic error*". But a well-sampled data set, free of random error, may be grossly inaccurate if there is some systematic error or bias in the data. Resampling from a biased set of data can only lead to some assessment of the precision in those biased data; it is mute on the accuracy of those data. The notion that bootstrap values (whether they be phylogenetic or otherwise) have some bearing on truth is not defensible. As to precision, Sanderson ([12]: 301) explains that "bootstrapping is aimed at generating a representation of the sampling error that would be obtained on repeated sampling of characters," and yet, it is never specified what is meant by repeated sampling. It is unproblematic that in a phylogenetic analysis of, for example, cytochrome oxidase, from some set of specimens, repeated sampling of these cytochrome oxidase characters from the same specimens will yield the same result each time. This introduces the question of what is the so-called universe from which items are ostensibly being sampled. Sanderson ([12]: 308) offered the 155, 844 nucleotides in the tobacco chloroplast. Surely, though, when armed with such a parametric universe there is no longer any meaning to (or need of) resampling. Any method, like a bootstrap tree, which renders a different solution than is obtained from the use of these data *in toto* must then be a very poor measure of the parametric universe. In short, because molecular data are obtained by sequencing a segment of DNA from one end to the other, the reference universe has been obtained from the start and resampling can only tell you how poor a tree might have been if it was based on a random sample of those data. If clades are not found to be well supported, you could then conclude that it is a good thing that you obtained the data by sequencing instead of randomly acquiring those data. If clades are well supported you could conclude that random acquisition of a portion of the sequence would have done a reasonable job. Insofar as obtaining sequence data randomly is impossible, all of this is uninteresting.

No proponent of bootstrapping has yet to adequately express what it means to have sampling error in a set of character information that is known in its entirety for all included taxa. This belies the central question of independent and identical distributions (i.i.d.) which would be required for any such method to be considered in a statistical framework. Sanderson ([12]: 302) admits that phylogenetic data are not so distributed, and suggests that they could be made so providing that the sampling routine was made random. He does not explain how we should randomly acquire data, nor why we should do so. Randomly sampling data would save the bootstrap, but it would do so at the expense of adequate data for phylogeny reconstruction. Finding well-supported phylogenies is more interesting to phylogeneticists than is bootstrapping. Sanderson ([12]: 305) concludes that "bootstrapping requires no assumptions about the evolutionary process," a bit of a straw man since this charge never has been leveled at the bootstrap, though "it does require assumptions about data gathering." Until someone can convince the systematic community that "random sequencing" is not an oxymoron, phylogenetic data remain nonindependent both in terms of their logical relations as well as the manner in which they are sampled. Sanderson ([12]: 315) also suggested that some of the limitations of the bootstrap are not unique to phylogenetic data, but are general to statistical methods. In particular, there is the problem that autapomorphies and other uninformative character distributions on remote parts of a tree can pose for the relative bootstrap proportions of clades that should be considered well supported (see also [13]). However, whereas it is clear that there are phylogenetic data that are uninformative of group membership (and that in many molecular data sets, these data are in the majority), there is no such thing as a fish length that is uninformative about its population mean. Sanderson's ([12]: 315) charge that "this is not unique" to phylogenetics then is just false (see [14]). Sanderson ([12]: 307), conceding non-random sampling, suggests that there might be something useful in the bootstrap, but does not actually specify what that might be. Of course, this alone denies the premise of the bootstrap in the first place, which was to assess confidence. The remainder of the objections that Sanderson [12] responded to are objections to statistics and probabilism in general, not objections to bootstrapping phylogenies. All of the reasons why bootstrapping cannot be interpreted statistically were described in detail by Felsenstein [4] when he proposed the method. Sanderson's [12] consideration of the issue offers neither a defense of the statistical interpretation of the phylogenetic bootstrap, nor any new way around the well-known [4] limitations to such an interpretation.

In an attempt to more thoroughly assess the behavior of bootstrap values, I have applied Felsenstein's [4] method to 30 datasets included in Platnick's [15] benchmark study. For the bootstrap to be taken as a meaningful measure of character support for clades in a phylogenetic hypothesis, they must not only be free of influences from uninformative data [14], but also from effects that are distinct from the focus of the sampling regime, that is, character information. Reassuringly, bootstrap values were found to be positively correlated with the number of informative characters in an analysis ($r = 0.378$; p = 0.0001). However, much of this appropriate relationship to character information is confounded by negative relationships between bootstrap values and various parameters relating to taxonomic composition, such as number of taxa in an analysis ($r = -0.288$; $P = 0.0001$), number of taxa in a clade ($r = -0.235$; $P = 0.0001$) and tree balance ($r = -0.111$; $P = 0.009$). The most troubling phenomenon was a significant autocorrelation of bootstrap values obtained for clades with the clades they are within and with clades included within them (residuals $dL = 1.44 < 1.65$ dU, Durbin-Watson serial autocorrelation for n > 100 clades). This is not surprising when one admits that the frequency with which one might find clade (D E) in bootstrap replicates never should have been thought to be independent of how frequently one finds clade (C D E) that includes it.

Because the bootstrap amounts to randomly including and excluding character information, and simultaneously randomly reweighting those characters that are included, it is impossible to separate out the effects of these two forms of data manipulation. If one is interested in the effect that character inclusion might have for a hypothesis, one should merely randomly include characters according to a sampling routine, but not reweight them. Parsimony jackknifing [16] appears to be appropriate to these ends. If one is interested in the effect that random weighting might have on a hypothesis, this too can be accomplished without much difficulty. The meaning of the latter is unclear because random weighting of characters is neither a common practice, nor one that is of much interest even to systematists who advocate weighting.

3   The Jackknife

Even though the bootstrap should not be interpreted as a confidence measure, consideration of its merits has unveiled an important issue in the use of randomization methods in systematics. That is, if there is to be some value to these methods, it will be found in terms of sensitivity of some analysis or its parts to certain kinds of perturbation of the data. The value of any method, then, will relate not to statistical interpretations, but to the relevance of the perturbation to some substantive issue in the practice of systematics. There are two basic dimensions to gathering data for phylogenetic analyses: the gathering of taxa and the gathering of characters from those taxa. A source of some angst for systematists relates to the interminable difficulty in deciding when to stop, be satisfied to some degree and then publish their results: a stopping rule. More taxa usually can be gathered and more characters usually can be obtained by histology or by DNA sequencing. Of course, if one could have done so, or had the resources, one surely would have. We can turn this question on its head and ask "What would have happened if I had fewer taxa or characters than I presently have, and then added the remainder?"

Jackknifing methods, originally employed for statistical purposes by Tukey [17], are appropriate for the purposes of clear and simple assessment of the "What if I had fewer and added the remainder of what I do have?" In phylogenetics, Lanyon [18] introduced the use of a taxonomic jackknife to assess the stability of clades to the removal of each of the taxa in succession (but replacing the removed taxon before the next is removed). Lanyon [18] suggested as well that the appropriate way to evaluate the trees found in the jackknifed replicate data sets (each lacking one of the taxa) was to construct a majority rule consensus of these replicates. Like the use of a majority rule 'bootstrap tree' (which is automatic in PAUP), this practice is regrettable. Whether these methods are taken to be statistical or not, their utility should be in assessment of variability, not as alternative optimality criteria for discovering a tree. Because they necessarily employ fewer than all of the available data, they are strictly verificationist in nature and lack any power to reject hypotheses of monophyly. No subsample of data has the force to reject an estimate based on all of the data as being the best estimate of some parameter. No amount of bootstrapping can cause one to reject the sample mean as the best estimate of the parametric mean, though it can give a more precise evaluation of the variance around that mean. Bootstrap trees, as such, are specious, and with them too go jackknife consensus trees.

Lanyon's [18] idea, though, still can be used as a means to determine the stability of a hypothesis to taxonomic inclusion, even though it lacks any statistical power for the acceptance or rejection of clades. Many current debates about sister group relationships can be reduced to issues of species sampling. Arguably more than the acquisition of additional characters, the inclusion of more taxa, and in particular a sample of taxa that is more dense, can do much for stabilizing a phylogenetic hypothesis. The relative monophyly of groups of whales [19], protists [20, 21], insects [22] and plants [23] has been shown to be profoundly influenced by the scope of taxonomic inclusiveness of analyses. As is perhaps made most clear in relation to inclusion or exclusion of fossil taxa [24–26], the addition of another taxon can incorporate a complexity of character correspondences that the simple addition of more characters cannot [27]. The use of taxonomic jackknifing, then, would appear to be a reasonable way to approach the question of how stable a current hypothesis is to the vagaries of these character complexities embodied in the species that comprise an analysis.

Homoplasy is not a feature of characters *per se*. After all, when furnished with any number of taxa and but a single character, or any number of characters and only three (or fewer) taxa, there will not be homoplasy. Homoplasy, as determined from a phylogenetic analysis, is embodied in the character correspondences of taxa in the tree. Removal of a taxon from an analysis can be expected to have not only local effects on tree topology (that is local to that taxon's most parsimonious placement) but also can be expected to have an influence on more remote parts of a tree [28]. Siddall [29] suggested two ways in which the relative effects of taxonomic composition can be assessed, both of which involve taxon jackknifing (Fig. 2). The effect that included taxa can have on overall stability can be determined simply by deleting each taxon once and, for each, determining the effect that this has on well-characterized measures such as number of equally parsimonious trees, consistency index and retention index. If the removal of a taxon results in a dramatic increase in the number of equally parsimonious trees, one might legitimately consider that taxon to be rather critical for stabilizing relationships. Though it may give some rough indication of the influences that individual taxa can have, this approach ignores the fact that in the event that the removal of a taxon still renders one tree, it may yet be a very different tree. The Jackknife Monophyly Index (JMI) [29], in contrast, is designed to evaluate the relative stability of the various clades in a phylogenetic hypothesis to the effects of inclusion or exclusion of taxa by calculating the frequency with which a clade appears in the most parsimonious tree(s) obtained when each taxon is removed and replaced in succession. This approach has never been included in versions of PAUP [3] so it has never caught on as a common tool for data analysis.

There is no particular reason why such analyses must be restricted to the removal of only one taxon at a time instead of two or more. However, the method as currently employed requires only as many analyses as there are taxa, and it is not clear how one should scale the effects of removing multiple taxa to be comparable with those obtained from removing only one at a time.

The character jackknife, or parsimony jackknifing [16] is an approach that, as noted above, is more in concert with an interest in the effect that character sampling may have, as opposed to the bootstrap which simultaneously re-weights and variously includes character information. The idea of removing a portion of characters as a way to address the issue of stability to this kind of information traces back to Davis [30] and indirectly, to Bremer [1]. Bremer's [1]

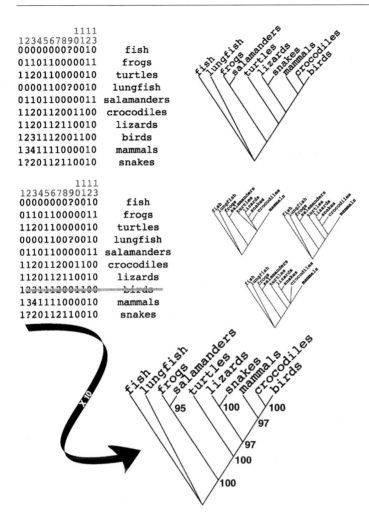

Figure 2 For a most parsimonious tree of t taxa (top), the taxonomic jackknife evaluates the stability of clades to the exclusion of each taxon once in t pseudoreplicates (middle). The frequency that a clade is found in the pseudoreplicates is interpreted as the relative stability of that clade to taxon sampling (bottom).

support index for a clade is merely the number of extra steps required to deresolve a clade found in a given tree (see next section). Davis [30] considered first a systematic approach to the problem of character removal in which each character is removed once in a manner identical to first-order jackknifing [17]. Any clade that disappears under this regime would have a clade stability index (CSI) of 1. To determine those clades having a CSI of 2, one need only evaluate all possible ways that two characters can be simultaneously removed. Perhaps "only" is an understatement. As Davis [30] noted, for datasets of even moderate size, the geometric increase in number of analyses required as more and more characters are removed renders this approach impractical. An alternative approach considers the relative effects of character removal. This approach employs a sampling strategy, like the bootstrap, but since characters are not randomly sampled in the first place, we are precluded from interpreting these values in a probabilistic sense. The simplest of these is random character deletion (the rcd; option HEYJOE from Random Cladistics, [31]) wherein characters are sampled for inclusion with replacement, but weights are not incremented. The resulting trees then can be evaluated in the same manner as the bootstrap by determining the frequency that clades found in the most parsimonious tree are found in the replicates.

Parsimony jackknifing [16] is similar in that the number of characters included or excluded varies stochastically according to a sampling protocol and the frequency of finding a clade is determined. Some may find it surprising that an individual who claims to eschew a probabilistic approach to phylogenetics [32–34] would express the virtues of employing what appears to be just such an approach. This misunderstands the intent of parsimony jackknifing. The size of datasets and their complexity have been growing at a rate that exceeds the limits of memory, processor speed and algorithms for evaluating all possible trees. A review of the literature on metaphytan relationships based on the nearly 4,000 available rbcL sequences is indicative of this (e. g., [35]) wherein the limits of heuristic search strategies have rendered revisions of relationships a common occurrence. The parsimony jackknifing algorithm has one crucial feature that was previously unavailable to phylogeneticists: the method, though heuristic, should only resolve clades that would also appear in all of the most parsimonious trees if one could actually find those trees. Although it may not necessarily resolve all of the clades in the most parsimonious tree(s), it will not resolve clades that are not in those trees. Moreover, and unlike bootstrapping for example, if there is even a single uncontroverted synapomorphy for a clade, that clade should be found in the sampling strategy. This does not impart a statistical framework to the method, and no such interpretation ever has been suggested. It does allow, though, for computation of large data sets in a reasonable length of time. Moreover, it provides a means for creating a partially resolved tree that can narrow the focus of the search for the most parsimonious solution by way of a constraint tree in PAUP [3] or NONA [36]. Figure 3 illustrates the difference in performance of the bootstrap, random character deletion (rcd), random character reweighting (heavyboot) and parsimony jackknifing (Jac) for a phylogenetic analysis of some mammals using cytochrome c oxidase subunit I. It is interesting that the artiodactylid clade (composed of the cow and the two whales), as well as the larger clade containing these, the horse and the primates, both have bootstrap values that are approximately the average between the effects of excluding characters and of reweighting them. In three other clades, the bootstrap overstates the effects of these two phenomena. It is also clear that parsimony jackknifing provides a considerably more conservative estimate of support.

4   Noise

Systematists have been troubled by the problems that random variation of characters can pose for accurate phylogeny reconstruction. Indeed, this is the substance of the long-branch attraction problem [37, 38], in that random

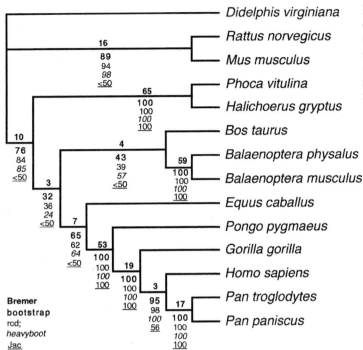

Figure 3 Comparison of nodal indices that evaluate character evidence in an analysis of cytochrome *c* oxidase subunit I for 14 mammals.

nucleotide data will suggest that two taxa are related with an expected frequency of 3/16ths. Systematists contemplating the issue of accuracy and homoplasy regularly suggest that this 'noise' somehow needs to be subtracted out either by avoiding certain kinds of data [39], by avoiding 'saturated' taxa [40–42], by arbitrarily weighting or merely discarding whole portions of data [43–48], or by keeping data sets of different 'kinds' separate in analyses [49, 50]. Unfortunately, more space has been devoted to worrying about noise than to measuring it, and it would seem that there should be some indication of just how important noise is before anyone should have worried too much in the first place. Helm-Bychowski and Cracraft ([51]; see also [52], as well as [53]) approached the issue of measuring noise somewhat indirectly. They suggested that either a majority rule consensus of the top 1% of most parsimonious trees, or the use of bootstrap values could give an indication of degree of noise in a data set. This may only be reasonable where character conflict is due to noise instead of mixed signals. A data set with two strong but conflicting signals, such as:

```
X    000000000000
Y    000000000000
A    111111000111
B    111111000111
C    111000111111
D    111000111000
```

will legitimately resolve two equally parsimonious trees. The top 1% (or any other arbitrarily chosen percentage) of shortest trees will render the majority rule tree ((A B) C D) and bootstrap values will only meaningfully exceed 50% for clade (A B). However, the failure to resolve a relationship for C or D here is due to patterns of character conflict, not to noise (which by all other definitions is not supposed to be pattern but a stochastic lack thereof). Hillis ([40]; see also [54]) suggested that the use of g1 skew statistics of all possible or randomly generated topologies was a meaningful way to assess noise in data, the supposition being that if there are many trees that are only a few steps away from the most parsimonious tree, this is an indication of noisy data (but see [55]). The same data set for X, Y, A, B, C and D above renders the following distribution of tree lengths when assessed on 1000 Markovian generated trees in Random Cladistics [31]:

Frequency distribution of random trees for:
'untitled'
        begin
        63   at 18 steps
        0   at 19 steps
        0   at 20 steps
        196   at 21 steps
        0   at 22 steps
        0   at 23 steps
        461   at 24 steps
        0   at 25 steps
        0   at 26 steps
        280   at 27 steps
        MEAN = 23.874001
        VARIANCE = 3.800000
        ST.DEV. = 1.949359
        G1 = −2.358787 (skewness) P < 0.001 for 1000 random trees
        G2 = 6.157025 (kurtosis) P < 0.001 for 1000 random trees

but the significance of the g1 statistic assumes a standard-normal model which cannot be applied here because of the discontinuous distribution of tree lengths: lengths of 19, 20, 22, 23, 25 and 26 are impossible to obtain from any tree. Nonetheless, if the g1 statistic is taken at face value, it would suggest that this data set is not noisy, which appears to be correct. Unfortunately, though, interpretation of the value of the g1 statistic is not straightforward. The behavior of g1 statistics in the face of increasing noise content on three empirical data sets [56–58], although revealing a positive relationship between noise content added and g1, also suggests a more complex relationship (Fig. 4). The primate data set (12 taxa) and the arthropod data set (26 taxa) both suggest a logistic relationship but the holometabolous insect data set (13 taxa) reveals a linear one. The absolute values of g1 suggest that the primate data set is more noisy than the arthropod data set, but the position and shape of their curves suggest that they are the same. Even after 50% of the matrix has been replaced with noise (with Mootoo in Random Cladistics, [31]), the holometabolan data set

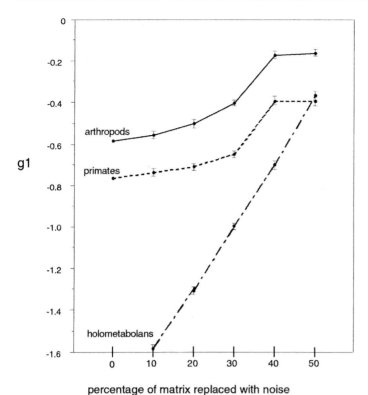

Figure 4 Behavior of g1 statistics on three different data sets in light of increasing proportions of those data sets being replaced with random noise.

percentage of matrix replaced with noise

still renders a significant skew ($P < 0.001$ on 1000 Markovian generated trees).

Wenzel and Siddall [59] developed a randomization technique to address the question of noise more directly than bootstrapping or g1 values will permit. The procedure (mojo) involves replacing a specified proportion of a data matrix with randomly selected states. In the case of nucleotide data, these states are simply randomly chosen from the set {A, C, G, T} to overwrite a pre-existing character state in the procedure. If this procedure is followed many times (e. g., 100 is practical), and trees are recalculated each time, the frequency with which clades in the most parsimonious trees are found in these noisy data sets can be determined. This has the added benefit, over g1 for example, of examining where on a tree additional noise might be said to be problematic. Notably, there is no bias by shape of the tree, and size of the clade has an effect only at extreme noise levels (> 50% overwrite). Simulations suggested that with 3 or more synapomorphies, nearly half of the data set would have to be noise before there would be a less than even chance of recovering any clade. Moreover, as noted by Wenzel and Siddall [59], "results show that signal is additive across different matrices, but that noise is averaged. Consider two data sets of equal size with s support and n% noise each. Their combination will produce 2s support, but still only n% noise. This can be expected to decrease the problem of noise by half, even though the analysis now includes twice as many noisy characters."

5   Direct Measures of Support

All of the preceding methods are indirect measures of support. That is, they examine the effects of sampling strategies or perturbations of the data on the recovery of optimal trees. Although in this chapter the methods are presented in the context of parsimony analysis, each of the bootstrapping, jackknifing and mojo data manipulations can be assessed in distance analyses, likelihood methods or even UPGMA if you find that method more suited to your phylogenetic philosophy. However, parsimony considers the support for a clade to be the balance of the number of synapomorphies in favor of that group *versus* those that deny it, and likelihood considers support in the context of how probable those changes are in light of things like branch lengths and other stochastic parameters. As such it should not be surprising that there are measures of support that are unique to each method.

Parsimony analysis is a non-stochastic, non-statistical approach to phylogenetic inference; so too the most widely used measure of support is an absolute value, not a relative one. The Bremer support index (b) for a clade is defined as the number of extra steps required to lose that clade in the most parsimonious tree. For example, if there are multiple equally parsimonious trees, those clades that are in dispute among those trees all have b = 0 because though some tree may have a clade, there is some other tree of equal length which does not. The difference in length between the tree with the clade and the one without is 0 since they are equally optimal. In other words, clades that are not found in all of the equally parsimonious trees have no unambiguous support (b = 0). Thus, for clades not found in any optimal trees, there must be more evidence against them than for them and their support values should be negative. Clades found in all most parsimonious trees must have more support for them than against (b > 0). Consider the following matrix:

```
(X)  AAAAAAAAAAAAAAAAAAAAA
(Y)  AAAAAAAAAAAAAAAAAAAAA
(A)  CCCCCCCCCCCCCAAAAAAAA
(B)  CCCCCCCCCCCCCAAAAAAAA
(C)  CCCCAAAAAAAACCCCACAC
(D)  CCCCAAAACCCCCCCCACA
(E)  CCCCAAAAAAAACCCCCCCC
```

There are three equally parsimonious trees (Fig 5) variously suggesting clades:

(A B C D E),
(C D E),
(D A B),
(A B),
(C E) and

(D E)
and the strict consensus (Fig 5) has two clades:
(A B C D E) and
(A B).

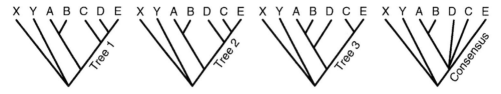

Figure 5   Three optimal trees and the strict consensus tree.

How much support is there, for example, for clade (C E)? The answer comes
from asking how many steps there are for the tree with this clade and how many
there are for the shortest tree that does not have clade (C E). This tree has 26
steps, but there is also a tree of 26 steps that does not have this clade; 26–26 = 0.
Bremer support for (C E) = 0. So too for (D E), (C D E), and (D A B). But what of (A
B) or (A B C D E)? In order to determine the Bremer support for these clades we
have to search for the shortest tree that does not contain them. In this case, a
next best fully resolved tree without (A B) has A with D or B with D at 30 steps.
The Bremer support for (A B) then is 30–26 = 4. For clade (A B C D E) a next best
tree without this clade also has 30 steps, again 30–26 = 4. Notice that in both
cases, the Bremer support for the clade is the same as the branch length. In fact,
the maximum value that b may take is the number of unambiguous changes
along the internode subtending that clade. To see how homoplasy affects
Bremer support consider the following matrix:

```
(X)  AAAAAAAAAAAAAAAAAAA
(Y)  AAAAAAAAAAAAAAAAAAA
(A)  CCCCCCCCCCCCAAAAAAA
(B)  CCCCCCCCCCCCAAAAAAA
(D)  CCCCAAAACCCCCCCCACA
(E)  CCCCAAAAAAAACCCCCCC
```

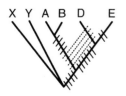

Figure 6   Optimal tree with character changes indicated (homplasies are dashed lines).

and clades (A B) or (D E) in Figure 6. There are 4 characters unambiguously supporting and no characters that conflict with clade (A B). Clade (A B) has a Bremer support of 4. However, although there are 6 characters supporting (D E), there are 4 characters that change homoplasiously, each suggesting clade (A B D). The most parsimonious tree with clades (A B), (D E) and (A B D E) has 24 steps. The next best tree without clade (D E) has 26 steps, b = 26–24 = 2. The next best tree without clade (A B) or (A B D E) has 28 steps and for each, b = 28–24 = 4. So, although the branch length for clade (D E) is 6, it has a Bremer support value of only 2. Typically it is advisable to report both values, Bremer support and branch lengths at internodes on a tree.

For PAUP aficionados to find the most parsimonious tree without clade (A B) the following commands will suffice:

> constraint cladeAB = (X, Y, C, D, E, (A, B));
> hsearch enforce converse constraint = cladeAB;

This widely used and highly intuitive measure of support is not supported in PAUP* [60], so unlike other point-and-shoot options that are available its calculation is cumbersome for those who are confined to Mac-OS. There are stand-alone programs such as TreeRot [61] that will help with this search by parsing tree descriptions and passing converse constraint options to PAUP.

Using Nona, however, Bremer support values are determined without recourse to external stand-alone software:

```
_NONA_____
|File/Data   f |
|Trees         _____
|Searches  |View trees            t |
|Options   |Make current          r |
|_____  |Consensus             c |
          |Diagnosis             d |
          |Evaluation            _____
          |Save (trees)          |Define as ref        r |
          |Save (parents')       |Compare to ref       c |
          |Tree file             |Check monophyly      m |
          |Edit tree             |_____
          |_____      |Branch support       b |
                                 |Swap clade           s |
                                 |_____
                                 |Indices              i |
                                 |Length               l |
                                 |Steps                x |
                                 |_____
```

For the latter, Bremer support values are written directly on nodes, ready for printing from a prespecified output file. For example:

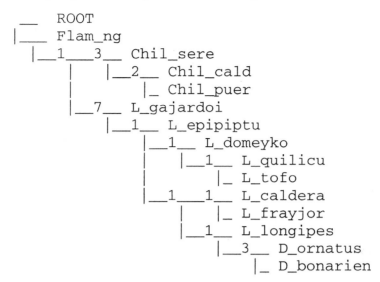

```
___    ROOT
|___    Flam_ng
  |__1___3__  Chil_sere
       |     |__2__  Chil_cald
       |          |_  Chil_puer
       |__7__   L_gajardoi
            |__1__   L_epipiptu
                 |__1__   L_domeyko
                 |     |__1__   L_quilicu
                 |          |_   L_tofo
                 |__1___1__   L_caldera
                 |     |_   L_frayjor
                 |__1__   L_longipes
                       |__3__   D_ornatus
                            |_   D_bonarien
```

Direct support measures for likelihood analyses are more complicated and have not yet been widely employed for all clades in an analysis. Since a direct clade support measure requires as many analyses as there are clades in an optimal tree, this would require many time-consuming likelihood analyses that few workers would seem to have the patience for. In any case, the appropriate measure is not so much the difference in likelihoods as it is the likelihood ratio between the optimal tree and the next-most-optimal tree without the clade using the Kishino-Hasegawa test. To calculate this for a clade, the following commands in PAUP will suffice:

```
set crit = likelihood;
hsearch;
constraint cladeAB = (X, Y, C, D, E, (A, B));
hsearch retain enforce converse constraint = cladeAB;
lscores all /khtest = yes;
```

The "retain" directive ensures that the optimal tree is held in memory for comparison with the best one found under the converse constraint. Since the objective is to only depict well-supported clades, a more rapid approach is available. In the analysis of holometabolan relationships conducted by Huel-senbeck [62], for example, the optimal tree using a Jukes-Cantor model had a -log likelihood of 2965.93089. By searching afterwards for a variety of sub-optimal trees with scores longer than this but less than 3000 using the commands:

hsearch retain = yes keep = 3000;
lscores all / khtest = yes;

and examining the output, it is clear that significant likelihood ratios are confined to scores of greater than 2985. Having so determined this, Huelsenbeck [62] could have determined the strict consensus of trees that are not significantly different from the best tree as follows:

```
hsearch; [to get best tree again];
hsearch retain = yes keep = 2985;
```

yielding the tree in Figure 7.

6   Remarks and Conclusions

Stability of a hypothesis can be justified in terms of a systematist's desire, for example, to not want to prematurely make drastic changes to their taxonomic system if there is some reason to believe that in the near future, as additional data are gathered, those changes will then be changed yet again, and so on interminably (but see [63]). Stability, though, is not a virtue in itself [64], and these methods are in no way ampliative. In an analysis of 30 taxa and 651 characters from cytochrome oxidase, Siddall and Burreson [65] found an unexpected sister group relationship for the piscicolid leeches with the jawed leeches, but this was not well supported (Bremer Support = 2) and yet, adding 10 more taxa and quadrupling the number of included characters continue to suggest this relationship [66], in spite of the previously low support. Other well-supported relationships (like monophyly of the medicinal leeches) have since evaporated. This then demands the question of what use is there at all for measures of stability, be they randomization-based or otherwise. They can say nothing beyond those data and taxa that are in the present analysis. They speak nothing to the future prospects of any particular group remaining monophy-

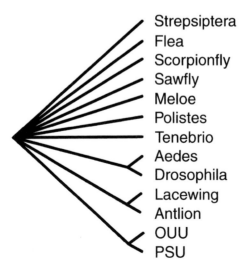

Figure 7   Likelihood tree from a small holometabolan dataset showing only those clades that survive the Kishino-Hasegawa test.

letic. But they are comforting. As dubious and unscientific a rubric as this may be, everyone needs a stopping rule. Moreover, most critically minded scientists desire a means by which they can second-guess their results, and by which they might evaluate the best use of their time and resources for the next project. A cladogram in which all of the clades have Bremer support values exceeding 20, 100% parsimony jackknife values and 100% JMI values may well be compelling, but it offers little direction for interesting new projects and hoped-for discovery of unexpected relationships. A cladogram in which one group is marked by low stability to character and taxonomic composition is interesting because it is this group that promises to reveal something new if a research focus is applied to it. Consider the relationships of leeches based on 18S rDNA, CO-I and morphology (Fig. 8) wherein Bremer support, JMI and mojo values have been determined. There is, as one would hope for if these measures mean anything in some general sense, an overall positive relationship among all of these measures (Fig. 9). It appears as though a Bremer support exceeding 5 roughly corresponds to being absolutely stable to taxonomic inclusion, and that those exceeding 10 correspond to a point at which signal strength is already 'loud' enough to be heard unambiguously over additional noise. However, these relationships are not absolute. That is, a Bremer support of 3, though sufficient to reflect considerable taxonomic stability (JMI = 100) and a better than even chance of overcoming noise (mojo = 55) in the medicinal leeches (species of *Hirudo* and *Limnatis*), is nonetheless insufficient in the glossiphoniids, where

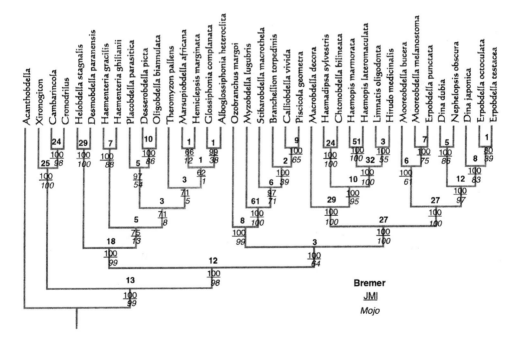

Figure 8   Relative support and stability of clades in the most parsimonious tree from combined 18S rDNA and CO-I data for 37 leech species.

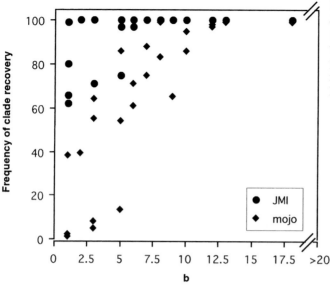

Figure 9 Bivariate plots of JMI (taxonomic stability) and mojo (noise stability) values *versus* Bremer support (b) from combined 18S rDNA and CO-I data for 37 leech species.

relationships appear to be extraordinarily labile to noise and where there is the greatest taxonomic instability. Why this should be so is obvious. The medicinal leeches, though themselves having low character support, are nestled in a clade (Hirudiniformes) marked by Bremer support indices that are all greater than (some considerably so) or equal to 10. The stability of these other clades confers stability on the medicinal leeches indirectly. Taxonomic instability of the medicinal leeches would require that they easily could group elsewhere, which then implies taxonomic instability of some other group (where they would otherwise group), which in turn is not possible because those other groups have so much intrinsic support. So the medicinal leeches are stable as a clade, not directly due to evidence in their favor, but for lack of cogent alternatives. In contrast, the overall instability of the glossiphoniids is a reflection of clades, nested one in the other, each having marginal support. So again, these nodal values are not independent measures any more than bootstrap values are, and cannot be interpreted statistically. Their value lies in what they say about the course of discovery. It does not seem likely that additional taxa or additional characters are going to quickly deny monophyly of the medicinal leeches, *Haemopis* species, the Haemadipsidae or the relative relationships amongst those taxa and I would, thus, feel comfortable in revising the systematics of the group, removing *Macrobdella* species from the family Hirudinidae where they do not group, and establish another family for them in the Hirudiniformes. In contrast, I think it would be somewhat premature to make drastic revisions to the Glossiphoniidae. JMI values suggest that I need a more dense sample of species. Mojo values suggest that the two genes used do not confer sufficient signal to noise ratio and that some other gene might be more useful in this regard. As such, work is continuing on an expanded set of glossiphoniid leech taxa (40 species in this one family) using CO-I and ND-1 in an attempt to better assess how these are related [67]. I see nothing particularly wrong with this approach to nodal values. No, there is nothing ampliative in this. No, stability is not a virtue in its own right. Neither does it have any bearing on unknowable truth. However, when the randomization technique has a clear and unambiguous meaning (which the bootstrap does not), then these values can provide clear and unambiguous direction for fruitful research directions. That this may be so, however, necessitates the use of more than one such measure, unless one can defend being interested in stability to taxonomic composition but not in noise, or to character support and not in the distribution of character conflict as it is variously embodied in the taxa.

References

1 Bremer K (1988) The limits of amino acid sequence data in angiosperm phylogenetic reconstruction. *Evolution*, 42:795–803

2 Popper K (1983) *Realism and the aim of science*. Routledge, London

3 Swofford D L (1991a) *PAUP Phylogenetic Analysis Using Parsimony, v. 3.1.1.* Champaign: Illinois Natural History Survey

4 Felsenstein J (1985) Confidence limits on phylogeneties: An approach using the bootstrap. Evolution 39: 783–791

5 Sanderson MJ (1989) Confidence limits on phylogenies: The bootstrap revisited. *Cladistics* 5: 113–129

6 Carpenter JM (1992) Random cladistics. *Cladistics*, 10: 215–220

7 Kluge AG, Wolf AJ (1993) What's in a word? *Cladistics*, 9: 183–199

8 Rodrigo AG (1993) Calibrating the bootstrap test of monophyly. Int. J. Parasitol., 23: 507–514

9 Brown JKM (1994) Bootstrap hypothesis tests for evolutionary trees and other dendrograms. P. Natl. Acad. Sci. USA, 91: 12293–12297

10 Hillis DM, Bull JJ (1993) An empirical test of bootstrapping as a method for assessing confidence in phylogenetic analysis. Syst. Biol., 42:182–192

11 Siddall ME and Kluge AG (1997) Probabilism and phylogenetic inference. *Cladistics*, 13:313–336

12 Sanderson MJ (1995) Objections to bootstrapping phylogenies: A critque. Syst. Biol., 44:299–320

13 Harshman J (1994) The effect of irrelevant characters on bootstrap values. Syst. Biol. 43:419–424

14 Carpenter JM (1996) Uninformative bootstrapping cladistics 12: 177–181

15 Platnick NI (1989) An empirical comparison of microcomputer parsimony programs, II. Cladistics, 5:145–161

16 Farris JS, Albert VA, Källersjö M, Lipscomb D et al. (1996) Parsimony jackknifing outperforms neighbor- joining. Cladistics, 12: 99–124

17 Tukey JW (1958) Bias and confidence in not quite large samples. Ann. Math. Stat. 29:614

18 Lanyon SM (1985) Detecting internal inconsistencies in distance data. Syst. Zool., 34:397–403

19 Adachi J and Hasegawa M (1995) Phylogeny of whales – dependence of the inference on species sampling, Mol. Biol. Evol., 12: 177–179

20 Spiegel FW, Lee SB, Rusk SA (1995) Eumycetozoans and molecular systematics. Can. J. Bot., 73: 738–746, Suppl. 1 E-H

21 Cavalier-Smith T, Couch JA, Thorsteinsen KE, Gilson P et al. (1996) Cryptomonad nuclear and nucleomorph 18S rRNA phylogeny. Europ. J. Phycol., 31:315–328

22 Mitchell A, Cho S, Regier JC, Mitter C et al. (1997) Phylogenetic utility of elongation factor-1 alpha in Noctuoidea (Insecta: Lepidoptera): The limits of synonymous substitution. Mol. Biol. Evol., 14: 381–390

23 Kadereit JW, Schwarzbach AE, Jork KB (1997) The phylogeny of *Papaver* s l (Papaveraceae): Polyphyly or monophyly? Plant Syst. Evol., 204:75–98

24 Gauthier J, Kluge AG and Rowe T (1988) Amniote phylogeny and the importance of fossils. Cladistics 4:105–209

25 Wheeler WC (1992) Extinction, sampling, and molecular phylogenetics. Columbia Univ. Press, New York

26 Brochu CA (1997) Morphology, fossils, divergence timing, and the phylogenetic relationships of *Gavialis*. Syst. Biol., 46: 479–522

27 Lecointre G, Phillippe H, V'n LÍ HL and Guyader HL (1993) Species sampling has a major impact on phylogenetic inference. Mol. Phylog. Evol., 2:205–224

28 Wheeler WC (1990) Nucleic-acid sequence phylogeny and random outgroups. Cladistics 6: 363–367

29 Siddall ME (1995) Another monophyly index: Revisiting the jackknife. Cladistics 11:33–56

30 Davis JI (1993) Character removal as a means for assessing stability of clades. Cladistics, 9: 201–210

31 Siddall ME (1996) Random Cladistics. Department of Biology, University of Toronto, Toronto, Canada

32 Farris JS (1983) The logical basis of phylogenetic analysis. In: NI Platnick, VA Funk (eds): *Advances in Cladistics, Vol. 2*. Columbia Univ. Press, New York, 7–36

33 Farris JS (1986) On the boundaries of phylogenetic systematics. Cladistics, 2:14–27

34 Farris JS (1997) "Who, really, is a statistician?" Sixteenth Meeting of the Willi Hennig Society. October 23–26, 1997. George Washington University, Washington, D.C

35 Chase MW, Soltis DE, Olmstead RG, Morgan D et al. (1993) Phylogenetics of seed plants – an analysis of nucleotide-sequences from the plastid gene. Ann. Missouri Bot. Garden. 80: 528–580

36 Goloboff PA (1996) NONA- Bastard son of PiWe with No Name

37 Felsenstein J (1978) Cases in which parsimony or compatibility methods will be positively misleading. *Syst. Zool.* 27: 401–410

38 Felsenstein J (1988) Phylogenies from molecular sequences: Inference and reliability. Ann. Rev.Genet. 22:521–565

39 Avise JC (1994) Molecular Markers, Natural History and Evolution. Chapman and Hall, New York

40 Hillis DM (1991) Discriminating between phylogenetic signal and random noise in DNA sequences. In: MM Miyamoto, J Cracraft (eds.) Phylogenetic analysis of DNA sequences. Oxford University Press, New York, 278–294

41 Hillis DM and Dixon MT (1991) Ribosomal DNA: Molecular evolution and phylogenetic inference. Q. Rev. Biol., 66:411–453

42 Graybeal A (1994) Evaluating the phylogenetic utility of genes: A search for genes informative about deep divergences among vertebrates. Syst. Biol. 43:174–193

43 Knight A and Mindell DP (1993) Substitution bias, weighting of DNA sequence evolution, and the phylogenetic position of Fea's viper. Syst. Biol., 42:18–31

44 Sidow A and Wilson AC (1990) Compositional statistics – an improvement of evolutionary parsimony and its application to deep branches in the tree of life. J. Mol. Evol. 31:51–68

45 Mindell DP, Shultz JW and Ewald PW (1995) The AIDS pandemic is new, but is HIV new? Syst. Biol., 44:77–92

46 Mindell DP and Thacker CE (1996) Rates of molecular evolution: Phylogenetic issues and applications. Ann. Rev. Ecol. Syst., 27:279–303

47 Hedges SB and Maxson LR (1996) Molecules and morphology in amniote phylogeny. Mol. Phylog. Evol., 6:312–314

48 Naylor GJP and Brown WM (1997) Structural biology and phylogenetic estimation. Nature, 388:527–528

49 Bull JJ, Huelsenbeck JP, Cunningham CW, Swofford DL et al. (1993) Partitioning and combining data in phylogenetic analysis. Syst. Biol., 42:384–397

50 Miyamoto MM and Fitch WM (1995) Testing species phylogenies and phylogenetic methods with congruence. Syst. Biol., 44:64–76

51 Helm-Bychowski K and Cracraft J (1993) Recovering phylogenetic signal from DNA sequences: Relationships within the corvine assemblage (class Aves) as inferred from complete sequences of the mitochondrial DNA cytochrome-b gene. Mol. Biol. Evol., 10:1196–1214

52 Cracraft J and Helm-Bychowski K (1991) Parsimony and phylogenetic inference using DNA sequences: Some methodological strategies. In: MM Miyamoto, J Cracraft (eds.): Phylogenetic analysis of DNA sequences. Oxford Univ. Press, New York, 184–220

53 Swofford DL (1991b) When are phylogeny estimates from molecular and morphological data incongruent? In: MM Miyamoto, J Cracraft (eds.): Phylogenetic analysis of DNA sequences. Oxford Univ. Press, New York, 295–333

54 Huelsenbeck JP (1991) Tree-length distribution skewness: An indicator of phylogenetic information. Syst. Zool., 40:257–270

55 Källersjö M, Farris JS, Kluge AG and Bult C (1992) Skewness and permutation. Cladistics, 8: 275–287

56 Wheeler WC, Cartwright P and Hayashi CY (1993) Arthropod phylogeny: a combined approach. Cladistics 9:1–39

57 Carmean D and Crespi BJ (1995) Do long-branches attract flies? Nature 373:666

58 Hayasaka K, Gojobori T and Horai S (1988) Molecular phylogeny and evolution of primate mitochondrial DNA. Mol. Biol. Evol., 5:626–644

59 Wenzel JJ and Siddall ME (1999) Noise. Cladistics 15: 51–64

60 Swofford DL (2000) PAUP*. Phylogenetic analysis using parsimony (*and other methods). Version 4. Sinauer associates, Sunderland, Massachusetts

61 Sorenson MD (1999) TreeRot version 2b. Department of Biology, Boston University, Boston, MA 02215

62 Huelsenbeck JP (1997) Is the Felsenstein zone a fly trap? Syst. Biol. 46: 69–74

63 Kluge AG (1997) Testability and the refutation and corroboration of cladistic hypotheses. Cladistics, 13, 81–96

64 Dominguez E and Wheeler QD (1997) Taxonomic stability is ignorance. Cladistics, 13:367–372

65 Siddall ME and Burreson EM (1998) Phylogeny of leeches (Hirudinea) based on mitochondrial cytochrome c oxidase subunit I. Mol. Phylogenet. Evol., 9:156–162

66 Apakupakul K, Burreson EM and Siddall ME (1999) Molecular and morphological phylogeny of leech familial relationships. Mol. Phylogenet. Evol., 12:350–359

67 Light JE and Siddall ME (1998) Phylogeny of the leech family Glossiphoniidae based on mitochondrial gene sequences and morphological data. J. Parasitol. 85: 815–823.

Partitioning of Multiple Data Sets in Phylogenetic Analysis

Patrick M. O'Grady, James Remsen and John E. Gatesy

Contents

1 Introduction

This chapter deals with the methodological aspects involved in phylogenetic analysis of multiple data sets within a maximum parsimony framework. Comparisons of character sets within a maximum likelihood paradigm are discussed elsewhere [1–6]. Below, we (1) review the three basic philosophies of data partitioning and combination in phylogenetic analysis (2) present some analytical methods available for examining the distribution of support and conflict among data sets in a combined analysis framework (3) use two empirical examples to illustrate different approaches to partitioned and combined phylogenetic analyses and (4) review the methodology required to perform some of these analyses using PAUP* [7] and ARNIE [8].

Advances in molecular techniques have enabled phylogeneticists to generate large numbers of systematic characters. Concurrent breakthroughs in computational biology have provided feasible means by which to analyze these data. However, much recent discussion has been focused on how one should analyze sets of characters from diverse sources (reviewed in [9–19]). Arguments for whether a given analytical procedure is valid or not hinge on philosophical arguments that, while undoubtedly fascinating, are not the focus of this chapter. We suggest that individual scientists read the primary literature (some of which is listed above, including references therein), weigh the pros and cons of different procedures, and proceed with caution (or confidence).

Methods and Tools in Biosciences and Medicine
Techniques in molecular systematics and evolution, ed. by Rob DeSalle et al.

Although most biologists would agree that different character sets are subject to unique phylogenetic constraints [11], some [10, 12] have argued that such classes of systematic characters are not "mind-independent," and questioned their "reality." Kluge and Wolf [12] stated that, even though systematists study "classes of characters, such as molecular and morphological, for reasons of tradition and technology, such subdivisions do not provide a reasonable basis for claiming the existence of classes of evidence." Nixon and Carpenter [17], however, countered that a boundary between data sets exists as long as "we choose to recognize it." Data partitions are, therefore, groupings of information that are quite "real," provided they are explicitly and clearly defined.

Some characters may be quite heterogeneous in their tempo and mode of evolution simply because the forces that govern their evolution are quite distinct. Molecular vs morphological vs behavioral characters, embryonic vs adult morphology, nuclear vs organellar genomes, one gene vs another un-linked gene, stems vs loops in ribosomal RNAs, introns vs exons, different codon positions, and hydrophobic vs hydrophilic amino acids are just some of the many partitions which describe systematic characters. These groupings may offer insights into congruence, conflict and support. For example, substitutions taking place in the third codon position are less likely to affect amino acid composition than those occurring in the first or second codon positions. Because of the higher rate of substitution, third position sites may become saturated more rapidly, undergoing multiple substitutions at a single site, and as a result have higher degrees of homoplasy than either first or second sites. Therefore, it might be useful to isolate the more rapidly evolving class of characters from the other codon positions to assess conflict and congruence between these char-acter partitions. Through this simple division of the data, the influence of homoplasious third codon positions, whether positive or negative, can be assessed [20–25].

With so many ways to define partitions and divide characters [18], the problem of how to choose between and summarize the results of multiple partitioned analyses becomes quite acute. The three primary strategies that are currently used to analyze several systematic data sets are taxonomic congruence, combined analysis, and conditional data combination.

The taxonomic congruence, or individual analysis, approach [26] requires separate phylogenetic analyses of each individual data set. Final systematic results are summarized in consensus trees (reviewed in [27, 28]) derived from the various separate analyses. Miyamoto and Fitch [13] supported taxonomic congruence and argued that "independent data sets should rarely be combined but should be kept separate for phylogenetic analysis because their indepen-dence increases the significance of corroboration." That is, the resulting tree topology is more "significant" when several independent lines of evidence support the same relationship than when that same relationship is supported in a single combined analysis. When conflicts among data sets exist, they can be scored as topological incompatibilities among the separate individual analyses.

In contrast to taxonomic congruence, all data sets are pooled and analyzed simultaneously in the combined analysis approach. Several arguments have been presented in favor of combined analysis. Kluge [10] pointed out that any phylogenetic analysis must attempt to explain all the data simultaneously. He claimed that, from a philosophical point of view, the combined approach is superior because it maximizes the explanatory power [29] and informativeness of all the data. Nixon and Carpenter [17] also advocated combined analysis, and pointed out that if each character transformation is assumed to be independent from all others (as in most phylogenetic methods), separating characters into subsets to accentuate the independence and significance of corroboration, as in taxonomic congruence [13], is nonsensical. Hillis [30] took a slightly different approach when arguing for combined analysis. He pointed out that while some partitions may be more effective at resolving terminal relationships on a tree, others may be better at elucidating basal relationships. Therefore, the combination of several character sets could yield a hypothesis that is better resolved overall. Finally, Barrett et al. [31] noted that various character sets may possess a weak phylogenetic signal that is masked by homoplasy in each data set. By combining data sets, and therefore increasing the overall number of characters in the analysis, the weak, but common, signal in the various data sets may emerge (also see: [17, 32–36]).

Although data sets are pooled in combined analysis, this does not preclude assessments of corroboration among data sets for particular relationships. Traditionally, the distribution of synapomorphies among data sets for a particular node has been used to assess the contribution of different character partitions to support for that node in a combined analysis (e. g., [10]). Recently, additional measures of data set influence in combined analyses have been proposed [22, 36, 37].

Others have advocated conditional data combination as an alternative position between the extremes of taxonomic congruence and combined analysis [11, 14, 32, 38]. Proponents of this view suggest that data partitions be subjected to a test of congruence and, if not significantly incongruent, combined. Data sets that sharply conflict with other data partitions are either excluded from combined analysis (e. g., [39]) or adjusted by either differential character weighting [35] or selective taxon removal [14] to improve congruence. This approach stresses congruence among data sets, but such an emphasis may, in some cases, hinder phylogenetic resolution and support by removing potentially informative characters.

Strong support for alternative relationships among data sets might be due to horizontal gene transfer (e. g.,[40]), incomplete sorting of ancestral polymorphisms [41–43], gene conversion [44], intense directional selection [45, 46], mutational bias [47], undetected gene duplications [48], hybridization [49], or other biological factors. Proponents of conditional combination argue that combined analysis of significantly incongruent data sets would only serve to hinder phylogenetic inference [11, 32].

Several tests for detecting significant incongruence have been suggested [38, 50, 51], although there is no consensus on which test is best fitted to the assumptions of conditional data combination [52]. Upon accepting a certain test of incongruence as valid, several criteria could be utilized to exclude data sets from analysis. Under a strict criterion, only those partitions that were not significantly incongruent with all other partitions, when compared in pairwise tests, would be included together in a combined analysis. According to a more permissive approach, all data sets that are not significantly incongruent with at least one other data set in pairwise tests could be included in combined analysis. Other conditions for exclusion of data sets could be formulated. For example, each individual data set might be compared, in statistical tests of congruence, to the sum of characters in all other data sets. A sound philosophical choice for choosing one exclusion criterion over another has not been presented.

2   Statistical Tests of Data Set Incongruence

Several tests for determining whether different data partitions are not congruent have been devised. Probably the most widely used measure of congruence among data partitions is the incongruence length difference (ILD). The ILD is a measure of the extra steps that come with the merging of different data sets in combined analysis [53]. Farris et al. [51, 54] have designed a simple procedure to test the extremity of an empirical ILD. This procedure is referred to as the partition homogeneity test, or PHT, in PAUP* 4.0 [7]. The null hypothesis of the PHT is that different data partitions are as congruent as randomly generated partitions of equal size to the originals. The null distribution is established by generating ILDs from a pre-set number of random data partitionings. The empirical ILD is compared to this distribution to assess significance [51, 54]. In addition to the PHT of PAUP* 4.0, the ILD is also available as the ARNIE program in Siddall's [8] Random Cladistics package (see Protocols 1 and 2).

Protocol 1: Using Hennig86 and random cladistics

Hennig86 [70] is a MS-DOS program that uses a parsimony algorithm to estimate phylogenetic trees. Here we include some basic instructions for creating and running a Hennig86 matrix. Further information is available online at http://www.vims.edu/~mes/mes/henhelp.html. The Random Cladistics software package [8] contains several MS-DOS applications that are meant to interface with Hennig86 [70]. ARNIE, which is used to perform the ILD test [51, 54], is most relevant to this review. Documentation for all Random Cladistics programs may be found online at http://www.vims.edu/~mes/mes/rchelp.html.

Constructing Hennig86 Data Matrices

The data matrix may be created with virtually any text or with a word-processor, and it must be saved as "text-only," "unformatted," or "DOS text." A DOS file suffix is not required at the end of the filename. Hennig86 recognizes the command "xread" to read in a data matrix. Immediately following "xread" can be a title enclosed in single quotes (this is optional) and then the number of characters followed by the number of taxa. It is vital when specifying certain characters or taxa in subsequent analyses to remember that in Hennig86, characters and taxa are treated as numbered beginning from zero, not 1. Thus, if you specify 7 characters, the program recognizes the first as "character zero" and the last as "character 6." The semicolon that closes each data matrix must be present for ARNIE to run.

```
xread
'test file' (do not put apostrophes or semicolons in your title)
7 6
zero 0000000
one 0001110
two 0111211
three 0111111
four 1012111
five 1012110
;
```

Running Hennig86 Data Matrices

Start Hennig86 by typing "ss" at the DOS prompt. You will be met with the generic Hennig86 prompt *>. To read in your matrix, type "proc" and your filename followed by a semicolon (*e. g.*, *> proc datamat.hen;). If you receive a procedure> prompt it means the semicolon is missing; enter it at the prompt. Next you must enter appropriate Hennig86 commands, which will tell the program how to treat your data when it performs the test. Complete definition of Hennig86 commands is not possible here. The Hennig86 web site should be consulted for a complete list of the commands available in this program. You must be able to construct appro-priate command lines in Hennig86 in order to use ARNIE.

Designating Partitions Using ARNIE

ARNIE allows two types of partitions to be designated, scopes and positions. Scopes may be partitions of any number, size, or character composition (molecular, morphological, etc.). Scopes are specified by using the command "scope" followed by number of scopes in the data matrix. For example, if you have 400 molecular characters followed by 100 morphological characters you should enter "scope 2." You will then be asked to identify the range of each scope. This tells the program at which character each scope ends. Once again, it is important when doing this to remember that in Hennig86, characters are numbered beginning from zero, not 1. Thus, in the example previously mentioned, the first scope will automatically begin at character 0 so you should input "399" for the first scope end point. The last scope end point will be predetermined.

The positions partition is only applicable to protein-coding sequence data. Using the positions command allows one to test whether codon positions are congruent. Positions are designated by inputting two numerals, the position to partition and the total number of positions. To determine if third positions are congruent with first and second, one would specify "position 3 3."

Protocol 2: Using PAUP*

PAUP* 4.0 [7] is a computer program that uses parsimony, maximum likelihood and a variety of distance-based algorithms to construct phylogenetic trees. PAUP* offers both menu-driven and command line interfaces. It can run on several platforms including Mac-OS, MS-DOS and UNIX. PAUP* recognizes the NEXUS file format. This is a modular file format that incorporates public and private "blocks" of information to house systematic data. Public blocks are recognized by several computer programs and contain information about taxa characters, assumptions, trees and other relevant attributes of the data set. The nine public blocks are TAXA, CHARACTERS, UNALIGNED, DISTANCES, SETS, ASSUMPTIONS, CODONS, TREES and NOTES [72]. Private blocks, such as PAUP and MACCLADE blocks, are used by a single program. The PAUP* User Manual [7] includes a full discussion of how to assemble and run NEXUS files.

Blocks

All blocks are essentially a series of commands starting with BEGIN command. The word directly after the BEGIN command defines the kind of block being executed. Commands within a block are separated by a semicolon (;). All blocks terminate with the END command. Here we will briefly review blocks that are essential when partitioning data for data exploration, the SETS and PAUP blocks.

SETS blocks can be used to define different sets of taxa, characters, trees, and character partitions. The general format for commands within the SETS block are described elsewhere [71]. The accepted format for defining TAXSETs, CHARSETs and CHARPARTITIONs is found in Maddison et al. [71]. TAXSETs and CHARSETs can be changed in PAUP by using the **Data** menu. CHARPARTI-TIONs can be accessed from the **Partition Homogeneity Test** dialogue box in the **Analysis** menu.

```
BEGIN SETS;
TAXSET analysisA=1–3 6 12–15;
TAXSET analysisB=1–5 7–12 14–15;
CHARSET geneA=1–610;
CHARSET geneB=611–1049;
CHARSET geneC=1050–1321;
CHARSET geneD=1322–1499;
CHARSET morph1=1500–1610;
CHARSET mtDNA=1–610 1050–1321;
CHARSET nuDNA=611–1049 1322–1499;
CHARSET allDNA=1–1499;
CHARPARTITION mtnu=mt:mtDNA,nu:nuDNA;
CHARPARTITION dnamorph=dna:allDNA,morph:morph1;
END;
```

PAUP blocks are useful for executing a suite of analyses without having to constantly check the progress of each search on the computer. When the file is executed the program will sequentially perform all the commands specified in the PAUP block. Output from all searches performed can be saved into a log file by using the **LOG FILE** command. Several other options can also be defined at this time using the **SET** command. These settings will ensure that multiple consecutive runs will not be interrupted by error messages after the file is executed. One should note, however, that any errors that do occur will be ignored and that this could affect the outcome of any searches. The **DELETE/ RESTORE** commands set which taxa sets are used in the analyses. The **INCLUDE/EXCLUDE** commands are used to define which character sets are used. The **HOMPARTITION** command defines which character sets will be subjected to a partition homogeneity test. The following PAUP block shows some of the commands available. The number of replications and the search type must also be defined in this step. Refer to the PAUP* 4.0 User Manual [7] for additional information.

```
BEGIN PAUP;
LOG FILE = 'LOG A' [APPEND/REPLACE];
SET [AUTOCLOSE, NOERRORSTOP, ERRORBEEP, NOKEYBEEP, NOWARNT-
SAVE, NOQUERYBEEP, NOWARNRESET, NONOTIFYBEEP, NOWARNTREE,
CRITERION=PARSIMONY];
```

```
HOMPART PARTITION=dnamorph NREPS=100 SEARCH=HEURISTIC;
DELETE ALL;
RESTORE analysisA;
EXCLUDE ALL;
INCLUDE mtDNA;
HSEARCH;
SAVETREES FILE= 'mtDNA.mpts' REPLACE;
CONTREE;
```

Templeton [50] devised a non-parametric test comparing character support for competing topologies. The trees selected as initial and alternative hypotheses can be most parsimonious trees from different data partitions or the combined data tree(s) and differing hypotheses of relationships implied by subsets of the combined data. The first step in the test is to select an initial phylogenetic hypothesis, and optimize characters from a data partition onto that topology by parsimony. Next, an alternative topology is selected, and characters from the data partition are optimized onto that topology by parsimony. The number of steps required for each character on each topology is compared using a Wilcoxon signed-rank test to determine whether character support is significantly better for initial or alternative topologies ([50]; see also [52, 55, 56]). This test has been implemented in PAUP* 4.0 [7].

Rodrigo *et al.* [38] have also suggested a test for measuring congruence between phylogenetic hypotheses from different data sets. Rodrigo's test has three steps. First, the symmetric distance (SD) is calculated between the most parsimonious trees (MPTs) from each data partition. Next, the distribution of this statistic for each data set is estimated using the bootstrap. For each partition, the distribution of SDs between MPTs is calculated from bootstrap pseudoreplicates taken from each data partition. If every bootstrap pseudoreplicate for a data partition supports the same tree, then the mean will be zero. A large amount of variation in trees supported by different pseudoreplicates will yield a wide distribution of SDs. Finally, the empirical SDs between partitions are compared to the null distributions within partitions to determine if they are significantly different. The utility and implementation of this test have been debated by Lutzoni and Vilgalys [57], Lutzoni [58], Rodrigo [59], and Lutzoni and Barker [60].

Cunningham [52] compared the PHT, Templeton's test and Rodrigo's test as assessments of data set combinability. He found that, in terms of congruence with well-supported phylogenetic hypotheses, the PHT performed the best of the three tests. The Templeton and Rodrigo tests were found to be too conservative and sometimes advised against data set combination when combination increased congruence with well-supported hypotheses of relationship [52]. It should be noted that this study is based on only a few empirical test cases and, as such, the generality of the above conclusions may not be widely applicable.

3   Measures of Character Interaction in Combined Analysis

Character support for clades favored by a combined analysis of multiple data sets may be attributable to one, some, or all of the individual data partitions. One way to assess the relative contribution of different data sets to the combined result is to quantify the numbers of character transformations from each individual data set that support each node (e. g., [10]). This type of analysis reveals the distribution of supporting evidence, synapomorphies, among different data sets, but does not highlight specific conflicts among data sets for particular relationships. So, several authors have derived alternative measures that attempt to quantitatively assess the interaction, conflict and influence of different data partitions in combined phylogenetic analyses.

Partitioned Bremer support (PBS, [22]) shows the contribution of each partition to the branch support [61] of a node supported by a combined analysis. The PBS score for a given node and a given partition in the combined data set is the minimum number of character steps for that partition on the shortest topologies for the combined data set that do not contain that node, minus the minimum number of character steps for that partition on the shortest topologies for the combined data set that do contain that node. If the partition supports a relationship represented by a node in the total evidence tree, then the PBS value will be positive at that node. If the partition better supports an alternative relationship, the PBS value will be negative, indicating conflict with the combined analysis. The magnitude of a PBS score indicates the level of support for, or conflict with, a node. The sum of all PBS scores for a given node will always equal the branch support for that node in the combined analysis of all data partitions.

The data set removal index (DRI, [36]) measures the stability of a given node in a combined analysis tree when subjected to the removal of individual or successively larger combinations of data sets. The index is defined as the minimum number of data sets that must be removed to collapse that node. For example, if a particular node collapses upon removal of one component data set, that node has a DRI of one. If the DRI and the number of component data sets are equal, this indicates that the node is present not only in the combined analysis tree but in all the individual partition trees and in all most parsimonious trees derived from all possible combinations of data sets. A node supported by one or more component data sets, but not supported in the combined analysis of all data sets, has a DRI of zero.

Nodal data set influence (NDI, [36]) quantifies the influence of removing a particular data set from a combined analysis upon the level of branch support at a particular node. For a given combined data set, data partition and node of a total evidence tree, NDI is equal to the branch support at that node for the combined data set minus the branch support for the combined data set excluding the given partition.

Reed and Sperling [37] suggested a method whereby differential weighting of data partitions might show "which partitions caused bias" and "where in the tree the bias occurred." In this approach, an initial bootstrap analysis is done for the total combined data set. Then, bootstrap scores are calculated for analyses in which different data partitions are downweighted relative to others. If bootstrap support for a node goes up with the downweighting of a particular partition, that partition is shown to have a negative influence at that node (in terms of bootstrap support). In contrast, if bootstrap support for the node goes down with the downweighting of a particular partition, that partition is shown to have a positive influence at that node. Similar results could be obtained by simply removing a data partition from the combined data set (the most extreme downweighting of a partition), calculating bootstrap scores, and comparing bootstrap scores before and after data set removal.

The interaction of characters in combined analysis may imply hidden character support and/or conflicts [31, 33, 34]. Hidden character support may be defined as increased support for a given node in the combined analysis relative to the sum of support for that node in the individual partitioned analyses. The most extreme demonstration of hidden support would involve the appearance of a node in a combined analysis that did not appear in any of the individual partition analyses (e. g., [33]). Hidden character conflict may be defined as decreased support for a given node in the combined analysis relative to the sum of support for that node in the individual partition analyses. Hidden branch support (HBS) measures hidden support and conflict in terms of stability to relaxation of the parsimony criterion [36]. For a given node, HBS is the difference between the branch support for that node in the combined analysis of all partitions and the sum of branch support scores for that node from each separate data partition. If the HBS is positive, then hidden support outweighs hidden conflict in combined analysis at that node. If the HBS is negative, then there is more hidden conflict than hidden support at that node in combined analysis (see [36] for other measures of hidden support and conflict). Indices of hidden support quantify the dispersion of homoplasy [31] that may occur with the merging of diverse data sets in combined analysis.

4 Congruence, Incongruence and Phylogenetic Inference

Many authors have discussed the implications of partitioning and combining diverse data sets in the context of empirical studies (e. g., [1, 10, 18, 19, 21, 22, 24, 25, 34–37, 39, 62]). We contrast two opposing viewpoints below.

Naylor and Brown [63] argued that judicious data partitioning is sometimes necessary to recover the historical signal across large phylogenetic distances. They used 13 mitochondrial (mt) protein coding genes to examine relationships among metazoans [63]. Their analyses of both nucleotide and amino acid se-

quences failed to recover the widely accepted phylogeny for chordate relationships (Fig. 1 A). They cited convergence in base composition across broad phylogenetic divergences and differences in rates of evolution among classes of nucleotide sites as factors leading to the obfuscation of the historical signal and the
resulting "incorrect" and robustly supported groupings (Fig. 1B). When they
partitioned their large data set and analyzed only first and second sites of codon
positions modally encoding a certain subset of amino acids, the "correct" historical
signal emerged (Fig. 1 A). Apparently, certain classes of characters present within
the existing data set could be used to obtain a more accurate phylogenetic
hypothesis. In essence, they defined a series of partitions that, at the level of their
analysis, were more highly conserved and less prone to homoplasy than were the
raw nucleotide or amino acid sequences. Exactly how these partitions can be
identified for any particular case study is a contentious issue (e. g., [15, 18]). The
selective exclusion of data certainly requires more assumptions than a total
evidence analysis of all relevant data (see [10, 12]), but these additional assumptions were simply considered a calculated risk by Naylor and Brown [63]. They
argued that rather than gathering additional characters or sampling additional
taxa, knowledge of molecular evolution could be used to partition the existing data
set and to winnow out a reliable set of characters.

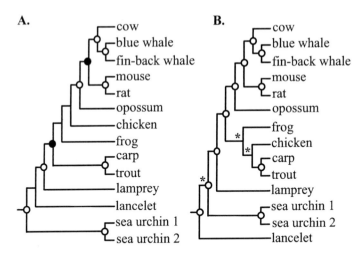

Figure 1 Different hypotheses of chordate relationships. The tree favored by a subset of
mtDNA nucleotides, first and second positions for sites whose modal amino acid is cysteine, glutamine, proline, methionine and asparagine [63], is shown in A. This is the traditional hypothesis of relationships. A nontraditional tree is shown in B. This topology is
recovered in an equally weighted analysis of all mtDNA nucleotide sites [63]. Open circles
at nodes indicate groups that were supported at the 100% bootstrap level in the analyses
of Naylor and Brown [63]. Solid circles mark traditionally recognized groups that have
lower bootstrap scores in the analysis of first and second positions of codons for cysteine,
glutamine, proline, methionine and asparagine relative to the analysis of all sites. Asterisks mark groups that are inconsistent with the traditional chordate tree. Note the position
of the non-chordate sea urchins in B.

Naylor and Brown [63] did not attempt a combined analysis of the 13 mt genes in their data set with the "wealth of other data" [63] that previously had been published in the literature. Such an analysis might have shown that the absurd relationships implied by the mt DNA data (Fig. 1B) are not supported by a more diverse collection of character data (i. e., a combined data set that includes mt DNA sequences, gross anatomical characters, nu DNA sequences, developmental information, fossil taxa, etc.). The tree supported by the mt DNA data has never been replicated in _any_ other analysis of metazoan phylogeny, but perhaps there is hidden support in the large mtDNA data set for the traditional hypothesis of chordate phylogeny (Fig. 1 A) that would reveal itself in a more inclusive analysis.

It should also be noted that bootstrap support scores for some "real" groups, such as Eutheria (cow, whales, rat and mouse), are lower in Naylor and Brown's [63] analysis of "reliable" characters (Fig. 1 A) than in the analysis of all nucleotides equally weighted (Fig. 1B). In an attempt to eliminate unreliable data, reliable data may be thrown out as well. It is not trivial to separate the bad from the good (see [64] for a criticism of compatibility methods).

An alternative approach is separating data to merge a variety of different data sets in combined analyses. With this approach, the potentially misleading quirks of any single data set may be diluted in an analysis of data sets that may each have their own unique biases (e. g., [62]). In a study of Cetartiodactyla (whales and even-toed hoofed mammals), Gatesy et al. [36] explored the interaction of characters within a combined analysis framework to assess the influence of different data sets. They showed that the combination of two significantly incongruent data partitions, nuclear (nu) DNA (κ-casein, β-casein, and γ-fibrinogen) and mt DNA (cytochrome b), increased both phylogenetic support and resolution (Fig. 2). The ILD for a division of the combined data set into nu and mt partitions is exteme (P_{ILD} = 0.0005). Nevertheless, when the nu and mt DNA data sets are combined, there is evidence of limited conflict, indicated by weak negative PBS scores for the mt DNA at only two nodes in the total data tree (Fig. 3, node A: PBS = –3, and node B: PBS = –0.5), weak negative NDI at one node (Fig. 3, node A: NDI= –1), and negative Reed and Sperling index (RS; [37]) at only one node (Fig. 3, node B: RS = –27). Closer examination of the data indicates that, at the two conflicting nodes, there is HBS from both the nu DNA and mt DNA data sets (Fig. 3), so disagreement between data sets in combined analysis is not as extreme as is suggested by either separate analyses (Fig. 2 A-B) or the ILD test. In fact, much of the apparent conflict between the individual data sets is offset by the HBS in the mt DNA data set (total HBS = +49). Because of HBS in the mt and nu partitions, BS is much higher in the combined analysis relative to either of the separate analyses (Fig. 2 A, B, and D). With the combination of significantly incongruent data sets, there is a total increase in BS of +30 relative to the separate analyses of nu and mt DNA partitions (sum of BS scores for the separate analyses = +306; sum of BS scores for the combined analysis = +336). Five more nodes are resolved in the combined analysis relative to the strict consensus of trees derived from the

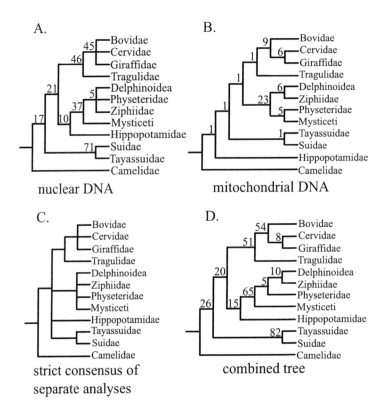

Figure 2 Trees derived from analyses of different data partitions for cetartiodactyl mammals [36]: strict consensus of shortest trees for the nuclear (κ-casein, β-casein, and γ-fibrinogen) data set (A), minimum length tree for the mitochondrial (cytochrome *b*) data set (B), strict consensus of trees supported by separate analyses of the nuclear DNA and mitochondrial DNA data sets (C), tree favored by a combined analysis of the nuclear and mitochondrial data sets (D). Branch support scores [61] are above internodes.

separate nu DNA and mt DNA data sets (Fig. 2C-D). Additionally, only five nodes are replicated in both of the separate analyses (Fig. 2 A-C), but PBS scores are positive for both mt and nu data sets at seven nodes in the combined analysis tree, and NDI scores are positive for both data sets at eight nodes (Fig. 3). Although the mt DNA and the nu DNA partitions are significantly incongruent, the combined analysis results in greater overall support, resolution and corroboration among data sets than is indicated by the taxonomic congruence approach.

Gatesy et al. [36] divided their combined data set (part of which is discussed above) into 17 partitions that separated different genes, exons from introns *within* each gene, morphology and transposon insertions, but further subdivisions of the data set could have revealed additional insights into support and conflict among different classes of characters. For example, in many empirical case studies of

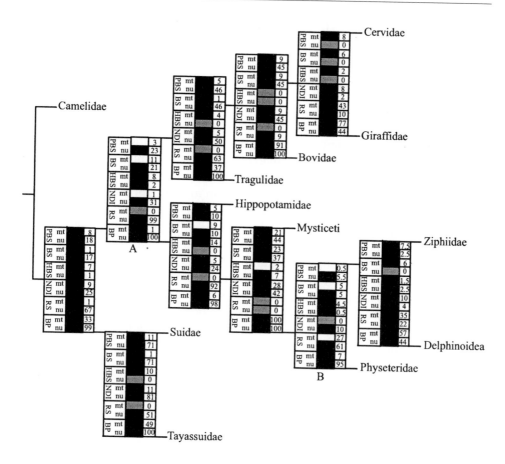

Figure 3 Topology supported by combined analysis of nuclear (κ-casein, β-casein, and γ-fibrinogen) and mitochondrial (cytochrome *b*) data sets for Cetartiodactyla [36] with indices of data set influence at nodes. Partitioned branch support (PBS), branch support in separate analysis (BS), hidden branch support (HBS), nodal data set influence (NDI), Reed and Sperling index (RS), and bootstrap percentage in separate analysis (BP) are shown for the nuclear DNA and mitochondrial DNA partitions. Positive, zero, and negative scores are indicated by black, shaded and white boxes, respectively. BS for separate data sets is defined as in Gatesy et al. [36], so some BS scores are negative or zero. Nodes labeled A and B are discussed in the text.

DNA sequences, second positions of codons evolve at a more conservative rate than do first or third positions. Similarly, transversion substitutions generally are more rare than transition substitutions, and nucleotide substitutions that cause radical amino acid replacements often are more scarce relative to those that cause conservative replacements. Therefore, it would be helpful to determine how these different classes of nucleotide changes conflict with or support the total evidence tree favored by Gatesy et al. [36]. Furthermore, weighting schemes based on these different classes of evidence could be used to assess the stability of an equally weighted analysis. For example, would the exclusion of rapidly evolving third codon sites increase support for the total evidence tree? Would an analysis of the third codon positions alone support the total evidence tree?

5   Remarks and Conclusions

The two examples above [36, 63] demonstrate that full exploration of the data, by the analysis of different partitions and through combined analysis, often provides information about the phylogenetic signal, support and conflict in diverse character sets. Extensive data exploration allows the researcher to fully examine the dynamics of character interaction and support.

Without such an approach, a strict proponent of conditional combination might always separate data sets that are significantly incongruent according to tests of data set congruence. Aside from philosophical arguments in favor of combining significantly incongruent data sets (e. g., [17, 18, 65]), there are often empirical reasons that incongruent partitions should be combined (e. g., [22, 52, 66, 67]). Most tests of incongruence measure the net character conflict between two data sets and not the conflict at particular nodes (e. g., [51]). Incongruence between two data sets may be due to differences at only a few nodes in a tree, rather than widespread disagreement among characters (e. g., [19, 68, 69]). Conflict among data sets does not have to be extreme to yield a significant PHT result [18]. So, it is possible that weak incongruence at only one or a few node(s) in a tree can yield a significant result with an incongruence test. Furthermore, the taxonomic congruence approach and, to a lesser degree, conditional combination ignore the hidden support that is commonly observed in combined analyses. Within a combined analysis framework, the interaction of characters from two significantly incongruent partitions may actually yield higher resolution, support and corroboration among data sets than that observed in separate analyses of individual data sets (e. g., Fig. 2). This cryptic information cannot be uncovered unless data sets are merged and analyzed simultaneously.

Alternatively, a strict proponent of combined analysis may gain insights into congruence, conflict and support by employing separate analyses of individual data sets. Statistical tests of data set congruence require separate analyses of data sets (e. g., [51]). Such additional calculations may not influence final systematic results, but could give clues as to what types of characters to sample in the future and which characters in the present data base should be scrutinized more closely. Furthermore, different subsets of data may be more informative or more internally consistent than others (e. g., [63]). These partitions or weighting schemes based on the consistency of individual characters (e. g., [70]) could be utilized to assess the stability of total evidence analyses.

Systematic characters should be examined thoroughly by performing both separate and combined analyses. Regardless of the philosophical basis for choosing between taxonomic congruence, conditional combination, or combined analysis, a better understanding of the data at hand will likely emerge through a variety of analyses.

Acknowledgements

JEG was supported by NSF DEB-9985847.

References

1 Cao Y, Adachi J, Janke A, Paabo S et al. (1994) Phylogenetic relationships among eutherian orders estimated from inferred sequences of mitochondrial proteins: Instability of a tree based on a single gene. *J. Mol. Evol.* 39: 519–527

2 Yang Z, Wang T (1995) Mixed model analysis of DNA sequence evolution. *Biometrics* 51: 551–561

3 Huelsenbeck JP and Bull JJ (1996) A likelihood ratio test to detect conflicting phylogenetic signal. *Syst. Biol.* 45: 92–98

4 Huelsenbeck JP and Rannala B (1997) Phylogenetic methods come of age: Testing hypotheses in an evolutionary context. *Science* 276: 227–232

5 Yang Z (1997) PAML: A program package for phylogenetic analysis by maximum likelihood. *CABIOS* 13: 555–556

6 Nishiyama T and Kato M (1999) Molecular phylogenetic analysis among Bryophytes and Tracheophytes based on combined data of plastid coded genes and the 18S rRNA gene. *Mol. Biol. Evol.* 16: 1027–1036

7 Swofford DL (1999) PAUP: Phylogenetic analysis using parsimony, v. 4. Smithsonian Institution, Washington DC

8 Siddall M (1995) ARNIE computer program. Random Cladistics Software Package

9 Miyamoto MM (1985) Consensus cladograms and general classifications. *Cladistics* 1: 186–189

10 Kluge AG (1989) A concern for evidence and a phylogenetic hypothesis of relationships among *Epicrates* (Boidae, Serpentes). *Syst. Zool.* 38: 7–25

11 Bull JJ, Huelsenbeck JP, Cunningham CW, Swofford DL et al. (1993) Partitioning and combining data in phylogenetic analysis. *Syst. Biol.* 42: 384–397

12 Kluge AG and Wolf AJ (1993) What's in a word? *Cladistics* 9: 183–199

13 Miyamoto MM and Fitch WM (1995) Testing species phylogenies and phylogenetic methods with congruence. *Syst. Biol.* 44: 64–76

14 de Queiroz A, Donoghue MJ and Kim J (1995) Separate versus combined analysis of phylogenetic evidence. *Ann. Rev. Ecol. Syst.* 26: 657–681

15 Brower AVZ, DeSalle R and Vogler A (1996) Gene trees, species trees and systematics: A cladistic perspective. *Ann. Rev. Ecol. Syst.* 27: 423–450

16 Huelsenbeck JP, Bull JJ and Cunningham CW (1996) Combining data in phylogenetic analysis. *TREE* 11: 152–158

17 Nixon K and Carpenter J (1996a). On simultaneous analysis. *Cladistics* 12: 221–241

18 Siddall M (1997) Prior agreement: Arbitration or arbitrary? *Syst. Biol.* 46: 765–769

19 Wiens JJ (1998) Combining data sets with different phylogenetic histories. *Syst. Biol.* 47: 568–581

20 Allard MW, Carpenter JM (1996) On weighting and congruence. *Cladistics* 12: 183–198

21 Milinkovitch MC, LeDuc RG, Adachi J et al. (1996) Effects of character weighting and species sampling on phylogeny reconstruction: A case study based on DNA sequence data in Cetaceans. *Genetics* 144: 1817–1833

22 Baker R and DeSalle R (1997) Multiple sources of character information and the phylogeny of Hawaiian drosophilids. *Syst. Biol.* 46: 654–673

23 Hastad O and Bjorklund M (1998) Nucleotide substitution models and estima-

tion of phylogeny. *Mol. Biol. Evol.* 15: 1381–1389

24 Bjorklund M (1999) Are third positions really that bad? A test using vertebrate cytochrome b. *Cladistics* 15: 191–197

25 Kallersjo M, Albert V and Farris J (1999) Homoplasy increases phylogenetic structure. *Cladistics* 15: 91–93

26 Nelson G (1979) Cladistic analysis and synthesis: Principles and definitions, with a historical note on Adanson's Familles des Plantes (1763–1764). *Syst. Zool.* 28: 1–21

27 Swofford DL (1991) When are phylogeny estimates from molecular and morphological data incongruent? In: MM Miyamoto and J Cracraft (eds.): *Phylogenetic Analysis of DNA Sequences*. Oxford Univ. Press, New York, 295–333

28 Nixon K and Carpenter J (1996b). On consensus, collapsibility and clade concordance. *Cladistics* 12: 305–321

29 Farris JS (1983) The logical basis of phylogenetic analysis. In: NI Platnick and VA Funk (eds.): *Advances in Cladistics*, Vol. 2. Proceedings of the Second Meeting of the Willi Hennig Society. Colombia Univ. Press, New York, 7–36

30 Hillis DM (1987) Molecular versus morphological approaches to systematics. *Annu. Rev. Ecol. Syst.* 18: 23–42

31 Barrett M, Donoghue MJ and Sober E (1991) Against consensus. *Syst. Zool.* 40: 486–493

32 de Queiroz A (1993) For consensus (sometimes). *Syst. Biol.* 42: 368–372

33 Chippindale PT and JJ Wiens (1994) Weighting, partitioning and combining characters in phylogenetic analyses. *Syst. Biol.* 43: 278–287

34 Olmstead RG and JA Sweere (1994) Combining data in phylogenetic systematics: an empirical approach using three molecular data sets in the Solenaceae. *Syst. Biol.* 43: 467–481

35 Cunningham CW (1997) Is incongruence between data partitions a reliable predictor of phylogenetic accuracy? Empirically testing an iterative procedure for choosing among phylogenetic methods. *Syst. Biol.* 43: 464–478

36 Gatesy JE, O'Grady PM and Baker RH (1999) Corroboration among data sets in simultaneous analysis: hidden support for phylogenetic relationships among higher level artiodactyl taxa. *Cladistics* 15: 271–313

37 Reed RD and Sperling FAH (1999) Interaction of process partitions in phylogenetic analysis: An example from the swallowtail butterfly genus *Papilio*. *Mol. Biol. Evol.* 16: 286–297

38 Rodrigo AG, Kelly-Borges M, Bergquist PR and Bergquist PL (1993) A randomisation test of the null hypothesis that two cladograms are sample estimated is a parametric phylogenetic tree. *N.Z. J. Bot.* 31: 257–268

39 Miyamoto MM (1996) A congruence study of molecular and morphological data for eutherian mammals. *Mol. Phylogenet. Evol.* 6: 373–390

40 Clark JB, Maddison WP and Kidwell MG (1994) Phylogenetic analysis supports horizontal transfer of *P* transposable elements. *Mol. Biol. Evol.* 11: 40–50

41 Nei M (1987) Molecular Evolutionary Genetics. Columbia Univ. Press, New York

42 Pamilo P and Nei M (1988) Relationships between gene trees and species trees. *Mol. Biol. Evol.* 5: 568–583

43 Hudson RR (1990) Gene genealogies and the coalescent process. *Oxford Surv. Evol. Biol.* 7: 1–44

44 Rozas J and Aguade M (1994) Gene conversion is involved in the transfer of genetic information between naturally occurring inversions of *Drosophila*. *Proc. Natl. Acad. Sci. USA* 91: 11517–11521

45 Stewart C, Schilling J and Wilson. AC (1987) Adaptive evolution in the stomach lysozymes of foregut fermenters. *Nature* 300: 401–404

46 Gauthier J, Kluge AG and Rowe T (1988) Amniote phylogeny and the importance of fossils. *Cladistics* 4: 105–209

47 Collins T, Wimberger P and Naylor G (1994) Compositional bias, character-state bias and character-state reconstruction using parsimony. *Syst. Biol.* 43: 482–496

48 Goodman M, Czelusniak J, Moore G, Romero A et al. (1979) Fitting the gene lineage into its species lineage, a parsimony strategy illustrated by cladograms constructed from globin sequences. *Syst. Zool.* 28: 132–163

49 McDade L (1990) Hybrids and phylogenetic systematics. I Patterns of character expression in hybrids and their implications for cladistic analysis. *Evolution* 44: 1685–1700

50 Templeton A (1983) Phylogenetic inference from restriction endonuclease cleavage site maps with particular reference to the evolution of humans and the apes. *Evolution* 37: 221–244

51 Farris JS, Kallersjo M, Kluge AG and Bult C (1994) Testing significance of congruence. *Cladistics* 10: 315–320

52 Cunningham CW (1997a). Can three incongruence tests predict when data should be combined? *Mol. Biol. Evol.* 14: 733–740

53 Mickevich M and Farris J (1981) The implications of congruence in *Menidia*. *Syst. Zool.* 30: 351–370

54 Farris JS, Kallersjo M, Kluge AG and Bult C (1995) Constructing a significance test for incongruence. *Syst. Biol.* 44: 570–572

55 Kishino H and Hasegawa M (1989) Evaluation of the maximum likelihood estimate of the evolutionary tree topologies from DNA sequence data and the branching order in the Hominoidea. *J. Mol. Evol.* 29: 170–179

56 Larson A (1994) The comparison of morphological and molecular data in phylogenetic systematics. In: B Schierwater, B Street, GP Wagner and R DeSalle (eds): *Molecular Ecology and Evolution: Approaches and Applications*. Birkhäuser Verlag, Berlin, 371–390

57 Lutzoni F and Vilgalys R (1995) Integration of morphological and molecular data sets in estimating fungal phylogenies. *Can. J. Bot.* 73: 649–659

58 Lutzoni F (1997) Phylogeny of lichen- and non-lichen-forming omphalinoid mushrooms and the utility of testing for combinability among multiple data sets. *Syst. Biol.* 46: 373–406

59 Rodrigo AG (1998) Combinability of phylogenies and bootstrap confidence envelopes. *Syst. Biol.* 47: 727–733

60 Lutzoni F and Barker FK (1999) Sampling confidence envelopes of phylogenetic trees for combinability testing: A reply to Rodrigo. *Syst. Biol.* 48: 596–603

61 Bremer K (1988) The limits of amino acid sequence data in angiosperm phylogenetic reconstruction. *Evolution* 42: 795–803

62 Wheeler W, Cartwright P and Hayashi C 1993. Arthropod phylogeny: A combined approach. *Cladistics* 9: 1–39

63 Naylor GJP and Brown WM (1998) Amphioxus mitochondrial DNA, chordate phylogeny and the limits of inference based on comparisons of sequences. *Syst. Biol.* 47: 61–76

64 Farris J and Kluge A (1985) Parsimony, synapomorphy and explanatory power: A reply to Duncan. *Taxon* 34: 130–135

65 Kluge AG (1997) Testability and the refutation and corroboration of cladistic hypotheses. *Cladistics* 13: 81–96

66 Sullivan S (1996) Combining data with different distributions of among-site rate variation. Syst. Biol. 45: 375–380

67 Wiens JJ (1999) Polymorphism in systematics and comparative biology. Annu. Rev. Ecol. Syst. 30: 327–362

68 Poe S (1996) Data set incongruence and the phylogeny of crocodilians. *Syst. Biol.* 45: 393–414

69 Goloboff P (1993) Estimating character weights during tree search. Cladistics 9: 83–91

70 Farris JS (1988) "Hennig86." Port Jefferson Station, New York

71 Maddison DR, Swofford DL and Maddison WP (1997) NEXUS: An extensible file format for systematic information. *Syst. Biol.* 46: 590–621.

7 Complex Model Organism Genome Databases

Charles G. Wray

Contents

1   Introduction

Until recently, molecular systematics was dominated by analyses of DNA sequences or derived peptide sequences generated by traditional cloning or PCR techniques. Many of these early studies were dependent upon only a single form of biological information, primary sequence. With the advent of genomic biology, multiple layers of biological information are available for comparative inquiry. It is possible to compare genomic DNA sequence and structure, mRNA sequence (via cDNAs), alternate transcript mRNAs, and peptide sequence with or without post-translation modifications. In the future it should also be possible to generate comparative, phylogenetically relevant data from 3-dimensional protein models, as well as from protein interaction assays, and possibly even microarrays. With the accelerating pace of data collection, it may be more time-consuming to assemble publicly available data than to generate one's own complement of pertinent data.

This contribution attempts to discuss the different complex eukaryotic organism databases now available on the world-wide web (Protocol 1). In so doing we are not attempting to compile information encompassing all known databases but rather highlight the differences between types of relevant biological data available (see Protocol 2 and Protocol 3). Several recent reviews provide URLs for a wide variety of databases (see Nucleic Acids Research 28(1)

Methods and Tools in Biosciences and Medicine
Techniques in molecular systematics and evolution, ed. by Rob DeSalle et al.
© 2002 Birkhäuser Verlag Basel/Switzerland

January, 2000). Proprietary data and software tools will not be discussed due to their rapid appearance and restricted availability to users.

Today the comparative biologist faces the task of sifting through tremendous amounts of information in a variety of public databases. Many database interfaces are user-friendly, while others require more complex queries or use of SQL (Structured Query Language) approaches. Databases continue to grow and multiply at unprecedented rates and many of the genomic or sequence-oriented databases are moving towards automation of records. For the comparative biologist the pace of accumulation and structure of information in such databases poses several basic and sometimes more insidious questions. Where did this information come from? Are published reports available? Is gene function based on inference or is this information known from a direct assay? How does information in one database compare to homolog (paralog or ortholog) information in a different database? How reliable are two different data sources? As each database functions slightly differently, these questions do not have unique answers; however, it is these questions (and others) that need to be resolved if one is to be confident about the comparisons or analyses dependent upon these resources.

Database reliability is an extremely important issue. Most sequence databases are only as accurate as the submissions they receive. As pointed out by Pennisi [1], there are numerous cases of contamination and/or vector DNA being inadvertently incorporated into sequence databases. Recently, trivial but actual human contamination of the *Drosophila* genome sequence has been identified [2]. While it is difficult to derive meaningful accuracy statistics, the larger sequence databases contain a wide variety of errors that can only be corrected by in-house curation or correction from the original submitting parties.

The large DNA sequence repositories remain as, virtually, write-only databases. Once a community member has submitted information to a repository it is rare for the record to be edited or corrected. The National Center for Biotechnology Information (NCBI) has recently made available a contamination BLAST tool (http://www.ncbi.nlm.nih.gov/VecScreen/VecScreen.html), entitled VecScreen, and this utility should become useful for filtering out vector-derived segments of sequence.

The advent of RefSeq (LocusLink) databases [3] at NCBI (www.ncbi.nlm.nih.gov:80/locuslink/refseq.html) will alleviate some contamination problems. The RefSeq initiative intends to provide a set of standardized sequences representative of naturally occurring genes. These sequences are semi-automated compilations which will provide a reference for polymorphism investigations, mutation studies, and gene expression work. Integration efforts between the RefSeq effort and the Mouse Genome Database (MGD) have uncovered inevitable nomenclatural difficulties and incorrect assignments; however, active, professional curation clears up many of these errors.

Database integration is a second fundamental issue. Some genomic databases may seem to be catch-alls of large amounts of data, while other databases may seem out-of-date but instead represent curated, boutique-style data sources. Integration attempts to connect these disparate sources of information. The NCBI is the largest source of public sequence information; however, it is not completely integrated with other relevant databases. Macauley et al. [4] modeled database integration and make several important points. Macauley [4] recognized that genome databases reflect their own origins. MGD and GDB were originally databases of mapped genes, genes which in many cases were not supported by nucleotide or amino acid sequence data. As a result many gene entries in MGD and GDB remain segregated from their logical NCBI records. Yet Macauley also points out that it is through integration efforts between databases that mistakes are recognized and corrections can be made.

2   Database Foundations: A Simplistic View

From a user's perspective three basic types of databases are now common. In general there are automated databases, community collaborative databases and centralized, curated databases. Databases built by the NCBI staff represent elegant examples of automated systems for the storage and easy retrieval of information concerning molecular biology, biochemistry and genetics. In particular LocusLink and COGs (Clusters of Orthologous Groups) [5] are two NCBI projects which depend on automation. While records are edited by the NCBI staff, the sheer bulk of data has driven the use of comparative algorithms for data sorting and binning. AceDB, Zfin (Zebrafish Information Network) and several other model organism and human databases are examples of community collaborative efforts. In collaborative schemes the scientific community submits their own data into a defined data structure which is freely searchable and edited by both community and in house efforts. The most labor-intensive databases, for example Flybase, SGD (*Saccharomyces* Genome Database) and MGD, work primarily from peer-reviewed literature and use a centralized curation staff.

For a wide range of reasons the three types of genome databases necessarily provide different types of information and different levels of potential data accuracy for the comparative biologist. Automated systems can and do contain the most current and comprehensive datasets. Yet use of this data should not come without some reservation. The data found in automated databases are only as accurate as the author or lab who submits the information. Furthermore, automated clustering, for example Unigene clusters or COGs, employ first-pass types of algorithms that should only be used to describe the scope but not the structure of the data. Automated databases also lack direct linkage to clear empirical supporting data. NCBI's PubMed does integrate publications and protein or nucleotide sequences but it places the onus on the user to retrieve the original paper and judge the fine details of the initial report.

Community efforts provide a vast array of data for particular organisms. Data are directly submitted to a database within certain standard formats. As such, community databases combine reviewed and unreviewed information. On the other hand, centralized, curated databases have been built from published data. Today direct submissions are becoming more common in curated databases; however, the architecture of these databases was framed upon careful annotation of peer-reviewed publications.

Curated databases provide the most exhaustive links to direct assays, but curation leads to a lag in data availability relative to automated databases. Within a curated database it is possible to track down specific information concerning gene expression, allelic variants and types of experiments from which a curatorial staff has established nomenclature and gene reports.

Standardized nomenclature is a benefit of curated databases. The literature is filled with colorful gene names representing an assortment of genomic features for many organisms. While gene names and symbols occasionally have some important historical context, the plethora of nomenclature systems confound the comparative biologist. Several mammalian databases (Human, Mouse and Rat) use standard conventions and established semantics for identifying genes. In so doing, Map3k1 (mitogen-activated protein kinase kinase kinase 1) in mouse is represented by MAP3K1 in human and so on. Standardization and integration of gene nomenclature simplify any search through large genome databases; however, unified nomenclature is not a sign of gene orthology. In some cases genes may be orthologs, but in other cases two genes with the same symbol may be paralogs or simply functionally related.

3   Genome Databases and the Comparative Biologist

The wide range of species-specific and general repository databases provide resources that would have been unimaginable to an earlier generation of comparative biologists and systematists. Yet the question remains, what is the immediate value of these data sources? There are four obvious uses of genome databases for comparative studies: 1) acquisition of sequence data for design and implementation of experiments 2) acquisition of data for phylogenetic analysis 3) investigation of anatomy, gene expression and pathway data for character development and 4) broad-scale multilayered surveying for elucidation of shared biological phenomena. The acquisition of data for either experimental design or for phylogenetic analysis shares enough similarity that it will be discussed together; however, it should be remembered that the strength of any study will be dependent upon both the judgment used in collecting data and the intelligence relied upon during data analysis.

There are a variety of on-line tools now available to aid in data acquisition. Primer and probe design is very specific to the scope of an investigation, so the following discussion will be limited to general aids to data acquisition. The similarity searches encompassed by the numerous BLAST [6] tools are a good starting point for identifying sequence data from any taxon which might be of interest. Nonetheless, there are numerous examples [7] of probable gene orthologs which will not be recognized by simple similarity searches.

Recently numerous advanced tools have emerged. In particular, SMART [8, 9], GHOST (Gene Homology On-line Search Tool; [10]) and the FAST_PAN strategy [11] seem appealing. SMART allows the user to specify the architecture of a gene prior to database querying. The user defines the identity and order of protein domains and then the SMART tool searches for other similar proteins. GHOST allows the user to input a protein and specify an organism of interest prior to searching via BLAST for similar proteins. By filtering the copious amounts of related sequences in ENTREZ, GHOST limits the time spent searching and offers the user the ability to select only those sequences (by a checkbox selection system) from which pairwise alignments can be presented. The FAST_PAN strategy is a technique designed to search for and identify newly broadcast and difficult to find (i. e., within EST database downloads) paralogs. If a user identifies a set of sequences from a gene family, FAST_PAN will query and collate results from nightly GenBank updates. With filtered results, FAST_PAN can be used to identify other protein superfamilies or can be prompted to winnow out high-identity sequences in order to narrow a search for new paralogs. The COGs resource is another way to search for probable paralog sets. At present 21 complete genomes (mostly bacterial) can be queried in multiple ways. Starting with a protein domain, it is possible to look for phenetically defined clusters which may or may not contain the target domain. For proteins within the clusters, COGs presents links to relevant sequences as well as functional classification and metabolic pathway information. All four of these resources are available either on-line, via on-line registration (free) or as source code from the authors (see Protocol 1 for more information).

The use of on-line genome database information for character development is a complex issue. While there is a wide range of anatomical and gene expression data now becoming available on-line (see Protocol 2 and 3) it is unclear how these data should be used by the comparative biologist. In very simple terms it is extremely naive to compare orthologous gene expression patterns, except in the most general way, between taxa. Despite this fact, the combination of anatomical information and gene expression data may expedite the refinement and understanding of morphological structures in many taxa. For instance using the Mouse Gene Expression Database (GXD), it is possible to query for gene expression data during limb development in a series of embryological stages. With such data, along with a controlled anatomical vocabulary, it should be possible to identify homologies (*sensu* Muller and Wagner [12]) that link individual structural units of the phenotype to their genotypic antecedents. These resources remain nebulous but could become a valuable dataset for study. Microarray and protein interaction data are an untested source of information. It appears that microarray data will be problematic until standard, iterative empirical studies and established data analysis tools are widely used. Here again, protein interaction and pathway data could yield a wealth of highly informative characters; however, a great deal of routine biochemical analysis and advanced mathematical modeling will need to be in place before the use of functional protein data becomes common.

Protocol 1:   Electronic tools and databases

Tools

COGs:	http://www.ncbi.nlm.nih.gov/COG/
Unigene clusters:	http://www.ncbi.nlm.nih.gov/UniGene/
GHOST: gene homology online search tool registration at:	http://www.hgmp.mrc.ac.uk/
FAST_PAN:	http://www.uvasoftware.org
SMART:	http://smart.embl-heidelberg.de/
GO:	http://www.geneontology.org

Databases

Mouse:	http://www.informatics.jax.org/
Zebrafish:	http://zfin.org/ZFIN/
Drosophila:	http://flybase.bio.indiana.edu/
	http://www.fruitfly.org/
Canine:	http://www.fhcrc.org/science/dog_genome/dog.html
	http://mendel.berkeley.edu/dog.html
Acedb:	http://www.acedb.org
Yeast:	http://genome-www.stanford.edu/Saccharomyces/
Plants:	Mendel database: http://www.mendel.ac.uk/
	Database of EST and STS sequences annotated with gene family information
Arabidopsis:	http://www.Arabidopsis.org/
Maize:	http://zmdb.iastate.edu/
Rice:	http://rgp.dna.affrc.go.jp/giot/INE.html
Microbial Genomes:	http://pbil.univ-lyon1.fr/emglib/emglib.html
Mitochondrial Genomes:	http://bio-www.ba.cnr.it:8000/BioWWW/
Genome Lists:	http://www.genome.ad.jp/dbget-bin/get_htext?Genome_Projects/

Protocol 2: Features of FlyBase (http://flybase.bio.indiana.edu) relevant to systematics

This database is a fully searchable tool on the biology, genetics and development of *Drosophila melanogaster*. Some of the important classes of information on the site are:
- Genome Annotation
- Cytologicae maps
- Gene Products
- Genome Projects
- Stocks (Tucson, Arizona)
- Anatomy (body part images and terminology)
- People

Anatomy

The basic anatomy of the fly is well outlined and documented on this website with drawings and images of the various body parts that are critical to understanding the genetics of the fly. The fly body plan is nearly exhaustively displayed in this part of the website. More importantly for systematists, the anatomy can be used as a comparative starting point for examination of other drosophilids, diptera and insects. This superbly constructed tool has drawings of the anatomical part of the fly set next to a column of anatomical terms for the localized body part. When each of the anatomical terms is "touched", the figure is highlighted for that anatomical term.

Genes and gene products

Data classes allow for access to GenBank for all accessed *Drosophila* sequences. The cytological maps class allows researchers to examine cytological locations of particular loci. These options are largely superseded by the addition of the Genome Annotation Class (see below).

References

Thousands of references on the genetics, development and biology of *Drosophila* are searchable by author, journal and title in the Bibliography class of data.

Genome annotation

The genome annotation option lists by location all sequences of the genome. For each location on the four chromosomes of *Drosophila* the annotation, gene symbol, map position, function, transcript, exons, and length of sequence in the genome are listed in a grid. The entries are listed by location on the four chromosomes by chromosomal location. Some of the entries in the grid lead to other parts of the FlyBase website. The most important internal link here is from the gene symbol link. By clicking on the various gene symbols one can obtain important information about each of the known genetic loci. Each of these loci has the following kinds of information listed: synopsis, full report, recent, expression and phenotypes, proteins and transcripts, stocks, references, similar genes and sequences, and synonyms. The similar genes and sequences link is probably the most useful link for systematists, as this link lists gene sequences from other organisms that show similarity to the *Drosophila* sequence. We have used this link extensively to design primers for PCR and to locate potential areas of the *Drosophila* genome for phylogenetic analysis.

Protocol 3: Features of Mouse Genome Informatics relevant to systematics *(http://www.informatics.jax.org/)*

The Mouse Genome Informatics (MGI) database provides integrated access to data on the genetics, genomics and biology of the laboratory mouse. Selected resources include:

Standardized gene/marker characterizations

Multiple genomic features can be rapidly queried in MGD. These features include: Genes, DNA segments, QTLs, Cytogenetic marker, Pseudogenes, BAC/YAC ends and gene Complex/Cluster/Regions. MGD employs accepted nomenclature conventions to identify each gene in the database. Nomenclature guidelines allow for: standard and other name query abilities, rapid searching within groups of related genes, and significant 1:1 correspondence with human gene names.

Homology information

MGD contains putative orthology data for mouse, human, rat and 16 other mammalian species. Homology data are supported by published references and standard evidence codes inform users on the empirical or computational criteria used to establish orthology.

Database integration

Curated links to NCBI (GenBank) connect genes to their representative sequence features, including genomic sequence (promoters, introns, where available), coding sequences, ESTs and allelic variants. Genes are also linked to their Swiss-Protein records, Enzyme (E.C.) classifications, Unigene clusters, as well as orthologous GDB records and pertinent On-line Mendelian Inheritance in Man (OMIM) records.

Expression data

As a sub-project fully integrated within Mouse Genome Informatics, the Gene Expression Database (GXD) provides endogenous gene expression data during mouse development. GXD includes a literature index as well as RT-PCR, *in-situ* hybridization, immunochemistry and blot data for mouse. Expression data are coded using the Edinburgh Mouse Atlas: The Standard Anatomical Nomenclature Database (http://genex.hgu.mrc.ac.uk/Databases/Anatomy/).

Comparative mapping data

MGD contains linkage, cytogenetic and physical map data. In addition it is possible to query and construct comparative maps between mouse and other mammal species. Therefore it is possible to form gene order comparisons across mammalian taxa.

GO consortium

MGD gene records are annotated using the controlled vocabularies of the Gene Ontology consortium. Users are able to query gene records for molecular functions, biological processes and cellular components. Use of the hierarchical GO vocabularies facilitates queries across yeast, mouse and fly (and in the near future other eukaryotic) databases.

Primer and probe information

For molecular probes and segments, MGD holds information on primers and probes used for gene expression studies.

Mouse inbred strains

MGD provides a genealogy of inbred mouse strains and other strain-specific data.

With the advent of full genome sequencing it has become apparent that many biological phenomena are shared by all. While the ability to rapidly sort through similar sequences is not new, our conceptual understanding of eukaryotic systems is beginning to expand broadly. In as little as two years relatively complete genomes from two fungi, a nematode, the fruitfly, a flowering plant, human and possibly mouse should be available. New tools are emerging which will be able to query these large datasets from a multilayered, systems-oriented approach. In particular, the GO consortium [13] is developing three hierarchical biological ontologies in order to describe the roles of genes and gene products in eukaryotes. Still in a developmental stage, it is possible to query for shared molecular function, biological process (one or more ordered assemblies of molecular functions) and/or cellular component (location of an active gene product) for mouse, yeast and fruitfly (with more consortium members ready to add other taxa). Members of the GO consortium are using structured, shared vocabularies in order to unify query capability across taxa. The GO consortium is not concerned with homology in the strict, evolutionary sense; however, the GO browser will serve a useful function for the comparative biologist. For example, rather than look at single genes across taxa, or at a single gene family, the GO browser should allow for analysis of genes which share ordered roles in biological processes. In the near future and with a great deal of careful analysis, it should be possible to investigate whether biological pathways are made up of networks of proteins which display sympleisiomorphic, homoplastic, or any other combination of phylogenetic relationships. The amount of intriguing comparative data which might be analyzed with GO as a starting point is exciting, and will be a testament to the collaborative nature of the consortium members.

4   Homology and Genome Databases

As proposed by Fitch [14], homologous genes that diverge through speciation are considered orthologous, while homologous genes that diverge after gene duplication are considered paralogous. Fitch's definition [14] articulates the fact that genes which are shared by a common ancestor will be orthologs, and genes that diverge after speciation from the most recent common ancestor will be paralogs. Molecular biologists, in particular, have occasionally confused the usage of homologous sequences to encompass both paralogous and orthologous comparisons (see [15]). The ability to align and compare nucleotide or amino acid sequences represents an empirical observation of similarity; in contrast, homologous similarity infers inherited similarity as a result of common ancestry. In many cases extremely similar sequences may not be homologous but instead are paralogous. On the other hand, low sequence similarity but clear phylogenetic footprints of motifs or domain structure between orthologs is expected when investigating genes under divergent selective pressure.

There appear to be no major on-line resources which identify paralogous sets of genes as determined by phylogenetic analysis. Most of the resources discussed herein use similarity search techniques which identify non-exclusive sets of potential paralogs. Many of these resources are useful for making a first pass analysis of those sequences which are publicly available, but it remains the job of the investigator to undertake phylogenetic analysis. Therefore systematists should be duly warned about any claims of orthology in many of the genome databases.

In the MGD the test for homology is not empirically defined similarity but the congruence of many forms of evidence for the gene under investigation. Each homology assertion is handled on a case-by-case basis and no similarity thresholds or cut-off points are employed. As a reference-driven database, the author's assertions and the combination of empirical assays are the most important information used for establishing a relationship of putative orthologs. This type of 'homology' assignment is powerful; however, these relationships of mammalian genes remain speculative until phylogenetic analyses are completed.

5   Remarks and Conclusions

It is a remarkable time in all of biology. The breadth and number of genome-related databases are ever-changing and in the next several years gigabytes of data will be available. Comparative biologists will be faced with the reality that vast amounts of information will exist for the 'superb six': humans, mice, fruitflies, worms, yeast and *Arabidopsis* [16], while other phylogenetically important taxa will lag behind in support and available data. The role of comparative biology will include using genome data to design investigations as well as finding and molding genome data into useful information. In this contribution several web tools and current databases are discussed. There are no state of the art databases and no online data are infallible. It is important for the user to recognize that automated and curated databases will always contain some degree of error. An exciting future development should include large scale phylogenetic analysis of genomic sequence data; however, inevitable computational assumptions will remain the absolutely critical factor in analytical success and subsequent information value for our field.

References

1   Pennisi E (1999) Keeping genome databases clean and up to date. Science 286: 447–450

2   Nature, 404, 916 (2000) News in Brief

3   Pruitt KD, Katz KS, Sicotte H, Maglott DR (2000) Introducing RefSeq and Locus-Link: curated human genome resources at the NCBI. Trends Genet. 16(1): 44–47

4   Macauley J, Wang H, Goodman N (1998) A model system for studying the integration of molecular biology databases. Bioinformatics 14 (7): 575–82

5   Tatusov RL, Koonin EV, Lipman DJ, (1997) A genomic perspective on protein families. Science 278(5338): 631–7

6   Altschul SF, Madden TL, Schaffer AA, Zhang J et al. (1997) Gapped BLAST and PSI-BLAST: a new generation of protein database search programs. Nucleic Acids Res 25: 3389–3402

7   Brenner SE, Chothia C and Hubbard TJ (1998) Assessing sequence comparison methods with reliable structurally identified distant evolutionary relationships. Proc. Natl. Acad. Sci USA 95: 6073–6078

8   Schultz J, Milpetz F, Bork P and Ponting CP (1998) SMART, a simple modular architecture research tool: Identification of signalling domains. Proc. Natl. Acad. Sci. USA 95: 5857–5864

9   Schultz J, Copley RR, Doerks T, Ponting CP et al. (2000) SMART: A Web-based tool for the study of genetically mobile domains. Nucleic Acids Res 28: 231–234

10 Curwen VA, Williams GW and Bard JBL (2000) GHOST: a gene homology online search tool. Trends in Genetics. 16(7): 321–323

11 Retief JD, Lynch KR and Pearson WR (1999) Panning for genes- a visual strategy for identifying novel gene orthologs and paralogs. Gen Res 9: 373–382

12 Muller GB and Wagner GP (1996) Homology, hox genes, and developmental integration. Amer. Zool. 36: 4–13

13 Ashburner M, Ball CA, Blake JA, Botstein D et al. (2000) Gene Ontology: tool for the unification of biology. Nat Genet. (1): 25–9

14 Fitch WM (1970) Distinguishing homologous from analogous proteins. Syst. Zool. 19: 99–113

15 Reeck GR, de Haen C, Tller DC, Doolittle RF et al. (1987) "Homology" in proteins and nucleic acids: A terminology muddle and a way out of it. Cell 50: 667

16 Murray AW (2000) Whither Genomics? Genome Biology 1(1)003.1–003.6

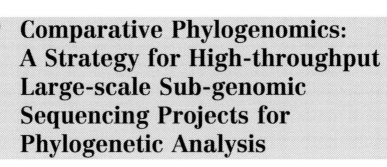

Comparative Phylogenomics: A Strategy for High-throughput Large-scale Sub-genomic Sequencing Projects for Phylogenetic Analysis

Aloysius J. Phillips

Contents

1 Introduction

Since rapid sequencing of bacterial genomes is currently feasible, it has been suggested that sequences for entire genomes of complex organisms will be commonplace in the future. We do not anticipate this happening within the near future, although the technology advances rapidly: A systematics thesis from the late 1980s would commonly be based on 5,000 base pairs of sequence, a thesis of a student of the early 1990s was commonly based on about 50,000 bases of sequence and students currently doing systematics PhDs are commonly basing their theses on 500,000 base pairs of data. Comparative genomics with just ten higher eukaryotic taxa might require a thesis to hold over 50,000,000,000 bases of sequence if entire genomes are utilized. This jump in information content would require the technology to implement a 10,000-fold increase in efficiency.

Methods and Tools in Biosciences and Medicine
Techniques in molecular systematics and evolution, ed. by Rob DeSalle et al.
© 2002 Birkhäuser Verlag Basel/Switzerland

Systematists interested in applying genome-level techniques to their systems must work under the assumption that it is not practicable to sequence the entire genome of a large number of eukaryotic taxa. Genome projects are expensive; they require a large commitment of resources and take several years to complete. In general, genomics labs are not interested in phylogenetics or at least it is certainly not their primary goal. Nevertheless, molecular phylogenetics has much to gain from the data and the techniques that are being produced by the genomics community. Megabase sequencing technologies and genome sequencing initiatives make available a large collection of molecular characters unavailable at this point in time. Given the emerging technologies from the genome projects now under way, how do we maximize the amount of data from as many taxa as possible in the most cost-effective manner using lessons learned from the genome projects? This chapter briefly describes and examines genome cloning and sequencing technology and attempts to point out where the technology can help systematists advance their data collection methods.

The paradigm of molecular systematics has been to focus on relatively few candidate genes that are easy to PCR or clone, easy to sequence and contain slowly evolving regions for which robust PCR primers can be designed (Fig. 1). Often these loci are present in high copy number in whole cell DNA preps (mtDNA or rDNA) and hence can be retrieved from poor-quality tissues. Two questions should arise when examining Figure 1 – how many taxa are needed to make a systematic study meaningful? And how much sequence is needed to obtain robust results? The first question about number of taxa is a matter of

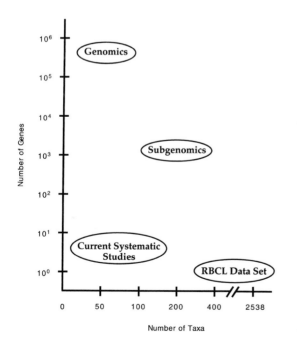

Figure 1 Schematic diagram showing the relative positions of the various approaches to obtaining sequence information in organismal studies. The X axis represents the average or relative number of taxa in studies. The Y axis represents the number of genes or gene regions targeted by particular studies.

level and depends on the phylogenetic question being asked. The second question is important in the sense that if current molecular character sets (one to a few separate gene systems) are adequate to make robust statements about phylogenetic relationships of organisms then there is no need to add 100,000 times more characters to resolve the question. We suggest that specific larger subsets of genomes can be targeted for future systematics sequencing projects. The targeted subsets can generate anywhere from 100 to 1000 times more sequence for a taxon than is currently generated.

There are only a few taxa from which we have a large amount of sequence data. These taxa belong to the venerated group of what are called "model" organisms (*Drosophila melanogaster*, *Caenorhabditis elegans*, *Mus musculus*, *Arabidopsis thaliana*, etc.). Most studies concerning "model" organisms seek to establish unifying biological principles. These organisms were chosen because they were amenable to genetic crossing and screening of mutant phenotypes, and they have been of enormous value in establishing the mechanisms of many human genetic diseases. Genomic studies of model organisms are only now being approached within a comparative framework.

Although we feel the current genome projects and their interpretation could benefit greatly from the infusion of systematic techniques and theory [1], how systematics can contribute to genomics is not the goal of this paper. Rather, in the post-genomic era the techniques used to investigate these "models" can be put to use by phylogeneticists to further the exploration of organismal diversity and molecular evolution. We suggest that there are methodologies in place today that can be co-opted and combined to establish a novel strategy for large-scale, high-throughput DNA sequencing projects for phylogenetic analysis.

2   Genomics and Systematics

There are two major ways in which we see genomics technology making an immediate contribution to systematics and phylogenetics. First we can simply continue to obtain whole genome sequences from a new set of model organisms. This demands the question of which taxa should be chosen for the next generation of genome projects. The criteria that should be used to make decisions on which organisms are chosen should be phylogenetically based. In order to do this we need more metazoan genome projects from lineages not yet represented, as well as mapped, sequence-ready reference libraries from related taxa. These projects can be used to access sequence data from many different organisms in a high-throughput framework. Basal chordates such as tunicates, cephalochordates, cyclostomes and invertebrate phyla not yet sequenced, such as Annelida and Platyhelminthes, are notable candidates for second-generation genome projects based on their intermediate phylogenetic position relative to current projects (Fig. 2). The information obtained from these second-generation genomic model systems in the form of both clone-based physical mapping studies and raw sequence can be used to focus large-

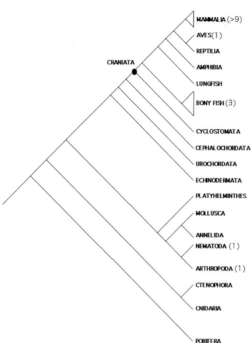

Figure 2 Current widely accepted phylogeny for the major lineages of metazoan animals.Numbers in parentheses after the major taxonomic name indicates the number of whole genome projects in progress

scale sequencing projects for many other taxa on a subgenomic scale. Whole-genome sequences from these organisms would then be acquired using the current genomic technology. Below we outline the workings of genome projects to describe in detail the complexity of the technology utilized and the requirements needed to complete a full-scale genomic project. As we will see, some of the organisms systematists work with are not even amenable to current genomics techniques.

This first approach of broadening our base of model organisms has an important ramification – the majority of biodiversity will be left by the wayside during its implementation if this approach is the only one we take. At the same time we feel strongly that genomics technology should be applied to modern systematics. From the systematists' point of view it is more efficient in the phylogenetic framework to subsample homologous regions from hundreds or even thousands of genomes simultaneously than to approach individual genomes as unitary projects. This approach poses special problems to which the genome projects can offer solutions that are discussed below. We call this the "subgenomic" approach.

3   Genomics Techniques

In order to place the generation of characters from genomics technology into a proper context, we present here in brief, some methodologies developed for genome level sequencing projects. There are several different strategies for carrying out a genome project. At one extreme is the shotgun sequencing procedure. This approach simply shears the DNA into small pieces, all fragments are cloned and a vast number of random clones are sequenced until the entire genome (or close to it) is complete. At the other extreme, one can construct physical maps based on overlapping large scale clones from artificial chromosomes. These clones are then subcloned into smaller libraries, mapped again and then sequenced using a directed approach. Each strategy has its advantages and disadvantages. In reality most projects use a combination of several techniques.

3.1   Cloning and library construction:

The ability to clone DNA fragments in excess of 100 Kb is crucial to physical mapping, positional cloning, and many types of molecular analysis of eukaryotic genomes. Conventional Cosmids hold 40–50 Kb of DNA and are not usually sufficient for analysis of very large genomic regions. Vectors have been developed which have chromosome-like properties that can accommodate longer DNA sequences. These artificial chromosome vectors originally were derived from chromosomal elements from *Saccharomyces cerevisiae*, the common baker's yeast that is used as a powerful eukaryotic genetic model organism. These vectors are known as Yeast Artificial Chromosomes (YACs). There are many YAC libraries available. Unfortunately, YACs have some technical drawbacks. Spontaneous deletions often occur, as well as concatamerization of multiple noncontiguous DNA fragments during cloning. The occurrence of chimeric and deleted inserts in YAC libraries can be on the order of 50%. Verification of clone integrity requires a substantial amount of overlap with other clones. This decreases the efficiency gained by the large size of the insert. YACs are also difficult to purify. Due to their large size they can easily shear and they can also be contaminated with native yeast chromosomal DNA.

Other artificial chromosome vectors have been developed. Bacterial artificial chromosomes (BACs) are based on the *Escherichia coli* fertility plasmid or F factor; P1-derived artificial chromosomes (PACs) combine features from both the P1 bacteriophage and F-factor plasmid. These vectors are propagated in *E. coli*. The average size of a BAC clone is about 120 Kb and the average size of a PAC clone is about 150 Kb. Since it is easier to clone DNA fragments into these vectors and they have a lower occurrence of chimeras, BACs and PACS are replacing YAC-based cloning in the construction of contig and physical maps. In the future, long-range physical mapping, positional cloning and whole-genome sequencing projects will probably rely more and more on BACs and PACs than on YACs.

3.2 Megabase DNA isolation

In order to construct DNA libraries in either YAC, BAC or PAC libraries, methods must be developed which isolate very high molecular weight DNA. To isolate the DNA cells, nuclei or protoplasts must be embedded in low melting point (LMP) agarose or agarose microbeads. The agarose acts as a semisolid porous matrix, which allows for diffusion of various reagents for DNA purification and subsequent enzymatic manipulations while preventing the DNA strand from being broken by shearing forces.

Obtaining sufficient quantities of very large DNA is problematic and would rule out many samples currently used in systematics studies. The best way to obtain a sufficient quantity of high molecular weight DNA is with tissue culture. A suspension of intact dividing cells can be produced which is amenable to the manipulations required to obtain high quality DNA. However, culture conditions need to be worked out for each organism. This is usually done through trial and error and can be quite time-consuming. Most non-model organisms do not have cell lines nor has there been much primary tissue culture. Primary tissue culture requires intact living tissue for propagation and will produce a limited number of cell divisions. Fresh tissue can also be used to generate high-quality DNA but it requires special treatments, such as enzymatic digestion of the extracellular matrix, and can involve shearing forces to free the cells.

In order to have cohesive ends for cloning, the DNA must be digested with a restriction enzyme. The most common method is to partially digest the genome. Since restriction sites are not randomly dispersed throughout the genome, complete digestion with a rare cutter would result in certain regions of the genome not being represented in the library. If the DNA is only partially digested at a relatively common restriction site, then the library will be more representative. The proper conditions for partial digestion to generate a population of fragments of the correct size range must be determined empirically. These restriction digests are typically performed while the tissue is embedded in the LMP agarose blocks. These incomplete digests can be produced either through timed digestions or through partial methylation of common restriction sites using DNA methyltransferases. Timed digestions create a broader range of fragments which need be size selected. Methyltransferase/endonuclease combinations create a more discrete size class of fragments [2].

The digested DNA must then be analysed by size using pulse field gel electrophoresis (PFGE). PFGE requires specialized apparatus that specifically alternates the direction of the electric field, while varying its intensity and duration. PFGE allows separation of DNA molecules as large as 10 Mb [3].

Conventional electrophoresis methods can only resolve up to 50 kb. Cells or tissue previously embedded in agarose can be loaded directly into the gel without pipetting which will damage the DNA. Size selection of larger target fragments from the smaller fragments ameliorates the occurrence of concatenated chimeric clones in the library. PFGE is not only crucial in isolating DNA fragments of the proper size for library construction but it is essential in the isolation, purification and analysis of clones from the library.

3.3 Physical mapping

In light of the recent success in whole-genome shotgun sequencing [4, 5], why bother constructing physical maps? A genome that has a complete or extensive contig map does not necessarily need to be sequenced completely. Annotated clones can be outsourced to labs with specific questions that are not equipped to do large-scale genomics work. These sequence-ready maps are of great utility in large-scale sequencing. They decrease the amount of redundant sequencing and the organization of the clones assists in distributing the effort between labs. In the context of sub-genomic projects these maps are necessary in order to avoid whole-genome sequencing. Arrayed libraries of different taxa provide an invaluable tool for isolating orthologous and paralogous regions for comparison.

Genetic mapping requires isolation of mutations and sufficient genetics to observe crossover events. Most organisms do not fit the requirements of genetic models. Physical mapping is more direct and represents the actual physical distance between genes on the chromosome. Physical mapping requires unique (single copy) markers that are derived from known sequence(s). Sequence tagged sites (STS) are short sequences of DNA (200–500 bp) that are usually obtained by sequencing both ends of a cloned, sequence using vector sequence as a priming site. When STSs are obtained from end sequencing of clones, they are also referred to as sequence tag connectors (STC). From these sequences, PCR primers are constructed. If any two clones yield the same PCR product it is assumed that they contain overlapping sequence(s). With many STS primer sets a large number of clones can be ordered into what is called a contig map. The goal is to construct a complete contig map for every chromosome with the minimum number of clones. A minimally overlapping set of clones (or tiling path) reduces the amount of redundant sequencing between contigs.

There are alternative markers for generating sequence-ready maps. Restriction digest fingerprinting can be used to generate maps, but can be troublesome to construct, because most enzymes are not randomly distributed within the genome and are sometimes blocked by methylation. Microsatellites are thought to be relatively randomly distributed throughout the genome and can be useful as markers as well. The expressed sequence tag (EST) technique consists of isolating mRNA from a specific tissue, making cDNA and cloning the cDNA for high-throughput sequencing where the orphan sequences are collected in a database. An immediate problem with this approach is that many of the tissues available to the systematist are not preserved for isolation of RNA. Library walking uses probes derived from the clone to find contiguous clones in the library. This method can detect gaps in the coverage of the library. Hybridization-based screening protocols can be vulnerable to difficulties caused by repetitive sequences. Most mapping strategies require some combination of these methods. The end result is a "sequence -ready" physical map of minimally overlapping clones for a single chromosome.

3.4 Shotgun sequencing

DNA sequencing strategies fall into two distinct categories, directed and random. In the directed strategy, individual reads are ordered and overlapped in relation to some starting sequence using methods such as subcloning, nested deletion, transposon-facilitated sequencing and primer walking. Each round of sequencing is based on information derived from the previous round. These methods reduce the amount of redundant sequencing but require a certain amount of effort to generate primers and ordered templates. In contrast, the random or shotgun sequencing strategy picks clones at random and thus allows the fragments from the original clone to be sequenced without prior selection. This is accomplished by cleaving the original BAC or Cosmid randomly either by shearing, sonication or restriction enzymes and randomly cloning all fragments into an appropriate vector such as M13, pUC or pBluescript. DNA from the vector will also be incidentally cloned. The sequence from these clones can be filtered out from a contig-generating program. The random method is then dependent upon the capacity of computer assembly programs to overlap the reads to generate contiguous DNA sequences. The challenge in the random strategy is to minimize sequencing redundancy while generating overlaps that close gaps between sets of clones. The main advantage is that it is suitable for rapid high-throughput strategies, with the resulting high accuracy of the sequence due to multiple overlapping sequence reads.

3.5 Finishing

As a sequencing project nears completion, the number of novel clones sequenced diminishes. Due to nonrandomness of cloning preference and toxic sequences, some regions of the genome are not represented in the contigs. A significant effort is required to obtain these difficult sequences. This aspect of sequencing projects is called 'finishing'. Some of these sequence gaps can be filled by walking, using synthetic primers to sequence clones or genomic DNA.

This is not always feasible since many gaps are caused by repetitive DNA, where there are multiple priming sites and are therefore difficult to PCR and sequence. PCR using primers outside of the repeat region can solve this problem but sequences on both ends of the region are not always known. There are various other technical approaches to obtaining problematic sequences at the terminal stage of a genome project. The bottom line is that finishing is expensive with little return phylogenetically. With a diminishing rate of data accumulation, finishing is not cost-effective and informative in a comparative framework and effort is better spent on increasing taxon sampling.

4 Subgenomics

Some of the methods described above for whole-genome analysis of model organisms can be co-opted to produce large amounts of subgenomic information for systematists. Certain limitations of the technology from above should be discussed first. The quality of DNA in systematics studies is often an issue. Often times very poor quality starting material for DNA isolation is all that is available for critical taxa. In addition, the only available specimens for a study are museum preserved (pinned insects, ethanol-preserved specimens or dried skins). These types of specimens prohibit whole-genome analysis in the way that it is currently practiced.

4.1 Comparative maps and synteny

Useful adaptation of genomic techniques to systematics entails establishing syntenic groups between taxa. Comparative maps pinpoint regions of conserved similarity. Methods for obtaining large regions of synteny (such as HOM C regions in animals) exist that may or may not require physical maps (see ZOO-FISH in the last part of this chapter). A comparative map between human and mouse reveals 105 conserved autosomal segments and 8 conserved X chromosome segments (www.informatics.jax.org) with an average length of about 15 Mb. Over 60% of mouse chromosome 11 is syntenic with human chromosome 17 . Data from other species such as chicken and *Fugu* indicate that there is still significant synteny compared to human [6–8]. Given this degree of conservation, clones of syntenic regions can be sequenced and this comparative information could be used to generate a battery of PCR primers that are applicable to a wide diversity of related organisms. Once isolated, shotgun cloning of a large region of synteny is amenable to high-throughput sequencing analysis and has the potential to generate immensely more sequence data than possible at present. Small labs could do sub-genomic projects in a problem-directed manner as opposed to the hypothesis-free approach of standard genomics projects. The one constraint is that chromosomal-length

DNA is required to identify the region of synteny. Fortunately, it is likely that only a few comparative maps are required in order to generate the syntenic multiplex primer sets.

4.2   Primer batteries and multiplexing

With the completion of several eukaryotic organisms' genomes, regions of high similarity in a wide range of genes and gene regions can be detected and primers for these regions can be designed (see chapter 14, this volume). The sequencing of these multiple PCR primer systems is a problem that high throughput sequencing analysis and multiplex priming can solve. Multiplex PCR was first described by Chamberlain et al. [9] as an application to detect deletions in the Duchenne muscular dystrophy locus. The regions of the gene most prone to deletion are simultaneously amplified in the same PCR reaction using six primer sets. A missing or size shifted PCR product indicates a deletion. Multiplex PCR has become a popular diagnostic tool for genetic diseases. Multiplex PCR is capable of both amplifying many (~25plex) regions at once or, in the case of long-range accurate PCR (LAPCR), generating several (7plex) large (up to 24 kb) PCR products [10]. LAPCR [11] utilizes a combination of thermostable polymerases, at least one of which has 3'–5' exonuclease activity such as Pfu or Klentaq to remove mismatched nucleotides. Multiplex PCR products that are too large to be sequenced directly without internal primers can be sheared and subcloned into an appropriate vector such as M13, pUC or pBluescript and shotgun sequenced. Alternatively, the clone can be sequenced directly using a Tn5 transposon based insertion of bidirectional sequencing primer sites [12, 13].

Ultimately, multiplex LAPCR in conjunction with shotgun cloning and sequencing can generate very large amounts of sequence without the requirements of internal priming information. This entire procedure is amenable to high throughput robotic sequencing technology. As with any kind of PCR protocol a certain amount of troubleshooting is required and these parameter variations have been discussed in detail elsewhere [14]. Most importantly, this type of protocol can be performed on DNA samples that can not be used to generate megabase clones due to degradation. Good field samples of ethanol preserved specimens generally yield DNA fragments on the order of 30 kb [15]. Fortunately, this size class is adequate for current LAPCR protocols.

5   Remarks and Conclusions

One of the most interesting and challenging issues in molecular systematics is the analysis of large datsets [16, 17]. Although taxonomic sampling is perhaps the foremost limitation in current studies, increased sampling of sequence data from individual species should improve resolution of phylogenetic relationships, particularly in datasets with many taxa .

Comparative maps of key taxa will provide evidence of synteny. This synteny suggests large-scale genomic homology between taxa. These maps provide a window into genomic regions not yet explored and they have the potential to provide a bounty of phylogenetic signals. The information is not only in the form of nucleotide sequence data but large-scale genomic characters such as translocations, inversions, deletions and duplications. Rare large-scale genomic events are now being considered as characters for phylogenetics [18]

Advances in high-throughput robotics and computer algorithms should be brought to bear by the molecular systematics community to identify and sequence large syntenic regions conserved over particular taxonomic divisions. Comparison between broad-ranging taxa can be used to establish efficient procedures to acquire sequence data from intervening taxa. These advances in molecular phylogenetics should not be limited to large dedicated facilities. With some organization, a central repository of archived genomic resources of biological diversity in the form of comparative map data, artificial chromosome libraries and sequence-ready arrayed libraries of key taxa could be outsourced to serve the molecular systematics community. Sequence-ready arrayed libraries are of great utility in comparative phylogenomics. Centralized facilities can farm out specific gridded libraries that can be screened and used by small labs for many different questions. The results of the collective action of these projects can be entered into a publicly available database.

The data generated by a genome sequencing project provide immense opportunities for study and analysis. Once a genome is sequenced, the genome center that produced the data must turn to a new organism. How does one decide which is the next organism to be sequenced? We submit that it is more informative and has greater general applicability to choose taxa from Phyla not yet represented. As indicated in Figure 2, the majority of genomic biodiversity is considerably underrepresented.

Glossary

Annotation – Organization, databasing and graphical representation of both library and genomic information including, clone designation, physical map and sequence data together with predicted genes, introns, promoters, alternative splicing, local, inverted and tandem repeats, and centromere and telomere location from genome sequencing projects.

Arrayed Library – Individual clones, usually Cosmids, YACs, BACs or PACs, are ordered in two-dimensional arrays of microtiter dishes. Each clone has an address that can be identified by its location (row and column) on a specific plate. Microtiter arrays are then duplicated and stored for distribution.

Arrayed libraries of clones can be used for many applications, including screening for a specific gene or genomic region of interest as well as for physical mapping. Information gathered on individual clones from various linkage and physical map analyses can be entered into a database and used to construct annotated physical maps.

Bacterial Artificial Chromosome (BAC) – A bacterial cloning system based on the *E. coli* F factor. It is capable of maintaining fragments up to 300 kb. It has the advantage of being very stable.

Comparative Map – The analysis of chromosomal location of homologous regions in different species. These maps are used to study chromosomal changes over time, to identify hereditary disease genes and to facilitate mapping in other species.

Contig – A set of overlapping clones covering an uninterrupted stretch of the genome. These are constructed by detecting overlaps by means of shared restriction sites or shared sequence-tagged sites (STS).

Cosmid – A cloning vector that is a hybrid between a phage DNA molecule and a plasmid. The Cos sites of λ-phage DNA when inserted into a plasmid allow the plasmid to be packaged into a λ-phage head *in vitro*. The λ-phage can then be used to infect *E. coli*. Cosmids can package inserts up to 47 kb.

Expressed Sequence Tags (EST) – Partial sequence from the ends of cDNA clones chosen at random from a library. This allows rapid identification of expressed genes and is useful in genome mapping.

FISH – *In situ* hybridization of a fluorescently tagged DNA probe to a preparation of condensed metaphase chromosomes or less condensed somatic interphase chromatin.

Multiplex PCR – A PCR reaction that utilizes several primer sets simultaneously to generate multiple products from separate regions of the genome.

P1-based Artificial Chromosome (PAC) – A vector based upon the P1 bacteriophage. P1 vectors are capable of cloning from 70–100kb fragments.

Sequence Tagged Sites (STS) – A short (200–500 bp.) single-copy sequence that characterizes mapping landmarks on a genome and can be detected by PCR reactions. The region of a genome can be mapped by determining the order of a series of STSs.

Shotgun Cloning – A cloning strategy in which genomic DNA is cleaved into fragments; these fragments are then inserted into a vector, which is used to transform bacteria. The product is a large number of clones (library) that between them contain the original genome.

Shotgun Sequencing – A sequencing strategy that picks clones from a library at random without any concern for what the clone contains. These orphan sequences are then submitted to an alignment regime that generates overlapping contigs. The program can filter out vector sequence or sequence that does not relate to the contig.

Synteny – Conserved order of genetic loci between the genomes of different taxa.

Yeast Artificial Chromosome (YAC) – A cloning system used to clone large (500 Kb) continuous segments of DNA, which is then used to stably transform yeast cells. The basic vector is Yac2, a plasmid that is readily propagated in *E. coli*. It is cleaved into two fragments that constitute chromosome arms, one arm of a telomere, a centromere, an autonomously replicating sequence (ARS) and a selectable marker. The other arm has a telomere and a selectable marker. The chromosome arms are then mixed and ligated with large (>100 kb.) exogenous DNA. The reconstituted chromosomes are then introduced into yeast. With selection only yeast that contains a construct with both arms survives.

ZOO-FISH Recent advances in the mapping of the human genome have generated a wealth of data that may be applied to other "map poor" species. Detailed comparative maps will facilitate targeted identification and cloning of evolutionarily conserved segments or phylogenetically informative genes and genome regions. Since conserved synteny can be maintained despite the karyotype modification that can occur during evolution, this extensive synteny can be observed directly by the use of heterologous chromosome painting, or ZOO-FISH, where whole chromosome specific probes from one species are hybridized to metaphase spreads of other species.

Not all taxa can feasibly be physically mapped. One needs large clones and metaphase chromosomes. Many taxa will not yield sufficient high-quality DNA for library construction, or these may be extinct taxa (ancient DNA), or these may be taxa with very large genomes for which we don't have sufficient STSs to physically map the genome or clones. These are the taxa that need to be approached using alternative approaches to library construction and mapping [19, 20]

References

1   Thornton JW and DeSalle R (2000) Gene family evolution and homology: Genomics meets phylogenetics. *Annu. Rev. Hum. Genom.* 1: 41–73

2   Nelson M, McClelland M (1992) Use of DNA methyltransferase/endonuclease enzyme combinations for maegabase mapping of chromosomes. *Meth. Enz.* 216: 279–303

3   Birren B, Lai E (1993) Pulse field gel electrophoresis: a practical approach (San Diego: Academic Press)

4   Adams MD, CS, Holt RA, Evans CA et al. (2000) The genome sequence of Drosophila melanogaster. *Science* 87, 2204–15

5   Myers EW, SG, Delcher AL, Dew IM, Fasulo DP (2000) A whole-genome assembly of Drosophila. *Science* 287, 2196–204

6   Burt DW, B. C., Dunn IC, Jones CT et al (1999) The dynamics of chromosome evolution in birds and mammals. *Nature* 402: 411–13

7   Schofield JP, Elgar G, Greystrong J et al (1997) Regions of human chromosome 2 (2q32-q35) and mouse chromosome 1 show synteny with the pufferfish genome (*Fugu rubripes*). *Genomics* 45, 158–67

8   Trower MK, Purvis IJ, Sanseau P et al. (1996) Conservation of synteny between the genome of the pufferfish (*Fugu rubripes*) and the region on human chromosome 14 (14q24.3) associated with familial Alzheimer disease. *Proc Natl Acad Sci U S A* 93, 1366–9

9   Chamberlain JS, Gibbs RA, Ranier JE et al. (1988) Deletion screening of the Duchenne muscular dystrophy locus via multiplex DNA amplification. *Nuclei Acids Res.* Dec 9;16(23):11141-56

10 Sorokin A, Lapidus A, Capuano V, Galleron N et al. (1996) A new approach using multiplex long accurate PCR and yeast artificial chromosomes for bacterial chromosome mapping and sequencing. *Genome Methods* 6, 448–453

11 Cheng S, Fockler C, Barnes WM et al. (1994) Effective amplification of long targets from cloned inserts and human genomic DNA. *Proc Natl Acad Sci* USA 19: 5695–5699

12 Nag DK, Huang HV, Berg DE (1988) Bidirectional chain-termination nucleotide sequencing: transposon Tn5seq1 as a mobile source of primer sites. *Gene* 64: 135–45

13 Goryshin IY and Reznikoff WS (1998) Tn5 *in vitro* transposition. *J. Biol. Chem.* 273: 7363–74

14 Zangenberg G, Saiki RK and Reynolds R (1999) Multiplex PCR: Optimization and guidelines. In: M Innis, GDH and JJ Sninsky, (eds): *PCR applications.* Academic Press, San Diego, 566

15 DeSalle R and Bonwich E (1996) DNA isolation, manipulation and characterization from old tissues. *Genet Eng* (N Y) 18: 13–32

16 Chase MW, Soltis DE, Olmsted RG et al (1993) Phylogenetics of seed plants: An analysis of nucleotide sequences from the plastid gene rbcL. *Annals of the Missuri Botanical Garden* 80: 528–80

17 Kallersjo M, Farris JS, Chase MW et al. (1998) Simultaneous parsemony jackknife analysis of 2538 rbcL DNA sequences reveals support for major clades of green plants, land plants, seed plants and flowering plants. *Plant Syst Evol* 213: 259–287

18 Rokas A and Holland PWH (2000) Rare genomic changes as a tool for phylogenetics. *TREE* 15: 454–459

19 Breen M, Thomas R, Binns MM et al. (1999) Reciprocal chromosome painting reveals detailed regions of conserved synteny between the karyotypes of the domestic dog *(Canis familiaris)* and human. *Genomics* Oct 15;61(2):145–55

20 Scalzi JM and Hozier JC (1998) Comparative genome mapping: mouse and rat homologies revealed by fluorescence in situ hybridization. *Genomics* Jan 1;47(1):44–51

9 Comparative Methods

Richard H. Baker

Contents

1 Introduction

Over the past ten years the comparative method has taken a central place in evolutionary biology. Following the seminal paper by Felsenstein [1] and the general treatments of Harvey and Pagel [2] and Brooks and McLennan [3], it is now widely recognized that knowledge of phylogenetic relationships among species is critical to understanding the evolutionary causes of interspecific patterns. This important theoretical advance, combined with the advent of molecular sequence data, has brought forth a surge of interest in using the phylogenetic relationships among species to test various evolutionary scenarios. Several different approaches now fall under the rubric of the comparative method and have been used to examine such evolutionary phenomena as (co)adaptation, constraints, allometry, trends, coevolution and ecological transformations. The major advance of these procedures centers on the recognition that, because species are connected in a historical hierarchy, they do not represent independent data points in a statistical analysis and that inferences concerning the rate and direction of evolutionary change are dependent on relationships among species. In addition, evolutionary transitions, rather than individual species, are often the variables of interest for adaptive hypotheses. Numerous reviews of comparative methodologies already exist [2, 4–6] and a comprehensive treatment of the entire field is beyond the scope of this chapter. Therefore, this review will 1) briefly present the basic conceptual issues for the most commonly used methodologies 2) provide a primer to the comparative method literature that can guide investigators new to the field to studies using different approaches and methodologies, and 3) highlight some of the controversies and advances that are at the forefront of the discipline.

Methods and Tools in Biosciences and Medicine
Techniques in molecular systematics and evolution, ed. by Rob DeSalle et al.
© 2002 Birkhäuser Verlag Basel/Switzerland

2   Correlated Evolution and Independent Contrasts

The majority of comparative analyses are interested in the correlated evolution of two continuous traits on a phylogeny and independent contrasts is now the most widespread method for testing this pattern. Recent tests of correlated evolution include the relationship between testis mass and sperm length [7], body mass and home range [8], and sexual dimorphism and mating systems [9]. Figure 1 illustrates the potential effect historical hierarchy can exert on the interpretation of correlated change among species. On the first tree in this example, all of the correlated change between the variables x and y has occurred along the basal split of the tree. The other branches of the tree do not support a positive relationship between the variables. An ahistorical correlation using only the species values, however, would support a relationship between the variables (i. e., Type I error), whereas a historical analysis correctly identifies that the relationship is limited to a single data point and that there is no general pattern. Alternatively, in the second tree, a positive relationship exists between the two variables in each of the two major clades, but this relationship is hidden when the species values are plotted by themselves.

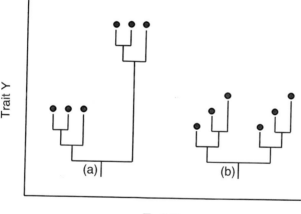

Figure1   The effect of historical hierarchy on interspecific correlation patterns. Each tree depicts hypothetical branching patterns and mean values for two continuous traits, X and Y, for six species. Tree (a) provides an example in which an ahistorical analysis suggests a positive relationship between the traits which is not supported when phylogenetic relationships are accounted for. In this case, all of the correlated change between the traits is limited to the basal split on the tree and not supported by the other branches. Tree (b) shows an alternative example in which a positive relationship between the variables would be rejected by an ahistorical analysis but supported by a phylogenetic analysis. In this case, the correlated evolution of the traits in each of the two clades is masked by the uncorrelated change in the basal branch when the species values are plotted by themselves.

Therefore, an ahistorical analysis would result in the rejection of a relationship between the variables when, in fact, one exists (i. e., Type II error).

Originally developed by Felsenstein [1] as a means for removing the statistical dependence created by phylogenetic relationships, independent contrasts examines evolutionary change horizontally along a tree by calculating the difference between the values at two derived nodes coming off from a given ancestral node (see Figure 2 for an example of how independent contrasts are calculated). For a tree with n species there will be n-1 contrasts. Statistical independence between data points is achieved because no two contrasts share the same tip or reconstructed trait value. Independent contrasts assumes a Brownian motion model of evolutionary change which is often used to describe evolution under random genetic drift. Under Brownian motion, the amount of divergence between two nodes is proportional to time, which is usually represented by the branch lengths separating these nodes. Therefore, contrasts are standardized by dividing each one by the square root of the sum of its branch lengths. Once standardized, the contrasts represent data points drawn from a normal probability distribution with the same mean and variance and can then be included in a standard statistical analysis. Garland et al. [10] stress the importance of visual inspection of the contrasts and their standard deviations

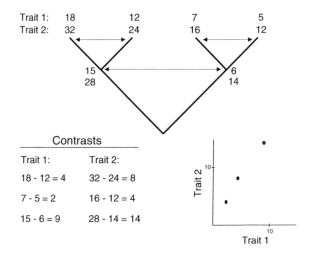

Figure 2 The calculation of independent contrasts. A hypothetical tree for four species shows the mean trait values for each of the species and the reconstructed ancestral states of these traits. Contrasts are calculated by taking the difference between two derived nodes coming off a given ancestral node. The graph on the right plots the corresponding values of each contrast for the two traits and reveals a positive correlation between the magnitude of change in the two traits. In this example, all the branch lengths have been set to one, but, in real analyses, contrasts are normally divided by the square root of the sum of their branch lengths before they are included in a statistical analysis.

to ensure appropriate standardization. With independent contrasts, the ancestral value at a given node is calculated using a weighted average of the two descendent values leading from that node and do not represent the most parsimonious reconstruction for a trait.

There are several alternative methods for calculating correlated evolution and the assumptions and implementation of most of these are described in Martins and Hansen [5]. A method that uses a parsimony criterion is the minimum evolution method [11]. Ancestral values are reconstructed using square-change parsimony which minimizes the sum of the squared changes along the tree. Unlike independent contrasts, the minimum evolution method examines evolutionary change vertically along the tree so that the data points are the difference between the ancestral and derived value for each branch of the tree. This leads to about twice as many data points as the independent contrasts method but with less independence among the data points. In addition, because the number of data points in any statistical test (i. e., 2n-1 evolutionary changes) is estimated from only n species, the degrees of freedom are inappropriately inflated [11, 12]. Therefore, it is more appropriate to use the number of species rather than the number of data points to compute confidence limits. Despite the statistical difficulties associated with the minimum evolution method, a few studies have suggested that it often produces estimates of evolutionary correlations similar to or superior to other methods [13–15].

Results from correlation analyses are often used to support hypotheses concerning adaptation. The validity of this approach has been questioned on the grounds that general evolutionary patterns reveal very little about the intraspecific mechanisms producing these patterns [16, 17]. Leroi et al. [17] have even suggested that phylogenetic patterns will often produce misleading results concerning the presence of adaptations. Correlations have multiple causes and factors such as pleiotropy and epistasis may be as important in shaping character coevolution as is adaptation. They stress the importance of obtaining genetic and developmental data for several species within a comparative analysis. This type of data, however, is often difficult to obtain for a single species, let alone several species, so it may be some time before comprehensive studies of this type emerge. In addition, because selection pressures are predicted to shape genetic and developmental relationships among traits [18], it may prove difficult to tease apart genetic and adaptive influences. In any case, it is important for researchers conducting correlation analyses to consider explanations alternative to adaptation when interpreting their results.

3   Importance of Topology

The first step in any comparative analysis entails reconstructing the phylogenetic relationships among the species of interest. Despite comparative methods being developed because of the recognition that ancestry affects evolutionary patterns, the importance of phylogenetic reconstruction has often been overlooked in the comparative literature. Recently, some authors [19] have suggested that accounting for phylogenetic information is not necessary in interspecific comparisons, while others have developed strategies for comparative analyses when robust phylogenies are not available [20, 21]. Both of these approaches, however, have been criticized.

Ricklefs and Stark [19] surveyed results from 13 comparative studies to demonstrate that there was a significant correspondence between interspecific correlations that accounted for phylogenetic relationships and the correlations between the same traits when phylogenies were ignored. Because non-phylogenetic correlations produced results similar to phylogenetic ones, they argued that it was "premature at this point to insist upon phylogenetically based approaches to comparative analysis". Their survey, however, included only 30 data points and more comprehensive analyses have highlighted areas where differences between phylogenetic and non-phylogenetic estimates can be more acute [22, 23].

Figure 3 shows the relationship between non-phylogenetic and phylogenetic (using independent contrasts) correlations for 32 morphological traits in 30 species of stalk-eyed flies. A nearly identical relationship was demonstrated by Ackerly and Donoghue ([22]; Figure 7) in a comparative analysis of morphological traits for 17 *Acer* species. As with Ricklefs and Stark [19], these graphs indicate that traits that have strong non-phylogenetic correlations also have strong independent contrasts correlations. This is not surprising given that both correlations are attempting to estimate the same relationship. However, there is also substantial spread in the data, particularly among the weaker correlations.

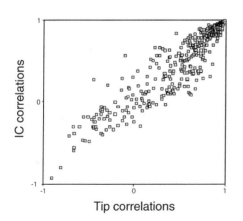

Figure 3 The relationship between the ahistorical correlation (Tip) and the phylogenetic correlation (IC) for all pairwise comparisons of 32 morphological traits for 30 species of stalk-eyed flies. The phylogenetic correlations were calculated using independent contrasts with branch length information. The tree used for these calculations was the single most-parsimonious tree derived from an analysis of six gene regions [81].

In fact, for the stalk-eyed fly measurements, of the 135 correlations weaker than 0.7, the use of independent contrasts affects the statistical interpretation of approximately one-third (42) of these. In these cases, ignoring phylogenetic information results in a different conclusion (e. g., the rejection of a relationship when in fact one is supported) than when relationships are accounted for. The difference between phylogenetic and non-phylogenetic correlations is also influenced by the degree of homoplasy in the traits measured. Ackerly and Donoghue [22] showed that this difference increases proportionally to the average amount of homoplasy of the two traits in the correlation. Therefore, accounting for phylogenetic relationships is likely to have a substantial impact on comparative results for traits that are relatively homoplasious and not strongly correlated in a non-phylogenetic analysis, but less so for traits that are strongly correlated.

Another strategy that has been suggested for comparative analyses when phylogenies are unavailable is the use of random trees [20, 21]. In this approach, if significant relationships are demonstrated on a large set of random trees, then it can safely be assumed to exist for the real topology and evolutionary parameters (i. e., allometric slopes) can be estimated by taking their average value from a set of random trees. This method, however, may be too conservative to detect many significant evolutionary patterns [23]. Abouheif [23] examined the difference in the allometric slopes and correlation coefficients for sexual size dimorphism between 1000 random trees and a tree derived from actual phylogenetic data. He found that many significant patterns of allometry that exist in the real analyses were rejected in the random analysis. Overall, estimates based on random trees appear to be good predictors of results generated from non-phylogenetic methods but poor predictors of results from actual phylogenetic relationships. This may result from the fact that even a relatively large subset of random trees is unlikely to contain much phylogenetic information in common with the actual tree. For instance, of 1000 random trees generated for 33 species of stalk-eyed flies, only 188 have at least one node that occurs on the most parsimonious tree for these taxa and only 23 contain two or more nodes in common with the most parsimonious tree.

One of the general criticisms of comparative method studies has been a lack of sensitivity to the difficulty of phylogenetic estimation [16, 24]. Taxonomic groupings or trees derived from cutting and pasting together results from different analyses are frequently used in comparative studies as a surrogate for real cladograms [9, 25–27]. Taxonomy, however, is extremely limited in its ability to accurately reflect phylogeny and should not be viewed as a reasonable substitute [28, 29]. Even when character data exist for all the species examined in a comparative analysis, they are often limited to a single data set. Despite the increasing volume of comparative studies, it is still rare to find results based on phylogenies derived from a single analysis of all the taxa using more that one data source (but see [30–32] for exceptions). Numerous molecular studies, however, have demonstrated the limitations of using a single gene region for resolving relationships [33–36].

Sensitivity analysis, in which results are examined also for near most-parsimonious topologies, has been advocated as a means for assessing the robustness of comparative results to uncertainties in phylogenetic estimation [15]. While this approach has heuristic value, there is no objective criterion for selecting the range of trees to examine and alternative topologies are still affected by biases in the specific data set. If a data set produces many erroneous results on the most-parsimonious topology due to poor taxa or character sampling, then many of these relationships will probably also be present in the near most-parsimonious trees. As with analyses derived from a set of random trees, similar comparative results from most-parsimonious and near most-parsimonious trees should not necessarily be viewed as strong corroboration for a hypothesis, especially when these trees are derived from a small data set and are weakly supported. There is no substitute for a strongly supported phylogenetic hypothesis generated from several diverse sources of character information and comparative studies need to place greater emphasis on this critical aspect of hypothesis testing.

4   Examining the Tempo and Mode of Evolutionary Change

In recent years considerable attention has focused on issues relating to rates and directionality of evolutionary change. The types of questions addressed include whether evolutionary change has occurred more often in one direction than another [37] and whether the rate of evolutionary change in a character differs between clades [38]. Figure 4 provides a simple example of how phylogenetic relationships among species are critical for calculating rates and direction of change. Both traits in this example have a similar range of values across species but differ considerably in their pattern of evolution. Trait 1 exhibits the minimal amount of change possible on the tree and all of the changes are in one direction (i. e., increases). Alternatively, trait 2 has undergone 50% more change on the tree than trait 1 (12 steps vs 8 steps) but this change has occurred in both directions.

Central to rate and direction calculations is the reconstruction of ancestral values on a phylogeny and numerous studies have examined the effect of different reconstruction methods on hypothesis testing [6, 39–42]. The majority of work has addressed the analysis of discrete traits [40, 41, 43] but many of the same concepts apply to continuous traits. While ancestral states have traditionally been reconstructed using parsimony (both linear and squared-change), maximum likelihood methods are increasingly being developed and implemented [41, 43]. These maximum likelihood methods are often preferred over parsimony methods because they provide a measure of estimation accuracy in the form of confidence limits for each reconstructed state [41, 42, 44]. Work so far in this area has indicated that the ancestral states and their confidence

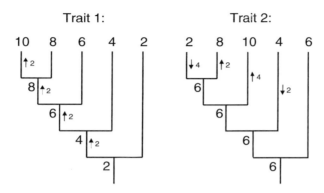

Figure 4 A simple illustration of how calculations of rates and direction of evolutionary changes are dependent on phylogenetic relationships among species. Each tree provides, for a given trait, the mean values of that trait for five species and the most parsimonious reconstructed values of these traits at the ancestral nodes of the tree. Arrows represent the direction of change in the trait on a given branch with the magnitude of the change shown next to the arrow. The two traits exhibit the same amount of variation across species (i.e., range 2–10) but very different patterns of change. The evolution of trait 1 relative to trait 2 is slower (8 steps vs 12 steps) but more trend-like (all gains and no losses vs equal numbers of gains and losses).

limits are dependent on several factors. For instance, the accuracy of reconstructions decreases as the rate of evolution in the character increases and as the placement of the node in the tree gets closer to the base [41, 45]. The number of taxa sampled, branch lengths, and the number of model parameters estimated can also significantly affect reconstructed states [46].

Maximum likelihood methods also provide a statistical framework for evaluating asymmetric rates of change [37, 47]. Numerous evolutionary theories (e.g., Dollo's Law, Cope's Rule) hypothesize that evolution is more likely in one direction than another, but rarely have these theories been tested using branching patterns among taxa. Recent examinations of directionality in evolutionary transitions using maximum likelihood tests include flower symmetry in angiosperms [37] and larval feeding in sea urchins [48]. This method assess trends by comparing the likelihood of the data on a given tree assuming equal rates of gain and losses in a character with the likelihood of the data allowing for unequal rates of gains and losses. In addition, rate tests have been developed that assess the directionality of continuous character evolution [43] and that distinguish between different modes of evolution such as gradual change, punctuated change and nonhistorical change [49]. Unlike parsimony methods, these maximum likelihood methods avoid specifying a particular transformation cost among character states by estimating rate parameters over all possible ancestral state reconstructions.

Proponents of maximum likelihood point to the incorporation of branch length information as a major advantage of these methods over parsimony methods [43, 50]. In these models, the amount of change in a character is proportional to time, with greater amounts of evolution occurring on longer branches. Results from these methods, therefore, are strongly affected by branch length calculations. For instance, analyses assuming equal branch lengths can provide significantly different results about ancestral states on the tree than analyses that estimate branch lengths [37, 48]. The reliability of branch length estimation has not received much attention in the literature, but it is clear that the rate of molecular evolution in a gene region can significantly affect their calculation. For instance, rapidly evolving genes will usually produce short internal branches relative to the tip branches. Therefore, given the diversity of evolutionary rates among different genes, branch lengths may vary considerably, depending on which genes are sampled for any given study.

As with estimates of topologies, single gene regions are generally inadequate for generating accurate estimates of branch lengths. Figure 5 compares the branch lengths generated from a data partition comprised of three nuclear gene regions with one comprised of three mitochondrial gene regions on the most-parsimonious tree for 34 species of stalk-eyed flies. As is evident from the scatterplot, the mitochondrial and nuclear data produce substantially different branch lengths. While the correlation (r = 0.380) between the estimates from the two data partitions is slightly significant, this is due mostly to the six shortest branches on both trees and the correlations between single genes from these partitions are as low as 0.197. The relationship is also not improved by calculating branch lengths using maximum likelihood. The correlation between the nuclear and mitochondrial data is 0.293 when branch lengths are generated from a HKY model, in which the ti/tv ratio, proportion of invariant sites and gamma distribution are estimated from the data. The relevance of these differences on comparative results will depend on the sensitivity of any particular analysis to branch length errors, but it highlights the substantial variation that exists in branch length estimation for different gene regions. Branch lengths are also strongly influenced by taxa sampling, such that clades which include more taxa will contain longer root to tip distances because the additional taxa help to reveal otherwise unobserved changes [51]. Overall, the enthusiasm for maximum likelihood methods based on their incorporation of branch length data should be tempered by the fact that accurate estimates of branch lengths are rarely available.

While the use of maximum likelihood methods for analyzing comparative data is sure to become more prevalent, detailed critiques of a model-based approach to studying historical events have been issued extensively [16, 24, 52, 53]. These authors point out that there is rarely strong justification for choosing among the array of parameters that can be incorporated into a model. Because results are so sensitive to differences in the model (e. g., one rate of change vs two rates of change), if the parameters are wrong, the conclusion can be incorrect, and it is difficult to determine when this has happened. In addition,

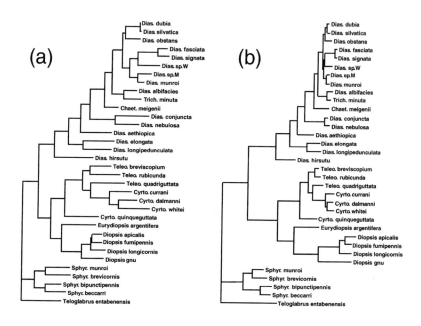

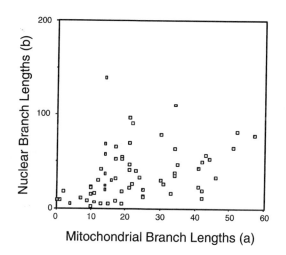

Figure 5 A comparison of branch length estimates between mitochondrial (a) and nuclear (b) data. The mitochondrial data include fragments from 16S, 12S and cytochrome oxidase II and the nuclear data include fragments from *white*, *wingless* and EF1-α. Branch lengths are calculated using an ACCTRAN reconstruction with the data constrained to the most-parsimonious tree from a combined analysis [81]. The scatterplot is between the corresponding branch lengths of the two trees.

examination of maximum likelihood in phylogenetic estimation has revealed conditions under which this method performs poorly [54], and analogous situations may exist for comparative analysis. For instance, some maximum likelihood models for reconstructing ancestral states impose rapid rates of change in a trait and produce equivocal reconstructions for nodes from which all descendent taxa share the same state [46]. Therefore, more research is needed to determine the performance of maximum likelihood methods under different evolutionary circumstances.

5   Future

The emergence of comparative methodologies was motivated by an interest in integrating historical and ecological data in a rigorous framework. Unfortunately, the popularity of these methods can promote relatively superficial analyses based on incomplete ecological data and taxonomic groupings that generally reveal little about the causes and effects operating within a system. Understanding the mechanisms of evolutionary change ultimately requires a synthesis of information from several different sources. One of the major challenges of evolutionary biology is to develop techniques and data that facilitate meaningful translation between disciplines. A particularly important area of research concerns the division between microevolutionary and macro-evolutionary phenomena and there has been increased interest in exploring this boundary [55–58]. The comparative method represents an important tool for this investigation and it is essential that studies continue this integrative approach by interpreting interspecific patterns in light of results from intraspecific ecological experiments [59], and genetic and developmental studies [60]. With a relatively robust phylogenetic hypothesis, the comparative method can provide valuable insights about the pattern of evolutionary change and identify particular taxa and variables deserving further investigation. For instance, Pitnick and his colleagues, using comprehensive comparative analyses among *Drosophila* species, have identified numerous associations among sexual traits such as sperm length, testis size, time to sexual maturity and seminal receptacle area [7, 62, 63]. Determining the evolutionary significance of these patterns requires intraspecific information, such as the genetic correlations and physiological relationships among traits [82], and the cost and benefits associated with variation in mating strategies. In general, our understanding of evolutionary change is best served by this type of multidisciplinary approach, of which comparative methods should be an integral part.

Table 1 A primer to the comparative method literature

This list is restricted to seminal papers, reviews and some significant recent additions. It is not an exhaustive survey but the majority of work on comparative methodologies, as well as many empirical papers, is cited within these papers. Discussions of primary assumptions, strengths and weaknesses of different approaches and practical suggestions are covered and the subjects are overlapping. Computer programs for various comparative methods are listed at http://evolution.genetics.washington.edu/phylip/software.html.

Comparative Method Subject	Issues Addressed	References
Correlated Evolution-Continuous Data	Examines whether evolutionary change in one character is associated with corresponding change in another character on a phylogeny. Characters can include morphological, molecular, behavioral, environmental or life-history traits. There are several different methods for calculating correlated change, although independent contrasts has become the most prevalent.	[1, 2, 5, 6, 10, 13]
Correlated Evolution-Discrete Data	Tests whether changes in a character are associated with a particular state of another character. There are tests which use both a parsimony (concentrated changes test) and maximum likelihood approach.	[1, 47, 64–66]
Adaptation	Most adaptive hypotheses require phylogenetic information on ancestral states and directionality of character change. The functional utility of a character state in a given selective regime is tested relative to the utility of its ancestral character state in that regime. This approach (the 'homology' approach) is often contrasted with other comparative methods ('homoplasy' approaches) because it focuses on a single evolutionary event within a clade rather than on statistical generalities across a clade. In addition, some authors have criticized the strictly historical view of adaptation and stressed the importance of stabilizing selection in the maintenance of character states.	[3, 29, 67–70]
Rates	Many models of evolutionary changes (i.e., punctuated equilibrium, adaptive radiations) make predictions about the relative rates of change for different characters. Commonly tested phenomena include whether two traits are evolving at different rates within a clade or if a single character is evolving at different rates on different parts of the tree. Approaches using independent contrasts, minimum evolution branch transformations and maximum likelihood have been used.	[12, 49, 71, 72]
Trends	Hypothesis concerning the reversibility (e.g., Dollo's law) of character change have always been prominent in evolutionary biology but rarely tested using phylogenetic data. Tests of trends examine the directionality of evolutionary change such as whether there are more gains or losses in a trait. Trends can be examined for both continuous and discrete data using either parsimony or maximum likelihood. The use of independent contrasts assumes there are no trends in the data.	[37, 42, 73]
Topological Uncertainties	Phylogenetic relationships used for comparative analyses are supported to varying degrees. In addition, many comparative data exist for species for which relationships are unknown. Therefore, attention has focused on the effect of incorrect topologies, as well as incomplete taxa sampling, on comparative results. Arguments supporting the use of random trees when phylogenetic data are unavailable have been made, but others have criticized this approach. Some authors have stressed the importance of sensitivity analysis in which the impact of alternative topologies is explored. The effect of polytomies on comparative results has also been investigated.	[21–23, 74, 83]
Branch Length Uncertainties	Independent contrasts and maximum likelihood analyses require calculations of expected amount of evolutionary change based on branch length information. The effect of errors in branch length estimation on comparative results has been examined. Various branch length transformations have been suggested to improve the fit between the model and the data.	[75]
Alternative Models of Change	Traditional methods assume a Brownian motion model of change in which rates are uniform across the lineages of a phylogeny and the amount of change is proportional to time (i.e., branch lengths). The impact of deviations from this model has been explored and alternative methodologies that allow rate heterogeneity and directionality have been developed. Other recent models allow the incorporation of microevolutionary assumptions and evolutionary change based on adaptive radiations.	[47, 72, 76, 77]
Confidence Limits on Ancestral Reconstructions	Ancestral character state reconstruction is a critical part of many comparative analyses. Recently, there has been increased attention to the significance of ancestral state uncertainty on hypothesis testing and methods have been developed to assess the robustness of ancestral reconstructions. These investigations have argued that reconstructions, particularly those nearest the base of the tree, generally have large confidence intervals.	[41, 42, 73, 78, 84]
Allometry	Evolutionary allometry describes the scaling relationships among traits across species. Allometric slopes are often used to identify physiological or genetic constraints. Use of phylogenetic information has been less common in studies of allometry than other fields, but evolutionary relationships can have a strong influence on allometric coefficients.	[2, 79]
Criticisms of the Comparative Method	While many papers discuss the strengths and weaknesses of particular techniques, a few authors have been more critical of the endeavor as a whole. Their objections include the assertion that statistical generalities derived from unique historical events are either inappropriate or not informative about intraspecific mechanisms, and that accommodating phylogenetic information in cross-species comparison is unnecessary.	[16, 17, 19, 29, 80]

References

1   Felsenstein J (1985) Phylogenies and the comparative method. *Am. Nat.* 125: 1–15

2   Harvey PH, Pagel MD (1991) *The comparative method in evolutionary biology.* Oxford University Press, Oxford.

3   Brooks DR and McLennan DA (1991) *Phylogeny, ecology, and behavior: A research program in comparative biology.* University of Chicago Press. Chicago, FL

4   Gittleman JL and Luh H-K (1992) On comparing comparative methods. *Annu. Rev. Ecol. Syst.* 23: 383–404

5   Martins EP, Hansen TF (1996) The statistical analysis of interspecific data: a review and evaluation of phylogenetic comparative methods. In: EP Martins (ed) *Phylogenies and the comparative method in animal behavior* Oxford University Press, Oxford, 22–75

6   Garland T Jr, Midford PE, Ives AR (1999) An introduction to phylogenetically based statistical methods, with a new method for confidence intervals on ancestral values

7   Pitnick S (1996) Investment in testes and the cost of making long sperm in *Drosophila. Am. Nat.* 148: 57–80

8   Garland T Jr, Dickerson AW, Janis CM and Jones JA (1993) Phylogenetic analysis of covariance by computer simulation. *Syst. Biol.* 42: 265–292

9   Owens IPF, Hartley IR (1998) Sexual dimorphism in birds: why are there so many different form of dimorphism? *Proc. R. Soc. Lond.* 265: 397–407

10 Garland T Jr, Harvey PH, Ives AR (1992) Procedures for the analysis of comparative data using phylogenetically independent contrasts. *Syst. Biol.* 41: 18–32

11 Huey RB, Bennett AF (1987) Phylogenetic studies of co-adaptation: Preferred temperatures versus optimal performance temperatures of lizards. *Evolution* 41: 1098–1115

12 Butler MA, Losos JB (1997) Testing for unequal amounts of evolution in continuous characters on different branches of a phylogenetic tree using linear and squared-change parsimony: An example using lesser Antillean *Anolis* lizards. *Evolution* 51: 1623–1635

13 Martins EP, Garland T Jr. (1991) Phylogenetic analyses of the correlated evolution of continuous characters: a simulation study. *Evolution* 45: 534–557

14 Martins EP (1996b) Phylogenies, spatial autoregression, and the comparative method – a computer simulation. *Evolution* 50: 1750–1765

15 Donoghue MJ, Ackerly DA (1996) Phylogenetic uncertainties and sensitivity analyses in comparative biology. *Phil. Trans. R. Soc. Lond. B* 351: 1241–1249

16 Wenzel JW, Carpenter JM (1994) Comparing methods: Adaptive traits and tests of adaptation. In: P Eggleton, R Vane-Wright (eds) *Phylogenetics and ecology.* Academic Press, London, 79–101

17 Leroi AM, Rose MR, Lauder GV (1994) What does the comparative method reveal about adaptation? *Am. Nat.* 143: 381–402

18 Cheverud JM (1984) Quantitative genetics and developmental constraints on evolution by selection. *J. Theoret. Biol.* 110: 155–171

19 Ricklefs RE, Starck JM (1996) Application of phylogenetically independent contrasts: a mixed progress report. *Oikos* 77: 167–173

20 Losos JB (1994) An approach to the analysis of comparative data when a phylogeny is unavailable or incomplete. *Syst. Biol.* 43: 117–123

21 Martins EP (1996a) Conducting phylogenetic comparative studies when the phylogeny is not known. *Evolution* 50: 12–22

22 Ackerly DD, Donoghue MJ (1998) Leaf size, sampling allometry, and Corner's Rules: phylogeny and correlated evolution in Maples (Acer). *Am. Nat.* 767–791

23 Abouheif E (1998) Random trees and the comparative method: a cautionary tale. *Evolution* 52: 1197–1204

24 Wenzel JW (1997) When is a phylogenetic test good enough? In: P Grandcolas

(ed): *The origin of biodiversity in insects: Phylogenetic tests of evolutionary scenarios.* Mem. Mus. natn. Hist. nat. 173, Paris, 31–45

25 Simmons LW, Tomkins JL (1996) Sexual selection and allometry of earwig forceps. *Evol. Ecol.* 10: 97–104

26 Prenter JR, Elwood W, Montgomery WI (1998) No association between sexual size dimorphism and life histories in spiders. *Proc. R. Soc. Lond. B* 265: 57–62

27 Morrow EH, Gage MJG (2000) The evolution of sperm length in moths. *Proc. R. Soc. Lond.* 267: 307–313

28 de Queiroz K, Gauthier J (1992) Phylogenetic taxonomy. *Ann. Rev. Ecol. Syst.* 23: 449–480

29 Coddington JA (1994) The roles of homology and convergence in studies of adaptation. In: P Eggleton, R Vane-Wright (eds): *Phylogenetics and ecology.* Academic Press, London, 53–78

30 Crespi B, Worobey M (1998) Comparative analysis of gall morphology in Australian gall thrips: the evolution of extended phenotypes. *Evolution* 52: 1686–1696

31 Vogler AP, Kelley KC (1998) Covariation of defensive traits in tiger beetles (genus *Cicindela*): a phylogenetic approach using mtDNA. *Evolution* 52: 529–538

32 Presgraves DC, Baker RH, Wilkinson GS (1999) Coevolution of sperm and female reproductive tract morphology in stalk-eyed flies. *Proc. R. Soc. Lond. B* 266: 1041

33 Cao Y, Adachi J, Janke A, Pääbo S, Hasegawa M (1994) Phylogenetic relationships among eutherian orders estimated from inferred sequences of mitochondrial proteins: Instability of a tree based on a single gene. *J. Mol. Evol.* 39: 519–527

34 Cummings MP, Otto SP, Wakeley J (1995) Sampling properties of DNA sequence data in phylogenetic analysis. *Mol. Biol. Evol.* 12: 814–822

35 Baker RH, DeSalle R (1997) Multiple sources of character information and the phylogeny of Hawaiian Drosophilids. *Syst. Biol.* 46: 654–673

36 O'Grady PM, Clark JB, Kidwell MG (1998) Phylogeny of the *Drosophila saltans* species group based on combined analysis of nuclear and mitochondrial DNA sequences. *Mol. Biol. Evol.* 15: 656–664

37 Ree RH, Donoghue MJ (1999) Inferring rates of change in flower symmetry in asterid angiosperms. *Syst. Biol.* 48: 633–641

38 Clobert J, Garland T Jr, Barbault R (1998) The evolution of demographic tactics in lizards: a test of some hypotheses concerning life history evolution. *J. Evol. Biol.* 11: 329–364

39 Schultz TR, Cocroft RB, Churchill GA (1996) The reconstruction of ancestral character states. Evolution 50: 504–511

40 Omland KE. (1997b) Examining two standard assumptions of ancestral reconstructions: Repeated loss of dichromatism in dabbling ducks (Anatini). *Evolution* 51: 1636–1646

41 Schluter D, Price T, Mooers AØ, Ludwig D (1997) Likelihood of ancestor states in adaptive radiation. *Evolution* 51: 1699–1711

42 Cunningham CW, Omland KE, Oakley TH (1998) Reconstructing ancestral character states: a critical reappraisal. *Trends Ecol. Evol.* 13: 361–366

43 Pagel M (1999) Inferring the historical patterns of biological evolution. *Nature* 401: 877–884

44 Martins EP (1999) Estimation of ancestral states of continuous characters: a computer simulation study. *Syst. Biol.* 48: 642–650

45 Maddison WP (1995) Calculating the probability distributions of ancestral states reconstructed by parsimony on phylogenetic trees. *Syst. Biol.* 44: 474–481

46 Mooers AØ, Schluter D (1999) Reconstructing ancestor states with maximum likelihood: support for one- and two-rate models. *Syst. Biol.* 48: 623–633

47 Pagel M (1997) Inferring evolutionary process from phylogenies. *Zool. Scr.* 26: 331–348

48 Cunningham CW (1999) Some limitations of ancestral character state recon-

struction when testing evolutionary hypotheses. *Syst. Biol.* 48: 482–496

49 Mooers AØ, Vamosi SM, Schluter D (1999) Using phylogenies to test macroevolutionary hypotheses of trait evolution in Cranes (Gruinae). *Am. Nat.* 154: 249–259

50 Swofford DL, Olsen GL, Waddell PJ, Hillis DM (1996) Phylogenetic inference. In: DM Hillis, C Moritz, BK Mable (eds) *Molecular systematics*, 2nd edition. Sinauer Associates, Sunderland, Massachusetts, 407–514

51 Omland KE (1997a) Correlated rates of molecular and morphological evolution. *Evolution* 51: 1381–1393

52 Farris JS (1983) The logical basis of phylogenetic analysis. In: NI Platnick, VA Funk (eds) *Advances in cladistics*, Volume 2. Columbia University Press, New York, 7–36

53 Siddall ME, Kluge AG (1997) Probabilism and phylogenetic inference. *Cladistics* 13: 313–336

54 Siddall ME (1998) Success of parsimony in the four-taxon case: long-branch repulsion by likelihood in the Farris Zone. *Cladistics* 14: 209–220

55 Schluter D (1989) Bridging population and phylogenetic approaches to the evolution of complex traits. In: DB Wake, G Roth (eds), *Complex Organismal Functions: Integration and Evolution in Vertebrates*. John Wiley & Sons, New York, 79–95

56 Futuyma DJ, Keese MC, Funk DJ (1995) Genetic constraints on macroevolution: The evolution of host affiliation in the leaf beetle genus *Ophraella*. *Evolution* 49: 797–809

57 Hansen TF, Martins EP (1996) Translating between microevolutionary process and macroevolutionary patterns: The correlation structure of interspecific data. *Evolution* 50: 1404–1417

58 Schluter D (1996) Adaptive radiation along genetic lines of least resistance. *Evolution* 50: 1766–1774

59 McLennan DA (1996) Integrating phylogenetic and experimental analyses – the evolution of male and female nuptial coloration in the stickleback fishes (gasterosteidae). *Syst. Biol.* 45: 261–277

60 Meyer A, Ritchie PA, Witte K (1995). Predicting developmental processes from evolutionary patterns: A molecular phylogeny of the zebrafish (*Danio rerio*) and its relatives. *Phil. Trans. R. Soc. Lond. B* 349: 103–111

61 Wilkinson GS, Taper M (1999) Evolution of genetic variation for condition dependent traits in stalk-eyed flies. *Proc. R. Soc. Lond. B.* 266: 1685–1690

62 Pitnick S, Markow TA, Spicer GS (1995) Delayed male maturity is a cost of producing large sperm in *Drosophila. Proc. Natl. Acad. Sci.* 92: 10614–10618

63 Pitnick S, Markow TA, Spicer GS (1999) Evolution of multiple kinds of female sperm-storage organs in *Drosphila. Evolution* 53: 1804–1822

64 Maddison WP (1990) A method for testing the correlated evolution of two binary characters: are gains and losses concentrated on certain branches of a phylogenetic tree. *Evolution* 44: 539–557

65 Pagel M (1994) Detecting correlated evolution on phylogenies: a general method for the comparative analysis of discrete characters. *Proc. R. Soc. Lond. B* 255: 37–45

66 Werdelin J, Tullberg BS (1995) A comparison of two methods to study correlated discrete characters on phylogenetic trees. *Cladistics* 11: 265–277

67 Coddington JA (1988) Cladistics tests of adaptational hypotheses. *Cladistics* 4: 1–22

68 Baum DA, Larson A (1991) Adaptation reviewed: a phylogenetic methodology for studying character macroevolution. *Syst. Zool.* 40: 1–18

69 Larson A, Losos JB (1996) Phylogenetic systematics of adaptation. In: MR Rose, GV Lauder (eds) *Adaptation*. Academic Press, San Diego, 187–220

70 Hansen TF (1997) Stabilizing selection and the comparative analysis of adaptation. *Evolution* 51: 1341–1351

71 Garland T Jr (1992) Rate tests for phenotypic evolution using phylogenetically

independent contrasts. *Am. Nat.* 140: 509–519

72 Martins EP, Hansen TF (1997) Phylogenies and the comparative method: a general approach to incorporating phylogenetic information into the analysis of interspecific data. *Am. Nat.* 149: 646–667

73 Sanderson MJ (1993) Reversibility in evolution: a maximum likelihood approach to character gain/loss bias in phylogenies. *Evolution* 47: 236–252

74 Garland T Jr, Diaz-Uriarte R (1999) Polytomies and phylogenetically independent contrasts: examination of the bounded degrees of freedom approach. *Syst. Biol.* 48: 547–558

75 Diaz-Uriarte R, Garland T Jr (1998) Effects of branch length errors on the performance of phylogenetically independent contrasts. *Syst. Biol.* 47: 654–672

76 Diaz-Uriarte R, Garland T Jr (1996) Testing hypotheses of correlated evolution using phylogenetically independent contrasts: sensitivity to deviations from Brownian motion. *Syst. Biol.* 45: 27–47

78 Omland KE (1999) The assumptions and challenges of ancestral state reconstruction. *Syst. Biol.* 48: 604–611

79 Abouheif E, Fairbairn DJ (1998) A comparative analysis of allometry for sexual size dimorphism- assessing Rensch's rule. *Am. Nat.* 149: 540–562

79 Price T (1997) Correlated evolution and independent contrasts. *Phil. Trans. R. Soc. Lond. B* 352: 519–529

80 Frumhoff PC, Reeve HK (1994) Using phylogenies to test hypotheses of adaptation: A critique of some current proposals. *Evolution* 48: 172–180

81 Baker RH, Wilkinson GS, DeSalle R (2001) The phylogenetic utility of different types of molecular data used to infer evolutionary relationships among stalk-eyed flies (Diopsidae). *Syst. Biol.* 50: 87–105

82 Pitnick S, Miller GT (2000) Correlated response in reproductive and life history traits to selection on testis length in *Drosophila hydei. Heredity* 84: 416–426

83 Ackerly DD (2000) Taxon sampling, correlated evolution, and independent contrasts. *Evolution* 54: 1480–1492

84 Oakley TH, Cunningham CW (2000) Independent constrasts succeed where ancestor reconstruction fails in a known bacteriophage phylogeny. *Evolution* 54: 397–405

10 Analyzing Data at the Population Level

Phaedra Doukakis, Kenneth D. Birnbaum and Howard C. Rosenbaum

Contents

1   Introduction

Molecular genetic data have become powerful sources of information for addressing population biology issues, particularly for conservation biologists. Probably the most popular of these applications concerns assessing the degree of genetic differentiation existing among populations, which has sparked an involved debate over the amount of genetic divergence necessary to warrant separate conservation and/or taxonomic attention [1–5]. A great deal of attention has also focused on the amount of genetic diversity residing within different populations, which can be another useful parameter for deciding conservation priority. Other applications utilize sequence information or co-dominant alleles to infer population size and history, examining whether a population has undergone a recent demographic reduction or expansion and whether two populations have experienced recent *versus* historical gene flow [6–9].

Methods and Tools in Biosciences and Medicine
Techniques in molecular systematics and evolution, ed. by Rob DeSalle et al.
© 2002 Birkhäuser Verlag Basel/Switzerland

On the individual level, knowing the specific genotypes of individuals within a population can help to guide reintroduction and captive management decisions [10–13], while also elucidating the breeding structure of wild populations [14]. Used in combination, these markers can be applied to regulate the harvest of endangered populations and/or subspecies through the development of forensic techniques. In many fisheries studies, such applications are used to quantify the contribution of individual stock populations to a large mixed-stock population in order to maximize harvest yield while minimizing the extinction risk of individual populations.

In this chapter, we discuss the use of both mitochondrial and nuclear DNA markers for population-level surveys. We provide more information on data handling and processing than on data collection, as this has been discussed elsewhere in this volume. We focus largely on DNA sequence, RFLP and microsatellite data as these techniques are widely used. The emphasis in this chapter is on specific applications of molecular genetic data, especially in conservation. For a more fundamental treatment of molecular evolution, numerous texts are available (e. g., [15–18]). Below we have structured our discussion such that analytical techniques are listed by the biological or ecological questions they can be used to address. Since much of the analysis of population-level information requires complex calculations and resampling procedures, we also provide suggestions on some useful computer programs. Our review of computer applications includes those we most commonly use but is by no means exhaustive. Some useful websites for finding additional programs are given in the Appendix (More information on computer programs). The choice of the gene region or set of loci used should be tailored to the question being addressed and several suggestions are given in "choosing markers" (see Appendix).

2   Sequence and Allele Frequency Data

This section assumes that one has already edited the sequence information or allele data and has linked each sampled individual in a population to a specific haplotype or genotype, either in the form of a DNA sequence or a designated allele. Generally, all of the following statistics apply to both sequence and haplotype data, although one should consult the manuals of each computer program listed for the specifics on input formats and data calculation nuances. At the end of each section, we will also cover techniques that are specific to the analysis of co-dominant alleles such as microsatellites and RFLPs. We organize this section around six commonly asked questions with respect to population level problems.

2.1 How diverse is the population?

A longstanding question in both evolutionary and conservation biology concerns the amount of genetic diversity characterizing wild populations and the amount of diversity necessary to allow organisms to adapt to changing environments. There are a variety of measures for assessing genetic diversity, and one should keep in mind that comparisons of genetic diversity among different organisms will depend upon the sample size surveyed and loci studied. A widely used measure is gene or haplotype diversity, which is equivalent to the expected heterozygosity for diploid data. It can be expressed as the probability that two randomly chosen haplotypes are different (see [15]). In cases where the sequence composition of the different haplotypes is known, it is also useful to consider the degree of genetic difference between different haplotypes. Haplotype diversity may be higher within one population as compared to another, but haplotypes might differ by only one or two base changes, so this should be considered. An understanding of this pattern is generally accomplished through analyzing the genetic distance among haplotypes within a given population (see the MEGA and Arlequin documentation for a good discussion of different distance algorithms). Furthermore, when the sequences of each haplotype are known, the nucleotide diversity of the population can also be quantified. This is simply the probability that two randomly chosen homologous nucleotides are different (see [15]). For allele data, heterozygosity (e. g., [19]) is a frequently used measure of diversity. Microsatellites have made allelic richness, or number of alleles detected at a locus, a sensitive measure of population-level diversity which may be more appropriate for some conservation questions [20]. However, making comparisons of allelic richness between any two sets of samples should include an adjustment for sample size (e. g., [21]) *Programs*: Both DNAsp ([22]; http://www.bio.ub.es/~julio/DnaSP.html) and Arlequin ([23]; http://lgb.unige.ch/arlequin/) will calculate these indices along with their respective variances. Additionally, these programs and others (e. g., MEGA: http://www.megasoftware.net/) will give you information on the number of polymorphic sites within the dataset.

Tips: Be sure to format your sequence data correctly, paying special attention to the treatment of missing (?, N) and polymorphic (R,Y, etc.) data as the programs will give you an error message if you don't consider these issues. In Arlequin, we have had the most luck when a haplotype file is created with the command file referencing the haplotype list (see the Arlequin manual for details).

2.2 What demographic parameters can be estimated using sequence or haplotype data?

Several applications have been developed that attempt to understand population demography and history. For sequence data, one way to infer whether a population is at demographic equilibrium or has undergone a recent expansion is through mismatch distribution analysis [25, 26]. This is an analysis of the distribution of the number of differences observed between pairs of haplotypes, with a graphical representation of this distribution indicating population history characteristics. A coalescence approach can also be used to estimate the growth rate of the population [27]. A description of the techniques for inferring population parameters can also be found in Avise [8]. Various programs will also give an estimate of theta that can give you information on the effective population size. For highly variable co-dominant markers such as microsatellites, statistical measures have also been developed to test for recent bottlenecks in a population by detecting an imbalance between allelic richness and heterozygosity from what is expected at mutation-drift equilibrium [9]. Goldstein et al. [28] used variance in genetic distance measures in microsatellites and coalescent-based simulations to infer population history. *Programs:* Both Arlequin and DNAsp will perform mismatch analyses and estimate theta, while the program Fluctuate within the LAMARC package (see documentation for references, http://evolution.genetics.washington.edu/lamarc.html) uses the coalescence approach. For microsatellites, the program, BOTTLENECK [29], tests for evidence of a recent population crash by detecting a significant excess in heterozygosity using allele frequency data.

2.3 Are two or more populations genetically distinct? How much gene flow has occurred among populations?

As stated earlier, an extensive debate has focused on the amount of genetic differentiation between populations that indicates that two populations are genetically distinct. Therefore, there are multiple ways to address this issue. For sequence data, one might calculate the genetic distance between populations using a measure such as D_{xy} (the average number of nucleotide substitutions per site between populations; [15]) or D_a, (the number of net nucleotide substitutions per site between populations; [15]). The major drawback of distance-based analysis is that there are no specific criteria for assessing what degree of genetic distance proves different populations are in fact distinct or the amount of genetic distance that corresponds to a specific taxonomic entity (e. g., population, subspecies, species) [3]. A more clear-cut analysis is to use diagnostic differences among populations, in the form of either specific nucleotide changes or distinct haplotypes (see [3, 5] for a discussion of diagnostic methodology). Frequency-based calculations are also widely used. One can analyze whether the frequency of shared alleles differs among populations. Caution should be exercised when using this approach as allele frequency estimation can vary greatly with sample size and can also vary temporally; therefore, such measures should only be used when large sample sizes taken over the course of many years are available.

Another approach to consider, especially when two populations do not display diagnostic distinctions, is to calculate the amount of gene flow among populations. A widely used statistic is the Fst [30, 30a], which is a measure with a value ranging from 0 (no isolation) to 1 (total isolation). There are many ways to calculate Fst values using rare alleles [31], intraspecific trees [32] and coalescence [33]. One measure routinely used in our laboratory is an application of analysis of variance techniques to sequence data called AMOVA [34]. One can calculate Fst values based on either sequence information or haplotypes. AMOVA is also suitable for microsatellite alleles. Some benefits of the AMOVA analysis include the consideration of the sequence distance among haplotypes and permutation procedures that test whether the Fst value is significant. Still another way to examine genetic difference among populations is through building intraspecific phylogenies [8, 35] or networks, and the newest version of Arlequin includes methodology to construct the latter. Methods developed by Templeton [36] also utilize network analysis methods and also attempt to assess the temporal nature of gene flow.

For determining population structure using microsatellites, a number of derivations of the Fst model have been developed. Recent theoretical analyses suggest that genetic distances and modified Fst statistics calculated from microsatellite data under the assumption of a Stepwise Mutation Model (SMM) provide better estimates than F statistics [37, 38]. These newer test statistics Hst, Rst, Gst, etc., were developed specifically to account for potential biases associated with mutation models for microsatellites (e. g., see [37, 39, 40]). Using simulations of a number of microsatellite datasets for well-characterized populations, several studies have shown that the frequency and distribution of microsatellite alleles follow a stepwise mutation model more closely than an infinite alleles model (e. g., see [41, 42].

Microsatellites can also be used to measure migration rates between populations. Pairwise measures of subdivision are calculated for populations based on a stepwise mutation model. Tests of significance are done by permutating genotypes or individuals between the populations. Assuming two populations of size N exchange a certain number of migrants per generation and the mutation rate is negligible, the resulting Fst values can be used to extrapolate the number of migrants between two populations. For diploid organisms (i. e., markers like microsatellites), the absolute number of migrants can be equalized to M=1-Fst/2Fst [31]. *Programs:* Calculations of the distance among populations can be accomplished using DNAsp. Searching for evidence of diagnosis is fairly straightforward and has been discussed elsewhere [3, 5]. Allele frequency analysis with small sample sizes should use the Monte Carlo simulation [43] in the ChiPerm program ([43a]; http://bioag.byu.edu/zoology/crandall_lab/

programs.htm). Coalescence based approaches are included in the LAMARC program [27]. Tree building can be accomplished using PAUP ([44]; http://www.lms.si.edu/PAUP/) and MEGA, while Templeton-based analyses can use a variety of specifically developed programs accessible through http://bioag.byu.edu/zoology/crandall_lab/programs.htm. AMOVA can be found within the ARLEQUIN package. The Genetic Data Analysis (GDA) program is also useful for examining population structure using discrete data, including routines for the traditional Fst (available at: http://lewis.eeb.uconn.edu/lewishome/gda.html). Two programs that calculate population subdivision based on a stepwise mutation model are: Rst Calc [40], which uses Slatkin's Rst, and GENEPOP, which is based on Rousset's [39] measure. In addition, GENEPOP has a useful routine to test for significant population differentiation using an estimation of an exact test.

2.4   Is the population in Hardy Weinberg equilibrium?

One standard question which can be addressed using diploid data is whether individual populations are at Hardy Weinberg equilibrium. Both Arlequin and GDA and GENEPOP will perform these calculations.

2.5   What is the contribution of individual populations to a mixed population? Can the source population be identified for an individual in a mixed population?

Several maximum likelihood-based analyses have been developed to address this question, with both empirical and theoretical examples in the literature [45–48].
    Programs: For fisheries stock assessment work, a useful program is the Statistics Program for Analyzing Mixtures (SPAM) available from Alaska Department of Fish and Game (http://www.cf.adfg.state.ak.us/). A similar program is available from http://www-bml.ucdavis.edu/imc/Software.html.

2.6   Is an individual in the population the parent of another individual in the same population (parentage analysis)? What is the likelihood of parentage/paternity?

The two main analytical methods used for parentage analysis are exclusion probabilities and likelihood-based approaches, both typically using co-dominant markers. In paternity analysis (where the mother's genotype is known), exclusion probabilities are based on the statistical power of the set of markers to rule out paternity given the genotypes of mother and offspring and population-level allele frequencies [20]. Methods that incorporate match probabilities have also been employed for cases where neither parent is known [49, 50]. Meagher [51] adapted an alternative approach using likelihood methods to assess paternity in natural populations and the technique has gained popularity in molecular ecology studies in recent years. Likelihood-based methods pose a different question, typically considering a ratio of probabilities: for example, a male is the father given all genetic evidence vs the male is not the father given the same evidence. The genotypes of mother, father and offspring are all considered. Likelihood-based methods are also available for cases where neither parent is known. One of the drawbacks of a likelihood approach is that it lacks a true confidence measure although Marshall et al. [52] have recently developed a simulation-based confidence estimator. In addition, Baye-sian statistics have been applied to paternity analysis [53] but this approach is not widely used in studies of natural populations. *Programs:* Several programs use likelihood-based parentage analysis (see chart). We have used CERVUS [52], which is based on the Meagher approach. Its advantages are that it incorporates population size, percent of population typed, genotyping error rates and other parameters into parentage likelihood and confidence estimates. Exclusionary analysis is more straightforward and the formulas can be formatted into a spreadsheet after genotype and summary data have been parsed.

3   Remarks and Conclusions

Genetic markers can provide detailed information about the population biology of many diverse organisms. Their application in conservation genetics and molecular ecology has led to a host of new questions being addressed in addition to more finely detailed analyses of population hierarchy. For instance, the widespread use of microsatellite markers has emerged within the last 10 years. With these markers, new and sensitive statistical and simulation techniques have been developed to reveal population history, such as recent bottlenecks. No doubt efforts into sequencing a variety of genomes will lead to new markers and new analytical tools. As we apply these new techniques, a word of caution is to become thoroughly familiar with their assumptions and their effect on the data being analyzed. A model that appears to answer the question at hand may have hidden assumptions that are inappropriate or over-reaching for a parti-cular dataset.

Appendix

More information on computer programs

Websites with a compiled list of programs: http://evolution.genetics.washington.edu/phylip/software.html, http://wbar.uta.edu/software/software.htm. Monitoring list servers such as the Evolution directory (http://life.biology.mcmaster.ca/~brian/evoldir.html) is also a good way to keep up to date on new programs.

In addition there is an extensive compilation of microsatellite programs listed in Luikart, G. and England, P.R. [54].

Choosing markers

The most widely used mitochondrial gene region for population surveys is the D-loop or control region in which certain sub-regions evolve 3 to 6 times faster than other mitochondrial genes [55]. Several studies have illustrated the utility of the NADH5/6 region, particularly in fishes . Yet restricting a survey to mitochondrial genes biases the study, as mtDNA is a nonrecombining, maternally inherited marker. If the species of interest has sex-specific differential homing fidelity, the observed results may not represent biological reality. In addition, mitochondrial markers may not always be informative at the population level because they generally evolve 5 to 10 times slower than nuclear DNA [35, 56]. Microsatellites are more useful than sequencing nuclear loci in this respect due to their more rapid evolution. These short DNA regions are comprised of repeated nucleotide patterns and tend to mutate to new lengths at a rapid evolutionary pace. They are generally the marker of choice when high levels of variation are needed, such as in paternity analysis (See chapter 16 for further discussion on where to find microsatellite markers).

References

1   Waples RS (1991) Pacific salomon, *Oncorhynchus* spp., and the definition of "species" under the Endangered Species Act. *Marine Fisheries Review* 53,11–22

2   Moritz C (1994) Defining "Evolutionarily Significant Units" for conservation. *Trends in Ecology and Evolution* 9, 373–375

3   Vogler AP, DeSalle R (1994) Diagnosing units of conservation management. *Conservation Biology* 8(2), 354–363

4   King TL, Burke T (1999) Special issue on gene conservation: identification and management of gene diversity. *Molecular Ecology* 8, S1-S3

5   Goldstein PZ, DeSalle R, Amato G, Vogler AP (2000) Conservation genetics at the species boundary. *Conservation Genetics* 14(1), 120–131

6   Hudson RR (1990) Gene genealogies and the coalescent process. *Oxford Surveys in Evolutionary Biology* 7, 1–44

7   Templeton AR, Routman E, Phillips C (1995) Separating population structure from population history: a cladisitc analysis of the geographical distribution of mitochondrial DNA haplotypes in the

Tiger salamander, *Ambrystoma tigrinum. Genetics* 140, 767–782

8   Avise JC (2000) *Phylogeography: the history and formation of species.* Harvard University Press, Cambridge, MA

9   Cornuet JM, Luikart G (1996) Description and power analysis of two tests for detecting recent population bottlenecks from allele frequency data. *Genetics* 144: 2001–2014

10 Amato G, Wharton D, Baker R, Ruvolo M. (1999) Molecular systematics for taxonomic placement of a Gorilla of uncertain origin. *Zoo Biology* 18(5), 429–432

11 Garcia F, Nogues C, Garcia M, Egozcue J et al. (1999) Characterization of constitutive heterochromatin in *Cebus apella* (Cebidae, Primates) and Pan troglodytes (Hominidae, Primates): Comparison to human chromosomes. *American Journal of Primatology* 49(3), 205–221

12 Houlden BA, Woodworth L, Humphrey K (1997) Captive breeding, paternity determination, and genetic variation in chimpanzees (*Pan troglodytes*) in the Australasian region. *Primates* 38(3), 341–347

13 Willis K, Wiese RJ (1997) Elimination of inbreeding depression from captive populations: Speke's gazelle revisited. *Zoo Biology* 16(1), 9–16

14 Nason JD, Herre EA, Hamrick JL (1998) The breeding structure of a tropical keystone plant resource. *Nature* 391: 685–687

15 Nei M (1987) *Molecular Evolutionary Genetics.* Columbia University Press, New York

16 Li W-H, Graur D (1991) *Fundamentals of Molecular Evolution.* Sinauer Associates, Inc, Sunderland Mass

17 Page RDM, Holmes EC (1998) *Molecular Evolution: A Phylogenetic Approach.* Blackwell Science, Oxford

18 Hartl DL (2000) *A Primer of Population Genetics.* 3rd Edition, Sinauer Associates, Inc, Sunderland Mass

19 Weir, BS (1996) *Genetic Data Analysis II.* Sinauer, Sunderland, Massachusetts

20 Frankel OH, Brown AHD and Burdon JJ (1995) *The Conservation of Plant Biodiversity.* Cambridge University Press, Cambridge

21 Hurlbert SH (1971) The nonconcept of species diversity: a critique and alternative parameters. *Ecology* 52, 577–586

22 Rosas J, Rosas R (1999) DnaSP version 3: an integrated program for molecular population genetics and molecular evolution analysis. *Bioinformatics* 15: 174–175

23 Schneider S, Roessli D, Excofier L (2000) *Arlequin: A software for population genetics data analysis.* Ver 2.000. Genetics and Biometry Lab, Dept. of Anthropology, University of Geneva

24 Kumar S, Tamura K Nei M (1993) MEGA: *Molecular Evolutionary Genetics Analysis,* version 1.01. The Pennsylvania State University, University Park, PA 16802

25 Slatkin M, Hudson RR (1991) Pairwise comparisons of mitochondrial DNA sequences in stable and exponentially growing populations. *Genetics* 129: 555–562

26 Rogers AR, Harpending H (1992) Population growth makes waves in the distribution of pairwise genetic differences. *Mol Biol Evol* 9: 552–569

27 Beerli P, Felsenstein J (1999) Maximum likelihood estimation of migration rates and effective population numbers in two populations using a coalescent approach. *Genetics* 152(2), 763–773

28 Goldstein DB, Roemer GW, Smith DA, Reich DE et al. (1999) The use of microsatellite variation to infer population structure and demographic history in a natural model system. *Genetics* 151: 797–801

29 Piry S, Luikart G, Cornuet JM (1999) BOTTLENECK: A computer program for detecting recent reductions in the effective population size using allele frequency data. *The Journal of Heredity* 90: 502–503

30 Wright (1921) Systems of mating. *Genetics* 6, 111–178

30a Wright S (1951) The genetical structure of populations. *Ann. Eugenics* 15, 323–354

31 Slatkin M (1985) Rare alleles as indicators of gene flow. *Evolution* 39: 53–65

32 Slatkin M (1989) Detecting small amounts of gene flow from phylogenies of alleles. *Genetics* 121, 609–612

33 Slatkin M (1991) Inbreeding coefficients and coalescence times. *Genetical Research* 58, 167–175

34 Excoffier L, Smouse PE, Quattro JM (1992) Analysis of molecular variance inferred from metric distances among DNA haplotypes: application to human mitochondrial DNA restriction data. *Genetics* 131: 479–491

35 Mortiz C, Dowling TE, Brown WM (1987) Evolution of animal mitochondrial DNA: relevance for population biology and systematics. *Annual Review of Ecology and Systematics*, 18, 269–292

36 Templeton AR, Routman E, Phillips C (1995) Separating populations structure from populations history: a cladistic analysis of the geographical distribution of mitchondrial DNA haplotypes in the Tiger salamander, *Ambrystoma tigrinum*. *Genetics* 140: 767–782

37 Slatkin M (1995) A measure of population subdivision based on microsatellite allele frequencies. *Genetics* 139: 457–462

38 Goldstein DB, Linares, AR, Cavalli-Sforza, LL, Feldman, MW (1995) An evaluation of genetic distances for use with microsatellite loci. *Genetics* 139: 163–471

39 Rousset F (1996) Equilibrium values of measure of population subdivision for stepwise mutation processes. *Genetics* 142: 1357–1362

40 Goodman SJ (1997) Rst Calc: a collection of computer programs for calculating estimates of genetic differentiation from microsatellite data and determining their significance. *Molecular Ecology* 6: 881–885

41 Shriver MD, Jin L, Chakraborty R and Boerwinkle E (1993) VNTR allele frequency distributions under the stepwise mutation model: a computer simulation approach. *Genetics* 134: 983–993

42 Valdes AM, Slatkin M, Freimer NB (1983) Allele frequencies at microsatellite loci: The stepwise mutation model revisited. *Genetics* 133: 737–749

43 Roff DA, Bentzen P (1989) The statistical analysis of mitochondrial DNA polymorphisms: χ^2 and the problem of small samples. *Molecular Biology and Evolution* 64(5), 539–545

43a Posada D (1998) *Chiperm version 1.0* Department of Zoology, Brigham Young University

44 Swofford DL (1999) *PAUP*. Phylogenetic Analysis Using Parsimony (*and other methods)*. Version 4.0b2. Sinauer Associates, MA

45 Miller RB (1987) Maximum likelihood estimation of mixed stock fishery composition. *Canadian Journal of Fisheries and Aquatic Science* 44, 583–590

46 Xu S, Kobak CJ, Smouse PE (1994) Constrained least squares estimation of mixed populations stock composition from mtDNA haplotype frequency data. *Canadian Journal of Fisheries and Aquatic Science* 51, 417–425

47 Wirgin II, Waldman JR, Maceda L, Stabile, J et al. (1997) Mixed stock analysis of Atlantic coast striped bass (*Morone saxatilis*) using nuclear DNA and mtDNA markers. *Canadian Journal of Fisheries and Aquatic Science* 54, 2814–2826

48 Wirgin I, Waldman JR, Rosko J, Gross, R et al. (2000) Genetic structure of Atlantic sturgeon populations based on mtDNA control region sequences. *Transactions of the American Fisheries Society* 129, 476–486

49 Dow BD, Ashley MW (1996) Microsatellite analysis of seed dispersal and parentage of saplings in bur oak, *Quercus macrocarpa*. *Molecular Ecology* 5, 615–627 (1996)

50 Westneat DF, Webster MS (1994) In: Schierwater B, Streit B, Wagner GP et al (eds): *Molecular Ecology and Evolution: Approaches and Applications*. Birkhäuser Verlag, Basel, 91–126

51 Meagher TR (1986) Analysis of paternity within a natural population of *Chamaelirium leuteum*. 1. identification of the most likely male parents. *The American Naturalist* 128, 199–215

52 Marshall TC, Slate, J, Kruuk, LEB, Pemberton JM (1998) Statistical confidence for likelihood-based paternity inference

in natural populations. *Molecular Ecology* 7, 639–655

53　Valentin, J (1980) Exclusions and attributions of paternity: practical experiences of forensic genetics and statistics. *Am J Hum Genet* 32: 420–431

54　Luikart G, England PR (1999) Statistical analysis of microsatellite DNA data. *TREE* 14: 253–256

55　Pesole G, Gissi C, De Chirico A, Saccone C (1999) Nucleotide substitute rate of mammalian mitochondrial genomes. *J Mol Evol.* 48: 427–434

56　Brown WM (1983) Evolution of animal mitochondrial DNA. In: M Nei, RK Koehn (eds) *Evolution of Genes and Proteins.* Sinauer, Sunderland, MA, 62–88

Part II
Laboratory Methods

Introduction to Part II

Part 2 of this manual is a collection of seven chapters that start with obtaining and archiving specimens and end with methods for visualizing ontogenetic information in organisms. Prendini, Hanner and DeSalle first summarize the essential and practical issues involved in acquiring, archiving and storing specimens for evolutionary studies. Nishiguchi et al. present several nucleic acid isolation procedures. Protocols for the isolation of nucleic acids from a wide variety of organisms are presented in this chapter and are based on protocols used over the past decade by researchers at the American Museum of Natural History in particular. Greenwood then describes techniques for isolation of DNA from ancient tissues by focusing on late Pleistocene tissues. The polymerase chain reaction (PCR) is widely used in molecular evolution studies to amplify target DNA for ease of analysis. Bonacum, Stark and Bonwich summarize PCR techniques and PCR primer design. Obtaining sequence data with automated gel and capillary sequencing methods are the major reasons for the leaps in order of magnitude of amounts of data and Whiting uses the fifth chapter in this part of the manual to describe in detail methods that implement the efficient and rapid throughput of template DNAs through the sequencing step. Population genetic studies have been aided greatly by the expansion of microsatellite techniques. Birnbaum and Rosenbaum examine several practical issues involved in this approach. This manual concludes with a description of the methods used to visualize the location of expressed genes in evolution and development studies. Gates et al. describe the various techniques involved in antibody staining and *in situ* hybridizations for a wide variety of organisms.

Rob DeSalle, Gonzalo Giribet
and Ward C. Wheeler

11 Obtaining, Storing and Archiving Specimens and Tissue Samples for Use in Molecular Studies

Lorenzo Prendini, Robert Hanner and Rob DeSalle

Contents

Methods and Tools in Biosciences and Medicine
Techniques in molecular systematics and evolution, ed. by Rob DeSalle et al.
© 2002 Birkhäuser Verlag Basel/Switzerland

Collection, storage and archiving of specimens and tissue samples are prerequisites for the successful acquisition of molecular data for any systematics or population genetics study. Field collections represent the most obvious source for tissues, but there are many ways in which tissues can be obtained without ever going into the field. Stock centers and culture collections, commercial supply companies, seed and spore banks, botanical gardens and zoological parks provide an array of live organisms, fresh tissues, cultures or cell lines, while vast quantities of frozen, desiccated or preserved tissues are stored in the collections of frozen tissue facilities, natural history museums and herbaria, among others. This chapter reviews the more important practical aspects of the selection of appropriate tissues for protein or nucleotide extraction, the preservation and temporary storage of freshly collected tissues in the field and the acquisition of specimens and tissue samples from alternative sources. Practical guidelines for the transportation, long-term storage and archiving of tissue samples are presented, emphasising the importance of identification, documentation and voucher specimens to scientifically validate the results of a study. Finally, legal and ethical issues concerning the collection, transportation and storage of specimens and tissue samples, the loan and deposition of samples in collections and the publication of results obtained from their analysis, are addressed from international and U.S. perspectives.

1   Introduction

The obvious first step in any systematics or population genetics study is to focus on a group of organisms or a level of interest. We assume that these issues are self-explanatory and the reader is referred to the accompanying volume (EXS 92) for examples of the application of molecular techniques to a wide range of questions and a summary of the problems that can arise during the course of such studies. Equally important in the initial stages of the study are sampling strategy, collection, storage, vouchering and archiving. These last two points are especially critical since no standard protocol currently exists for the disposition of tissue or DNA vouchers to scientifically validate the results of a study. Fortunately, this situation is changing as various museums and research universities establish biorepositories for the long-term storage of genetic resources.

There are several ways in which tissues can be obtained for analysis, each with its own peculiarities, nuances and requirements. The purpose of this chapter is not to give an exhaustive account of collecting techniques, but rather to focus on the more important aspects of preservation and storage for successful isolation of the molecules of interest. Other publications have reviewed the plethora of field collection methods that exist for organisms as diverse as plants, fungi, vertebrates and invertebrates and the reader is advised to consult literature relevant to the taxa of interest for specific details about collecting these organisms. In this chapter we focus on five aspects of sampling and storage: 1) selection of appropriate tissues for protein or nucleotide extraction 2) storage of freshly collected tissues in the field 3) obtaining tissues from other sources, e. g., museum collections, stock centers and commercial supply companies 4) transportation, long-term storage and archiving of tissue samples and voucher specimens and 5) legal and ethical issues involved in the collection, transportation and storage of tissues.

2   Selection of Appropriate Tissues for Molecular Studies

Before embarking on a field trip to collect fresh samples or requesting samples from an official repository, commercial company or informal source, the researcher must decide what tissues are required for the study. Depending on the size of the organism or the kind of material available, DNA or proteins can be extracted from tissue samples or from a homogenate prepared from the whole organism or "clonal" collections of organisms, such as isofemale lines [1]. Tissue selection for molecular isolation depends on the organism to be studied and on the type of molecular work to be conducted. The aim is to use parts of the organism that are relatively free of compounds potentially damaging to the protein or nucleic acid of interest or that may interfere with cloning, PCR, sequencing or restriction digestion. Using specific tissues has the added benefit of reducing the risk of contamination with host or parasite tissues, gut flora or recently ingested prey. For example, DNA extraction from parasitic *Cuscuta* required using only internode tissue to prevent DNA contamination from its host plant species [2].

Young, actively growing leaves or shoots are the best tissues for DNA extraction from plants, but seeds, roots, flowers, stems, pollen, spores and gametophytes have all been used successfully [2–4]. Brown [5] reported successful amplification of DNA from the charred remains of wheat seeds.

Among animals, successful amplifications have been achieved from bone, feathers, scales, hair, muscle, skin, whole blood (liquid and dried), serum, stomach contents and feces, among others. The following tissues, listed in order of desirability, are recommended for DNA extraction from vertebrates: brain, testes or ovary, liver, kidney, heart and skeletal muscle [1]. The nucleated erythrocytes of nonmammalian vertebrates provide a convenient source of DNA because they lack tough connective tissue [6], while milk provides an alternative source of DNA from mammals [7]. Toe clips are a common source of DNA for lizards and anurans [8] and scute notches provide a nondestructive alternative for chelonians [9]. Fecal and hair samples similarly offer noninvasive options for genetic sampling of mammals, especially endangered species [10–12].

Embryos are an exceptional source of insect mtDNA [1]. However, gonad or muscle tissues are generally suitable for both nuclear DNA and mtDNA isolation from arthropods and must be dissected out, avoiding the exoskeletal material if possible, since the presence of lipids and subcutaneous pigments may inhibit PCR. If specimens can be induced to spawn (for taxon-specific methods, see Strathman [13]), gametes may be obtained without the need for dissection. However, PCR may be inhibited if eggs or sperm are used, because of DNA viscosity [14]. Eggs are preferable to sperm for mtDNA isolation.

If gamete or muscle tissues are difficult to obtain, body tissues *sans* gut provide an alternative. For example, Koch et al. [15] extracted DNA from the heads of blackflies (*Simulium vittatum*) to avoid contamination by parasitic nematodes. Animals too small to be dissected should be starved for at least two days and homogenised whole. Even when all the above procedures are impractical, it is still possible to obtain good results with whole organisms, especially if PCR primers or DNA probes are sufficiently taxon-specific to circumvent possible contamination. For example, Dorn et al. [16] successfully preserved and amplified the DNA of *Trypanosoma cruzi* from human blood samples.

Irrespective of which tissues are selected, they should be used, frozen or preserved immediately, before they begin to degrade. Fresh tissue is preferable if purified mtDNA is required, as freezing may break mitochondrial membranes, reducing yield at the step in a protocol where mitochondria are pelleted. However, if genomic DNA or total cellular DNA with a mtDNA fraction is desired, frozen tissue can provide high yield and quality.

High molecular weight DNA is required in applications such as cloning and RFLPs. In handling tissues intended for such studies, care should be exercised not to allow warming of the tissue after initial storage. Similarly, if protein analysis is intended, care should be taken to maintain specimens at temperatures that will prevent protein denaturation.

Although high molecular weight DNA is preferred for PCR, degraded DNAs will often amplify if the target fragment is small enough and the primers specific enough. DNA intended for PCR is routinely used from both desiccated and ethanol-preserved museum specimens, which may yield fairly degraded DNA. As progress continues with protocols developed for ancient samples, such issues may become *passé*. Interested readers should consult the *Ancient DNA Newsletter*, *DNA Amplifications* (the newsletter of Perkin-Elmer Inc., Wellesley, MA) and the journal *BioTechniques* (Eaton Publishing Co., Natick, MA) for updates of such technological advances. The website of the Molecular Biology Techniques Forum (Appendix 1) also provides a useful platform for questions.

3   Storage of Freshly Collected Specimens and Tissue Samples in the Field

Field collections represent the most important source for tissues, because the origin of the material (including habitat or provenance data) is fully documented. Additional advantages include the ability to obtain rare or poorly collected species (which can seldom be obtained from other sources), to collect within and among populations, and to reduce the time prior to DNA or isozyme extraction. Disadvantages include the time and expense associated with expeditions to remote areas, together with problems of transportation and permits (addressed below). Nevertheless, many unusual and important molecular systematics and population genetics studies, particularly those addressing biogeographic issues and rare, range-restricted or endangered species, cannot be conducted without field collections.

The method of tissue storage or preservation in the field is determined primarily by the nature of the study – the hierarchical level (population, species, higher taxon) focused on, the molecules (DNA, RNA or isozymes) to be examined, what protocol will be effective for extraction of the molecules, and how pure or intact they must be. Ultimately, the methods of choice will be constrained by logistical factors, most importantly the duration of the field trip, the facilities available and the transportation options. Readers interested in pursuing molecular studies requiring visitation and collection in foreign countries with even limited logistical support are advised to consult the recommendations of Mori and Holm-Nielsen [17] for a botanical perspective. Additional useful addresses are contained in Davis et al. [18].

3.1   Aspects of preservation and molecular degradation

When obtaining tissue samples, regardless of the study or molecule of interest, the goal should be to acquire the freshest, or best-preserved, samples possible. Denatured proteins and degraded nucleic acids present the single greatest obstacle to a successful molecular research programme. For example, PCR amplification of DNA from old or poorly preserved tissues can be hindered because various forms of damage reduce the average length of intact template molecules for the polymerase enzyme [19, 20]. Damage to the templates may cause the polymerase to stall, thereby retarding the initial rounds of amplification [21]. The average length of the DNA from ancient soft tissue is less than 200 base-pairs (bp) [22]. This reduction in length is partially attributable to strand breakage caused by autolytic processes (e. g., DNAse activity) that occur rapidly after death [19, 23, 24]. However, equally important sources of damage are subsequent oxidative and hydrolytic effects that either break, or labilise, phosphodiester and carbon-nitrogen (sugar-base) bonds [21, 25, 26]. Although desiccation appears to place a limit on endogenous hydrolytic damage, oxidative attack continues with time.

In an effort to minimise the processes of denaturation and degradation, preservation methods must aim to maintain the tissue samples at low temperature, exclude light and other forms of radiation, remove water and oxygen and sterilise against micro-organisms. These objectives are accomplished by freezing, desiccation or the addition of preservative fluids, and by storing samples in the dark, at constant, low temperature.

3.2   Field storage of tissue samples for molecular studies

Live animal specimens
Fresh material consistently provides the highest yield and quality of DNA for amplification, restriction digestion and isozyme analysis. Field trips of short duration are amenable to the collection of live invertebrates and small vertebrates (e. g., herpetofauna), which can usually survive for several days if kept

cool, well ventilated and sufficiently humid (dehydration is the primary cause of death in terrestrial taxa). Containers should be kept away from bright sunlight, inside a Styrofoam box and specimens checked twice daily (morning and evening) throughout the duration of the trip. At this time, moisture sources should be replenished, excreta removed and individuals in poor condition preserved immediately in ethanol. Rare specimens should be preserved on collection, rather than kept alive, to safeguard against death going unnoticed during the course of the trip, resulting in inadequate preservation.

Fresh plant samples
Contrary to earlier reports (e.g., [3, 4]), most fresh plant tissue need not be immediately placed on ice or frozen. A diverse array of plant tissues can be kept in a stable condition in Ziploc bags, stored away from light and fluctuating temperatures, inside a Styrofoam box [2]. If collected as whole plants, leaf or shoot cuttings, and placed in bags, leaf tissue is maintained the longest, provides an immediate voucher when some of the leaf tissue is separated (see below) and allows for subsequent propogation in the greenhouse.

Ziploc bags are ideal for maintaining leaf or shoot tissues in good condition without the addition of moisture (since the tissues will transpire naturally). The addition of a damp paper towel to increase the moisture content of the atmosphere is not recommended, as it can promote waterlogging and rotting of delicate leaves [2].

A small amount of wet ice, placed inside the Styrofoam box containing the bagged samples, is recommended if the trip is of short duration. Once a continuous source of refrigeration is available, the tissue samples should be kept in the refrigerator until processed or placed in long-term ultracold storage (see below). To prevent DNA degradation of fresh tissues, it is important to avoid fluctuating heat exposure or warm up from cold temperatures. The sensitive nature of protein activity in isozyme analysis usually requires that fresh tissues be rapidly exposed to cooler temperatures. However, adequate isozyme activity and high molecular weight DNA have been obtained from samples collected 8 days before being placed in ultracold storage [27]. In some taxa, senescence occurs after collection, during which many proteins disappear from leaves with seasonal aging [28]. It is therefore important to record the age of the leaves at harvest. Seeds, pollen and fern spores should be harvested only when mature. The collector should also be aware that hot and dry weather prior to harvest might cause synthesis of "storage proteins" in seeds and tubers to cease prematurely.

Frozen samples
In comparison with plants, field-collected tissue samples from animals, especially vertebrates, have traditionally been stored frozen, using combinations of wet ice, dry ice and liquid nitrogen. An extensive discussion of methods can be found in Dessauer et al. [3, 4, 28], Simione and Brown [29] and Simione [30]. Although cryopreservation remains important for studies involving proteins

and RNA (see below), researchers are increasingly adopting alternative strategies for field-storage of animal tissues.

There are several reasons for this shift in protocol. First, cryopreservation in the field involves considerable planning and logistical support, since sources of dry ice and liquid nitrogen are required. Although dry ice is generally available from airlines, whereas liquid nitrogen can be obtained from medical and veterinary clinics, universities and hospitals, among others (refer to the list provided by Dessauer and Hafner [31]), field trips must be planned to intercept such sources at regular intervals. This may be impractical when trips take place in remote locations or are prolonged in duration. Second, air transportation of dry ice and liquid nitrogen is strictly regulated (see below). Finally, many investigators have obtained suitable yields of high molecular weight DNA, nearly comparable with those obtained from frozen samples, using less elaborate protocols (e. g., preservation in 95–100% ethanol).

If frozen tissues are to be collected, they should be sampled while the specimen is alive or as soon after death as possible and rapidly placed in the cold and away from light. Tissues should be packed in plastic cryotubes (e. g., Nunc), Ziploc bags, or wrapped tightly in aluminium foil, excluding as much air as possible. Small blood samples can be collected in heparinised hematocrit or microcentrifuge tubes, whereas larger samples are most efficiently collected from heart or caudal vessels using a heparinised syringe [3, 4, 28]. Blood cells should be separated from plasma, prior to freezing, by using a commercial hand centrifuge or a lightweight, plastic centrifuge (e. g., [32]).

As soon as possible after collection and packaging, tissues should be quick (snap) frozen by dropping directly into liquid nitrogen or covering with dry ice. However, since quick freezing often shatters hematocrit and microtubes filled with liquids, such tubes should be frozen slowly in a freezer (if available) before subjection to ultracold temperatures.

For cryopreservation of vegetative plant tissues (mainly leaves), material should be washed in distilled water and quick frozen in liquid nitrogen for subsequent storage at ultracold temperatures. Dessauer et al. [28] provided the following protocol for cryopreservation of seeds: remove fleshy portion of seed; dry seed; place seed in vacuum-sealed container; store in the dark, below 0°C. Seeds stored in this manner have remained viable for up to 10 years. Pollen and fern spores may be similarly treated (but must first be cleaned of debris and treated with organic solvents to remove lipids) and have yielded stable proteins after 4 years' storage at ≤ -30°C.

Cell lines

Cryopreservation of living cells requires special collecting, freezing and storage procedures if cells are to survive the freezing and thawing process [33]. Few cells will survive freezing and thawing without a cryoprotectant, e. g., glycerol or dimethylsulphoxide (DMSO). Each species, tissue and freezing system has an optimum cryoprotectant concentration and freezing rate, which must be ascertained to recover the maximum number of viable cells from the sample [34].

Cryoprotectant concentration must be sufficient to protect the cells from freeze damage, yet dilute enough to avoid chemical injury to cells, while the rate of freezing should also be precisely controlled, depending on the species, size of sample, cryoprotectant and freezing system. Rapid freezing induces death by the formation of crystals within the cells, while slow freezing causes death from the chemical consequences of solute concentration. For further guidelines on the field preservation of cell lines, refer to Hay and Gee [33] and Dessauer et al. [3, 4].

Desiccated samples

Rapid desiccation is one good alternative to cryopreservation for storing tissue samples in the field and may be preferred because it requires neither refrigeration nor flammable substances. Many proteins in carefully dried tissues are stable for short duration at ambient temperature and for longer periods in refrigerators or freezers. For example, acetone powders, solids precipitated from tissues with cold acetone, retain sufficient enzymatic activity for use in endocrinological studies [28]. Desiccated tissues are also suitable for subsequent extraction and PCR amplification of high molecular weight DNA. Specific protocols for the storage of desiccated samples are provided in Protocol 1.

Protocol 1   Methods of desiccation

1. *Insects.* Air-dried, pinned insect samples are convenient, require low maintenance and are lighter than fluid-preserved samples. Insects may also be preserved in ethanol or acetone and later desiccated using a critical point drier [35–37]. Both techniques are suitable for DNA extraction, but marked degradation may occur if tissue dehydration is prolonged (e. g., in humid climates or with large specimens). Placing samples in contact with silica gel for at least 12 hours (determined by the size of the specimen) is recommended in this case [38]. Removing the head, legs, wings, etc. enhances desiccation of large specimens.
Various methods of chemical desiccation, e. g., hexamethyldisilazane (HMDS), amyl acetate, xylene, methyl cellusolve and acetone vapour, provide an alternative to air desiccation, especially in humid climates, yielding high molecular weight DNA amenable to PCR [36]. Specimens can also be dried from ethanol using chemical techniques. Phillips and Simon [39] presented a protocol for the isolation of DNA from pinned insects without destruction of the exoskeleton.
2. *Parasites.* Toe et al. [40] evaluated methods for the field preservation of parasite and vector samples (e. g., *Onchocerca volvulus* and *Simulium damnosum*) for PCR. Preservation of desiccated tissues from parasites and their associated vectors on microscope slides yielded the greatest quantity of high molecular weight DNA.
3. *Vertebrates.* Blood samples may be spotted on Guthrie cards or on Whatman paper (3 mm) and left to dry in bright sunlight for 30 minutes. Makowski et al. [41] recommended usage of guanidium thiocyanate-impregnated filter paper

(GT-903), which binds PCR inhibitors and preserves DNA in an aqueous extractable form. If GT-903 is unavailable, consult Ostrander et al. [42] and McCabe et al. [43] on methods for extracting DNA from regular paper. Weisberg et al. [44] found that DNA extracted from lyophilised (freeze-dried) blood was similar in length to that extracted from fresh or frozen blood and suitable for PCR.

Sodium chloride (NaCl) can be used as a desiccant for solid vertebrate tissues, e. g., muscle. Approximately 1 g of tissue should be sliced off and centered in a 15 ml or 45 ml conical test tube containing NaCl. Upon return to the laboratory, the tissues need only be rinsed under distilled water before commencing with DNA extraction. PCR from such samples has resulted in mtDNA fragments up to 300 bp in length [45]. However, the success of this method for isozyme analysis has not been established and is probably unreliable.

4. *Plants.* Plant tissues are often difficult to preserve for molecular studies, due to the presence of phenols, polysaccharides and lipids. Methods of dehydrating plant tissue prior to transport include lyophilisation, dehydration with a food drier, air-drying and herbarium specimen (forced air or heat) drying [46–54]. However, these techniques cannot be universally applied to plants and should be tested before embarking on a field trip [55]. For example, lyophilisation often leads to decreased protein activity, although animal and fungal tissues are usually unaffected [56]. DNA extracted from dried plant tissues is often too degraded for use in restriction site analysis, although successful amplification may be achieved with PCR.

Desiccation of plant tissue in Dierite ($CaSO_4$) or silica gel offers the best alternative to collection of fresh or frozen plant tissue [55, 57, 58], having been successfully tested by extraction and sequencing of DNA, and by digestion with restriction enzymes, over a diversity of angiosperm taxa [2, 53]. It also appears to be less likely to damage proteins than other methods, e. g., lyophilisation [59]. Since the method is simple, the chemicals readily available and easy to transport, it is ideal for obtaining species that exist in remote areas or are to be collected by colleagues in other countries.

Silica gel is a more efficient desiccant than Dierite as it can be obtained in smaller mesh size (28–200, grade 12 from Fisher Scientific, Chicago, IL), allowing for greater surface coverage of leaf tissue. Usually 4–6 g of fresh leaves are placed in small Ziploc bags. Leaf tissue dries faster and thus with less DNA degradation, if first torn into smaller pieces. Subsequently, 50–60 g of silica gel (minimum 10:1 gram ratio of silica gel to leaf tissue) is added to the bag. Drying should take place within 12 hours, determined if tissue snaps cleanly when bent. Longer exposures to silica gel may be required, e. g., in monocots, but will usually result in DNA degradation. The dried tissue in Ziploc bags should be stored with trace amounts of indicator silica gel (6–16 mesh, grade 42), which changes colour from violet-blue to whitish pink when hydrated, to verify that rehydration has not occurred. The Ziploc bags should, in turn, be kept in tightly sealed plastic boxes. Schierenbeck [60] provided a modified polyethylene-glycol DNA extraction protocol for silica gel-dried plants.

5. *Marine algae*. Holzmann and Pawlowski [61] obtained high molecular weight DNA from air-dried samples of foraminifera stored at ambient temperature for up to 11 weeks. Positive results were also obtained with foraminifera stored for 2–3 years, although amplification products were < 500 bp long.

6. *Marine invertebrates*. Desiccation is generally not viable for marine invertebrates (e. g., Scyphozoa, Polychaeta) because their high water content is incompatible with the requirement that samples be dried within 12 hours to prevent DNA degradation [55].

Fluid-preserved samples

Tissue samples from invertebrates, intended for molecular studies, are routinely stored in 95–100% ethanol at ambient temperature (e. g., [62]). Preservation in methanol and propanol is contraindicated [38], although tissue samples preserved in 100% methanol will yield higher molecular weight DNA than samples preserved in formalin [63].

Greer et al. [64] demonstrated that only storage in 95–100% ethanol (v/v) results in PCR products of 1–2 kb after 30 days. Adequate preservation in 70–80% ethanol may occur if specimens or tissue samples are small and, in the case of arthropods, weakly sclerotised, but DNA isolated from larger specimens preserved in 70% ethanol is usually highly degraded. For large or heavily sclerotised samples, even 95–100% ethanol is no guarantee against degradation, because saturation of the tissues may be delayed by the size or impermeable nature of the tissues [65]. Such samples should be injected with ethanol, dissected into smaller pieces or, in the case of arthropods, cut in several places along the exoskeleton, to allow the ethanol to diffuse directly into the internal tissues. Excess ethanol (i. e., a high ratio of ethanol to tissue sample volume) should always be used when preserving samples for molecular studies, to minimise dilution that occurs with addition of the sample [66]. Ethanol should be replaced after the initial fixation and periodically thereafter. Initial fixation of fresh or frozen tissues at –20°C has been found to contribute significantly to the quality of tissue preservation in arthropod samples [67], but refrigeration is seldom available in the field.

The method of euthanasia may also affect the preservation of tissues, especially if whole organisms are placed directly in ethanol. Animals as diverse as nematodes and scorpions are prone to close their oral, anal and respiratory openings on placement in ethanol, thus further hindering ethanol diffusion into the internal tissues after death. Such animals should be frozen alive, placed in warm ethanol (unless being used for isozyme analysis), very dilute (e. g., 10%) ethanol, or dissected/cut (after euthanasia with ethyl acetate, chloroform or cyanide), prior to placement in 95–100% ethanol. Alternatively, lysis buffers may be used for their preservation (see below).

Ethanol is suited to the storage of vertebrate tissues and has been used successfully in DNA hybridisation and sequencing. Solid tissues, e. g., muscle, should be cut into pieces approximately 10 mm in diameter to allow rapid penetration of ethanol. After saturation in ethanol for at least two days, moist tissues may be transferred to plastic bags for storage or shipment [28].

Fungi and marine algae may also be stored in ethanol. Ethanol-preservation was widely considered ineffective for maintaining adequate yields of high molecular weight DNA from land plants [51] until Flournoy et al. [68] obtained excellent yields from samples stored in 95–100% ethanol or 100% methanol by addition of proteinase (Pronase E). Vacuum infiltration of ethanol resulted in better DNA preservation than passive infiltration [68].

In the absence of ethanol, most samples may be stored in saturated salt or buffer solutions until transported to laboratories equipped with appropriate resources. Even laundry detergent has proven to be a rapid and uncomplicated temporary storage solution for recovering high yields of DNA [69, 70].

Many of these buffer solutions are also used for DNA isolation and may be advantageous for the preservation of highly sclerotised organisms. For example, Sansinforiano et al. [71] used lysis buffer with a high urea concentration for preserving the impermeable mucopolysaccharide capsules of *Cryptococcus neoformans* and other pathogenic yeasts, and acquired high molecular weight DNA after storage at ambient temperature for up to 6 months. Details of these methods are listed in Protocol 2.

Protocol 2   Preservation in buffer solutions

1. *Blood.* Preservation of vertebrate blood samples in the field has traditionally employed buffer solutions, e. g., 2% 2-phenoxyethanol with glycerol or DMSO, 2-propanol with ethylene-diamine-tetraacetate (EDTA) or sodium dodecyl sulphate (SDS). Such buffers can preserve blood proteins (e. g., plasma albumin) and mRNA for up to three weeks without refrigeration [72, 73], but must usually be frozen within 24 hours of collection [28].

Quinn and White [74] recommended that blood samples from birds, injected into 5 ml vacutainer/EDTA tubes, should be frozen within 10 hours of collection to prevent significant DNA degradation. However, Cann et al. [45] obtained high molecular weight DNA from unrefrigerated blood samples of birds, bats and wallabies, using a scaled-down version of Quinn and White's protocol.

Cann et al. [45] reported that microhematocrit tubes of blood blown into 500 µl of TNE_2 (10 mM tris-hydroxymethyl amino methane, 10 mM NaCl and 2 mM EDTA, pH 8.0) are stable for months if refrigeration is impossible and still provide high molecular weight DNA. The use of mannitol-sucrose buffer in mtDNA studies is also reported to be adequate for storing samples up to 2 weeks without refrigeration [75]. Gelhaus et al. [76] provided protocols for the isolation of DNA from urea-preserved blood. For practical guidelines on the extraction of blood samples from vertebrates, refer to Dessauer et al. [3, 4].

2. *Vertebrates.* The lenses of vertebrate eyes, collected for sequence studies of alphacrystallin, have been preserved in saturated guanidine hydrochloride [28] and this solution is also purported to be effective for the preservation of proteins in other vertebrate tissues [45].

High molecular weight DNA has been extracted from vertebrate solid tissue samples, stored at ambient temperature in 4 M guanidium isothiocyanate (GITC) from 41 days [77] to 3 months [78]. Vertebrate tissue samples may be kept up to three years without refrigeration in TNES-urea (6 or 8 M urea; 10 mM Tris-HCl, pH 7.5; 125 mM NaCl; 10 mM EDTA; 1% SDS) and still yield high molecular weight DNA [79].

3. *Invertebrates.* Sperm has been stored in 0.01–0.02% sodium azide solution [14]. Laulier et al. [77] extracted high molecular weight DNA from field collected samples of viruses, bacteria, yeasts and invertebrates stored up to 41 days at ambient temperature, in 4 M GITC.

Dawson et al. [80] assessed the effects of five buffer solutions (70% ethanol, Queen's lysis buffer [0.01 M Tris, 0.01 M NaCl, 0.01 M disodium-EDTA and 1.0% n-lauroylsarcosine, pH 8.0], DMSO-NaCl solution, CTAB-NaCl solution, and a urea extraction buffer) on the preservation of marine invertebrate samples for DNA isolation. In accord with Seutin et al. [81], these authors concluded that dimethylsulphoxide and sodium chloride (DMSO-NaCl) was the best solution in which to store marine tissue samples. Reiss et al. [65] recommended that insect samples be thoroughly homogenised in order to achieve adequate DNA preservation in buffer solutions.

4. *Plants.* Samples of plant tissue preserved using cetyltrimethylammonium bromide (CTAB) yield high molecular weight DNA, thereby providing an alternative to standard desiccation methods [82].

Embryonic tissues

Developmental data are becoming increasingly important in modern evolutionary and systematics studies [83, 84]. Developmental studies utilise embryonic tissues that hold expression data (transcribed mRNA for *in situ* studies and translated proteins for antibody staining approaches), the preservation of which is essential. Accordingly, embryonic expression work is best conducted with live material. This can be obtained from embryos collected in the field or from sources of live organisms listed in the next section.

Knowledge of embryogenesis in the organisms of interest can aid in the collection and treatment of embryos in the field. For instance, if an organism displays brooding behaviour, large numbers of embryos can be obtained by locating reproducing individuals [85, 86]. Such individuals can be brought alive to the lab and manipulated for fixation.

If live material cannot be transported to the laboratory, embryos may be preserved in the field, provided that the preservation of mRNA and proteins can be assured. Although embryos can be frozen at −80°C with liquid nitrogen, fixation methods are easier and more efficient for their preservation in the field. However, knowledge of the embryology of organisms is required, because fixation of embryos without certain kinds of pretreatment may render them unusable. For example, *Drosophila* embryos require dechorionation prior to fixation. Such pre-fixation treatments may be simple to accomplish in the field. Alternatively, if embryonic tissues like imaginal discs in insects are the targets of molecular developmental research, whole larvae can be preserved in fixatives to allow later dissection of tissues for whole mount or sections. Specific protocols for preservation of embryos in the field are provided in Protocol 3.

Protocol 3   Storage and fixation of embryos

1. *Dechorionation*. This procedure is accomplished by treating embryos in sodium hypochlorite followed by thorough washing in water or 0.7% saline. This step can be accomplished in the field with test tubes and a small supply of Pasteur pipettes. Other pre-fixation treatments may be necessary to prepare the embryos or developing organisms for fixation or the researcher may determine that pre-fixation treatments are unnecessary and proceed directly to fixation of materials.

2. *Fixation*. Dechorionated embryos or untreated embryos can be fixed by several methods depending on whether antibody staining or mRNA *in situ* detection is the goal. Antibody staining requires that larvae be fixed in a solution of 4% formaldehyde in 50 mM cacodylic acid (pH 7.4) for 1 hour at 20°C, dehydrated in an ethanol series and stored in methanol at –20°C. *In situ* hybridization requires that larvae be fixed in a solution of 4% formaldehyde in PBS, dehydrated in an ethanol series and stored in methanol at –20°C.

3. *Storage*. Fixed embryos should be transferred as quickly as possible to a –20°C freezer, where they will remain for long-term storage.

Fecal samples
PCR amplification of DNA from fecal samples is dependent on preservation method, PCR-product length and whether nuclear or mtDNA is assayed. Storage in DMSO/EDTA/Tris/salt solution (DETs) is most effective for preserving nuclear DNA, but storage in 70% ethanol, freezing at –20°C, and desiccation using silica beads perform equally well for mtDNA and short (< 200 bp) nuclear DNA fragments [87, 88]. Protocols for the isolation of DNA from fecal samples stored by these methods are provided by Wasser et al. [87], Frantzen et al. [88], Shankaranarayanan and Singh [89] and Launhardt et al. [90]. Dowd et al. [91] described protocols for the extraction of DNA from formalin-fixed fecal samples.

3.3   Practical Considerations

Contamination
Throughout the duration of fieldwork, collectors should be aware of the importance of keeping their instruments, containers and reagents clean in order to prevent cross-contamination. Individual tissue samples should be

stored in separate containers, if possible, even when the level of analysis to be examined does not require it (e. g., DNA restriction site analysis among populations). Future studies involving single-copy genes, introgression, hybridisation or recombination require knowledge of specific sources of the molecules. If storage of samples in the same container is unavoidable, polyethylene sacks may be used to separate individual samples [92].

An attempt should be made at the outset to remove all dirt and contaminating organisms (e. g., epiphytes, fungi, ectoparasites) from the specimen or tissue sample. It may not be possible to remove all such organisms (e. g., endoparasites and gut flora), but this problem can be circumvented by judicious tissue choice (see above) and the use of taxon-specific primers in the PCR. Mites are a particularly problematic contaminant of field-collected fungal and algal samples. Not only can these arthropods destroy specimens, they can also cross-contaminate a culture collection as they move about [93]. Additionally, the removal of contaminating organisms is important for preventing the introduction of foreign organisms across international borders (see below).

Labelling and documentation
Investigators should meticulously label all materials they collect. Labels should include species (if identified), collection locality, brief habitat description (including reference to collection method used), date of collection, collector, voucher number, etc. [94]. When working in a team, the principle investigator should allocate documentation duties at the outset and assign a backup. All team members should be conversant with the system used to identify and inventory samples that have been collected, including unambiguous abbreviations, the order in which information should appear on the labels and other critical information about the provenance of the sample, which may disappear when field notes are transcribed.

Permanent ink markers should be used for labelling, but must be tested beforehand in water, ethanol and extremes of temperature [95, 96]. Labels should be written directly on the sample bag or container or, if affixed, tied as well as taped, since tape is liable to come loose during handling, freezing and thawing. Alternatively, labels can be etched into glass or plastic tubes with a diamond-tipped pen [97]. Permanent ink marker or heavy lead-pencil labels, written on roughened paper or card [98, 99] and tied to the specimen itself or placed inside the container, are the safest guarantee that collection and provenance data will remain with appropriate samples. The use of synthetic, polypropylene paper (e. g., Kimdura or Tyvek, available from Kimberly-Clark, Dallas, TX) prevents tearing of labels [92, 100, 101].

It is prudent to test all vials, tubes or other containers of unknown composition before departure, to ascertain if they are unbreakable, solvent-resistant and leakproof (for fluid-preserved samples) or can withstand ultracold conditions (for frozen samples). Fragile glass tubes (e. g., hematocrit tubes) can be inserted into the slots of corrugated cardboard for protection during storage [45].

Water-resistant pocket notebooks and pens with waterproof (and solvent-resistant) ink are as essential to successful collecting as sterile vials and bags [97]. The use of traditional collector's catalogues (e. g., [102, 103]) organised such that each specimen receives a unique number preceded by the collector's initials, is recommended [3, 4]. Catalogue entries should record the type(s) of tissues to be sampled and catalogue numbers should be recorded on the vials containing tissues.

Modern biorepositories often use standardised specimen vials with bar-coded labels printed on thermal- and solvent-resistant material suitable for long-term storage. Advance planning with regard to tissue voucher disposition may allow a researcher to obtain a series of numbered sample vials from the repository prior to the outset of fieldwork. The researcher may then refer to the vial number in the field notebook with regard to collection data, thereby circumventing the need to label vials in the field and facilitating the access to specimens in the repository and its associated database on completion of the study. Advance knowledge of the repository accession numbers associated with specimens also facilitates the preparation of manuscripts.

Primary documentation in the field will always be done by hand (if for no reason besides the fact that the backup systems for field computers are subject to failure). Nevertheless, notebook computers have provided an additional innovation to the documentation of field data. Battery operation for more than three hours, with one-hour recharge times from 110 V outlets or car batteries, together with the ability to link via internal or facsimile modems to remote laboratory computer facilities, has allowed the maintenance of up-to-date database files [2]. For further suggestions on methods of cataloguing in the field, refer to Baker and Hafner [97] and Dessauer et al. [3, 4].

Hazardous organisms
Preventative measures may be needed to protect the researcher working with certain organisms. Investigators should not attempt to work with these taxa without proper equipment and a thorough appreciation of the risks involved. For example, investigators working with venomous animals should wear protective clothing and carry antivenom or venom aspirators, in the event of accidental envenomation. These precautions are especially obvious for those involved in the "milking" of venom for their investigations [3, 4].

Investigators working with some mammals and their parasites should be vaccinated against rabies or tuberculosis. Animal necropsy should also include protective clothing and the containment of possible biohazards, in order to avoid contracting diseases such as psittacosis, erysipelas, rickettsial infections and brucellosis [104]. The National Wildlife Health Center (Madison, WI) has a Resource Health Team available to examine animal specimens for possible diseases (Appendix 2).

Some fungi or their secondary products can also cause virulent or chronic disease in animals and may require special procedures [105]. Culture collection staff (see below) may be able to advise on the necessary procedures for pathogenic fungi.

4   Other Sources of Specimens and Tissue Samples

There are many ways in which tissues can be obtained without even going into the field. Stock centers and culture collections, commercial supply companies, cell line centers, seed/spore banks, botanical gardens and zoological parks are the most frequently used sources for fresh tissues. The primary consideration in obtaining tissues from any of these sources is the reliability of the identification of the organisms that are used for the tissues. If the species designation is suspect, the specimen may not be worth the trouble. In addition, the exact origin of organisms from these collections is often unknown or ambiguous, such that caution should be exercised with specimens obtained from these sources if precise location data are needed for a particular study.

In addition to these sources of fresh samples, vast numbers of frozen, desiccated and fluid-preserved samples exist in natural history museums, herbaria and frozen tissue collections. Such samples are generally accurately identified, documented and, if collected in series, vouchered. However, some (e. g., desiccated and fluid-preserved samples) may provide considerable challenges to protein or nucleotide extraction, depending on the method and extent of preservation.

4.1   Sources of Fresh Specimens and Tissue Samples

Stock centers and culture collections
Stock centers exist for several animal taxa (e. g., *Caenorhabditis*, *Drosophila*, *Peromyscus* and *Tribolium*), from which particular strains or species may be ordered at cost (Appendix 3). The services of such centers, including preparation, testing, preservation, maintenance and shipping of the cultures, are expensive, hence user fees must defray costs [106].

Animal stock centers are often limited in the diversity of taxa available, although some hold a large variety. For example, the Jackson Laboratory (Bay Harbor, ME) maintains more than 2500 strains of genetically defined mice for biomedical research, while the *Drosophila* Species Stock Center (Tucson, AZ) can supply approximately 300 species of *Drosophila*. Samples of bacteria, fungi, algae and other unicellular protists may similarly be obtained at cost from culture collections, where they are maintained in pure culture. Lyophilised cultures of fungi may be shipped immediately on order, but algae must usually be grown out on agar or in liquid medium so that shipping may be delayed several weeks.

Detailed information on worldwide culture collections is available from the World Federation for Culture Collections (WFCC) which maintains the World Data Center on Microorganisms (WDCM) and is a component of the UNESCO Microbial Resources Centers (MIRCEN) network. As part of their mission to disseminate information on culture collections, the WDCM has produced two useful publications: the *World Directory of Collections of Cultures of Micro-organisms* [107], which includes 345 fungal collections, and the *World Catalogue of Algae* [108]. In addition, users may search the listings of all the collections for individual species on the WDCM electronic database (Appendix 4). According to this database, 472 culture collections in 62 countries are currently registered in WDCM. A guide to the database (*Guide to World Data Center on Microorganisms – A List of Culture Collections of the World*) is also available. The websites of other culture collections and germplasm repositories are listed in Appendices 3 and 4.

In addition to the major collections with broad general holdings, many smaller collections occur throughout the world, which might be of interest for providing samples for particular studies or geographic regions. Medical research institutes, which include departments specialising in parasitology, toxicology and medical entomology (e. g., Instituto Butantan, São Paolo, Brazil; South African Institute for Medical Research, Johannesburg, South Africa), often stock permanent cultures of common medically important local taxa (e. g., mosquitoes, spiders, scorpions, snakes and unicellular protists), from which samples may be supplied for molecular studies, on request. Insect fungal pathogens can be obtained from the specialist collection of entomopathogenic fungi maintained by the U.S. Department of Agriculture Agricultural Research Service (USDA-ARS) at the U.S. Plant, Soil and Nutrition Laboratory (Ithaca, NY). Cultures of many isolates of wood-rotting basidiomycetes are available from the USDA Forest Service at the Forest Products Laboratory and Center for Forest Mycology (Madison, WI), which maintains voucher specimens for every culture. The *Chlamydomonas* Genetics Center at Duke University (Durham, NC) houses a large collection of *Chlamydomonas reinhardtii* mutants and numerous strains of other *Chlamydomonas* species and provides a printout of file information on each strain. The Soil Microbiology Division of the International Rice Research Institute (Manila, Philippines) maintains a collection of prokaryotic nitrogen-fixing blue-green algae, and a collection of algae with high potential for use in biomass energy production is available at the National Renewable Energy Laboratory (Golden, CO).

Most curators of culture collections ensure that cultures are correctly identified, even returning transferred cultures to the depositing scientist for authentification before cataloguing [106]. Accordingly, the majority of samples obtained from such cultures will be correctly identified (although collection data may be unavailable). Where multiple strains of species are offered by culture collections, informed choices should be made, depending on the research question. Strains may differ in metabolism and other genetic characters, which may be important in selecting a culture. When strains are intended to be representative of the species, so-called type cultures (i. e., those isolated from the type collection of a species at the time of its description) should be ordered preferentially. Two useful publications of the American Type Culture Collection (ATCC) list processes and products associated with many isolates in their collection [109, 110] and additional information may be found on their website (Appendix 4). Similarly, some culture collection catalogues contain information, including literature references to strains [111, 112].

Ploidy number should be a further consideration when selecting fungal cultures. Highly variable DNA regions in single-copy or repetitive genes (e. g., ribosomal RNA genes) may vary at single base positions in the homologous chromosomes of diploid or dikaryotic isolates [113]. Haploid cultures from many fungi with gametic meiosis can be readily acquired from single spores, whereas isolates from basidiomata, ascomata or mass spore cultures of hetero-thallic basidiomycetes and ascomycetes provide strains with nuclear variation [93]. Diploid material is difficult to avoid in some organisms (e. g., oomycetes), but it is important to recognise the possibility of variation. DNA extraction from single spores [114] may eliminate some ploidy problems and provide a means for a variety of intraspecific studies. The ploidy level of most algae in culture collections is known or can be inferred from the life cycle (information available in general texts such as Bold and Wynne [115]).

Cell line centers
Cell lines offer an alternative source of fresh tissues for comparative research. A large selection of cell lines are held in storage by a variety of different centers (Appendix 4), e. g., viral infected cell lines, cell lines for large mammals and cell lines for human diseases [106, 116, 117]. Further examples of the breadth of taxonomic representation are provided on the websites of the ATCC (American Type Culture Collection) and the ECACC (European Collection of Cell Cultures). The CRES (Center for Reproduction of Endangered Species, Zoological Society of San Diego, CA) supplies a variety of endangered vertebrate cell lines and tissues (Appendix 5).

Many centers store multiple different lines derived from different tissues for a variety of taxa, thus allowing specific tissues to be selected for studies in which tissue specificity is a concern [106]. The major cell line centers such as the ATCC and ECACC are easily accessible via email or their websites, which provide online order forms to facilitate the rapid purchase of cell lines. These centers also offer advice on the materials and methods of cell culturing, once the cell lines have been shipped to researchers. Each cell line center has specific terms of usage that the researcher should consult before placing an order (see below).

Commercial Sources
Live representatives of most animal phyla and pure cultures of some protists can be obtained relatively cheaply from biological supply companies (Appendix 6). Ordering specimens is convenient insofar as they are shipped direct to the laboratory, thereby saving considerable time and effort. However organisms

should be ordered as far in advance as possible, since their appearance in the field may be seasonal or unpredictable. Identifications are usually correct (common, well-known taxa are usually stocked), but voucher specimens should still be preserved (see below).

The commercial pet trade provides yet another convenient source of specimens or tissue samples for some animal taxa, notably birds, fish, herpetofauna and selected terrestrial and marine invertebrates. Prices of commonly available species acquired from these sources are usually reasonable, but less commonly available taxa may be expensive. Moreover, although certain common or captive-reared species are available year-round, the availability of most field-collected species is highly unpredictable and requires constant vigilance on the part of the investigator. Apart from the commonly available species, identifications of most specimens acquired through the pet trade should be viewed with suspicion, especially where reptiles, amphibians and invertebrates are concerned. Besides general identification errors made by untrained individuals (especially common with sexually dimorphic taxa), deliberate misidentification is used by certain dealers to enhance the apparent "diversity" of taxa on offer and hybrids are often sold as "new species." Collection data, if available (seldom more than country of origin, which is acquired third-hand by the dealer from the importer, who received it from the collector!), should be mistrusted for the same reasons. Finally and most importantly, it behooves the investigator intending to use the pet trade as a source of specimens to ensure that the specimens to be purchased were acquired legally by the dealer (addressed further below).

Botanical and zoological gardens

Botanical gardens (including arboreta and university greenhouses) are an established source of land plant tissue, whereas zoological gardens (including aquaria, aviaries, butterfly gardens and snake parks) are just emerging as a source of tissue from vertebrates (and certain invertebrates). Currently, CRES is the only zoological institution to make their resources formally available. The large diversity of plant and animal families and genera maintained by botanical and zoological gardens is ideal for higher level molecular studies using exemplar taxa. More specialised collections (e. g., certain arboreta, butterfly gardens and snake parks) may also be useful for molecular studies at the genus or species level. Most directors of botanical or zoological gardens are very willing to provide access to their collections for molecular study as it enhances the role, and thus continued support, of the garden. However, since animal tissues may have to be sampled when animals are sedated for medical examination, investigators should request samples well in advance.

Despite the obvious benefits of obtaining tissue samples from botanical or zoological gardens (large diversity, ease of collection and transport, saving in cost and time), potential problems remain. Foremost is the lack of complete voucher information (collector, date and exact locality), attached directly or indirectly in the records, to the specimen. A more serious problem is the possibility of misidentifications and label switches, although these can be obviated by making vouchers for subsequent identification by specialists in each group. A further problem concerns collection and transport of the material. Many gardens have neither time nor personnel to oversee the multitude of requests for shipment of samples and should not be expected to bear the associated expenses. Personal contact with scientists and managers at the gardens is recommended.

Henderson and Prentice [118] provide a worldwide listing of botanical gardens. Various other sources, e. g., the teaching guide to the text *Biology* [119], also contain extensive lists of zoos and botanical gardens. Recent listings can be found at the websites provided in Appendix 7.

Seed and spore banks

Seed banks are becoming an increasingly important source of plant tissue in molecular studies, especially where agronomically important groups, including cultivated species and their wild relatives (e. g., *Brassica*, *Solanum*, *Triticum*, *Zea*, *Glycine*), are involved (Appendix 3). The U.S. National Plant Germplasm System (NPGS), coordinated by the National Germplasm Resources Laboratory (Beltsville, MD), comprises 22 repositories that are annually updated and planted out to maintain their collections. One of these, the U.S. Potato Gene Bank (Sturgeon Bay, WI), was essential for a large molecular research program on the cultivated potato and its wild relatives [2]. Besides maintaining an extensive collection of seeds, the National Seed Storage Laboratory (Fort Collins, CO), another repository of the NPGS, is investigating pollen storage as a means of storing germplasm of clonally held species [120].

In addition to such formal repositories, many botanical gardens maintain seed exchange lists (e. g., *Index Seminum* of the Modena Botanical Garden, Italy). The Millenium Seed Bank Project of the Royal Botanical Gardens, Kew has the most extensive listing of fully documented plant species, including many families of noncultivated plants.

Although fronds and rhizomes are most often used as tissue sources from ferns or fern allies, spores and resultant gametophytes are equally suitable. The American Fern Society maintains a spore exchange program and information can be obtained from the *Bulletin of the American Fern Society* or their website (Appendix 3).

Potential problems with seed or spore sources, whether from seed banks or university greenhouse collections, include contaminated seed, errors in handling and labelling, and misidentification. Such problems can be avoided by vouchering all plant tissue grown from seed.

Gifts or exchanges with scientific colleagues

A commonly used alternative source of field-collected material is through gifts or exchanges with colleagues in foreign countries, who are either specialists in the group under study or knowledgeable about specific collection sites. Although the use of local scientists can add greatly to the breadth of the study, several practical issues remain, including the provision of detailed instructions concerning shipment of material and the necessity for providing permits if crossing international borders (see below). A further issue to consider is whether the support in obtaining samples from colleagues has been so extensive and critical to the success of the project that nothing short of co-authorship would be adequate compensation.

4.2   Frozen Tissue Collections

Many natural history museums have created, and continue to support, frozen tissue collections or collections that harbor tissues in ethanol or lysis buffers [121, 122]. Two directories of frozen tissue collections ([4, 31], reproduced and updated in Appendix 5) list public and private institutions in the U.S.A. and abroad that have holdings of frozen tissues. Several frozen tissue collections are now online while others are in the process of producing websites for easy electronic access via the internet.

Some requests are easily fulfilled by these facilities, as with commonly available tissues of which large quantities exist in storage. Other requests are more difficult to fulfill, as with tissues of endangered species of which precious little remains. Consequently many frozen tissue facilities require a formal application from the researcher wishing to obtain loans, especially for the more difficult requests (for examples of frozen tissue collection policy, refer to the websites in Appendix 2). First, the loan request must be accompanied by some assurance that the applicant is proficient in isolating the molecules of interest from frozen specimens [97, 123]. Such assurance is usually demonstrated in a short proposal that the applicant submits to the repository for curatorial approval. Second, many frozen tissue facilities request that a proportion of the DNA or proteins isolated from the frozen tissue be deposited back in the frozen tissue facility in return for the loan.

Shipments of frozen tissues are costly to package and transport hence payment of tissue shipment costs is normally requested of the recipient [97]. Investigators planning to obtain samples from frozen tissue collections should provide for these costs in their grant budgets.

4.3   Sources of Desiccated or Fluid-preserved Specimens and Tissue Samples

Natural history museums
Museum collections represent a tremendous resource for molecular studies. The world's research collections contain vast amounts of material amenable to molecular analysis that may be otherwise unavailable due to extinction or

collection difficulties. This can be especially acute in higher level studies of geographically diverse taxa. Such collections can make possible, in time and resources, studies for which it would take years to gather material.

Zoological collections are composed predominantly of skeletal remains, dried skins, dried insects and fluid-preserved specimens. Skeletal material, which makes up a large proportion of vertebrate collections, is usually devoid of amplifiable DNA [19, 21, 22, 124]. Nevertheless, Hagelberg et al. [125] amplified DNA from human bone samples 300–5500 years old. More recently, methods developed for protein extraction from ancient bone were found to yield large amounts of DNA with molecular weights much higher than those seen from ancient soft tissue [126–128]. Dried, untanned skins routinely yield DNA sequences, as do dried, pinned insects, although the DNA may be extremely degraded [38, 64, 124].

The vast majority of helminth and arthropod specimens are preserved in ethanol, in which DNA preserves fairly well (depending on collection practices), although it is usually degraded to < 2 kb, especially if 70% ethanol was used, as is prevalent in older collections [65]. However, other fixatives, e. g., glycerin and 2-propanol (in arachnid collections), may have been added to the ethanol in an effort to harden, soften or maintain the colour of the specimens. For example, terrestrial planarians, stored long-term in 80% ethanol, must first be fixed in a formaldehyde calcium cobalt solution [62].

Fluid-stored insect collections may contain similar fixatives, including formamide, picric acid, formalin and glycerin. Any of these fixatives may likewise affect DNA, hence it is crucial to know how samples were collected and maintained. For example, DNA isolated from insects preserved in acetone, ethyl acetate, formal saline, Carnoy's solution (ethanol:acetic acid, 3:1), methanol, or propanol was highly degraded, compared with that extracted from insects preserved in liquid nitrogen, preserved in ethanol (stored at 4°C or at ambient temperature), sun-dried, or dried over silica gel [15, 38, 65].

Vertebrates and marine invertebrates (e. g., cnidarians, ctenophores and ctenostome bryozoans) have traditionally been preserved in solutions of 10% buffered-neutral formalin (BNF) (v/v) with implications for DNA isolation from such samples. PCR is greatly affected by the amount of time spent in formalin, the temperature at fixation and the method of buffering. For example, Greer et al. [64] found that fragments of DNA, approximately 1 kb in length, were impossible to amplify from formalin-fixed tissues after only 24 hours at ambient temperature. However, high molecular weight DNA was obtained from tissues fixed at 4°C in formalin buffered with 4 M urea [23, 129]. Similarly, Noguchi et al. [63] established that fixation in formalin at 4°C and in formalin containing 5 mmol/L EDTA at ambient temperature preserved significantly more high molecular weight DNA than fixation in formalin at ambient temperature. Savioz et al. [130] recently provided protocols for the isolation of high molecular weight DNA from tissues stored in formalin for nearly 50 years. A simple test for the identification of formalin *versus* ethanol-preserved specimens, based on a colour change in the acid-base indicators sodium sulphite and sodium metabisulphite, was reported by Waller and McAllister [131]. Tests for other fixatives were outlined by Moore [132].

Greer et al. [64] and Criscuolo [133, 134] evaluated the ability of various other fixatives, traditionally used in vertebrate collections, to yield DNA fragments of moderate size on PCR amplification. Most fixatives gave poor results, e. g., acetone, Zamboni's solution, Clarke's solution, methylated spirits, phenol, chloroform, glutaraldehyde, paraformaldehyde, formalin-ethanol-acetic acid (FAA) and metacarn. The worst results were obtained with highly acidic fixatives such as Carnoy's solution, Zenker's solution and Bouin's solution. Zenker's solution is a common tissue fixative (5 parts glacial acetic acid mixed with 95 parts saturated mercuric chloride and 5% potassium dichromate) which affects the stability of DNA in tissues preserved in it, owing to high heavy metal concentrations which promote phosphodiester breakage. Bouin's solution was formerly the fixative of choice for many studies of infectious or parasitic diseases in animals, where a large piece of tissue was preserved in excess solution. However, parasitic samples preserved even under these conditions may be useful if investigators design PCR primers to yield small (< 200 bp) fragments.

Overall, most of the wet and all of the dry samples in museum collections are potentially useful to molecular studies. Given the resources of some of the large institutional collections, some intractable problems may be resolved and some groups whose molecular biology is poorly known may be sampled more completely (e. g., [135]). Readers should consult the *Ancient DNA Newsletter* for updates on protocols suitable for the successful isolation and amplification of DNA from museum specimens.

Herbaria

As with museum collections, there is increasing interest in herbaria because they can potentially provide DNA from identified, fully vouchered species, including types and species that are rare or nonculturable [136–138]. Like museums, the role of herbaria will undoubtedly become more important as PCR and sequencing technology allows more rapid and extensive surveys at all taxonomic levels.

Herbarium specimens of fungi as old as 50 years [139] and nearly 100 years [140] have provided adequate templates for PCR of ribosomal genes. Several studies have also successfully extracted DNA from plant [48, 50, 51] and algal [141] specimens, which similarly proved to be good templates for PCR.

Douglas and Rogers [142] assessed the effects of seven cytological fixatives (3.7% formaldehyde at pH 3.0 and 7.0, FAA at pH 3.0 and 7.0, 1% glutaraldehyde at pH 3.0 and 7.0, and Lavdowsky's fluid, containing mercuric chloride, at pH 3.0) and one storage buffer (SED = NaCl-EDTA-DMSO, pH 7.0) on the DNA of nine plant and fungal species. DNA from untreated tissue and SED-treated tissue was of high molecular weight (> 50 kb). DNA from glutaraldehyde-treated tissues averaged 20 kb in length, while DNA from all other treatments averaged < 8 kb in length. Nearly all attempts to amplify from specimens treated with 3.7% formaldehyde (at pH 3.0 and 7.0) failed.

Although DNA from desiccated herbarium specimens may be partially degraded, primer-directed amplification, especially of multiple copy genes, can overcome the problem. However, the relatively poor quality precludes efficient restriction site analysis [55]. Generally, specimens fast-dried at moderate temperatures, or the edges of large specimens, provide the least-degraded DNA [93]. It is unlikely that isozymes will ever be routinely extracted in an active state from herbarium specimens, although certain desiccated tissues can provide good activity for some systems [58], discussed above. In addition to providing DNA from tissue, herbarium specimens of fungi and algae may contain viable propagules for establishing cultures [143] or extracting DNA [141]. Selected websites, containing recent listings of natural history museums and herbaria, are provided in Appendix 8.

Medical facilities
Archival collections of paraffin-embedded tissues prepared for histology [144–146] and collections of dried blood spots on Guthrie cards stored by state neonatal-screening laboratories [41, 147, 148] represent valuable resources for retrospective studies in clinical pathology and molecular epidemiology (Appendix 4).

Protocols for DNA extraction from paraffin-embedded tissues up to 40 years' old were provided by Frank et al. [149], Morgan et al. [150] and Pavelic et al. [151]. Pavelic et al. [151] assessed the effects of different fixatives on paraffin-embedded tissues, noting that 10% formalin caused irreversible DNA damage that was greater with prolonged fixation time, while tissues stained with Carnoy's, AmeX and Papanicolaou fixatives resulted in consistent yields of high molecular weight DNA. Makowski et al. [41], Ostrander et al. [42] and McCabe et al. [43] provided methods for extracting DNA from blood spots on paper.

5   Transportation, Long-term Storage and Archiving of Specimens and Tissue Samples

Whether specimens and tissue samples are collected in the field or acquired from other sources, they will have to be transported back to the laboratory for processing. Transportation methods must aim to retain the samples as intact as possible during transit, which may require rapid delivery and be subject to a variety of regulations, depending on the nature of the samples (e. g., live *versus* preserved) and the mode of shipping (e. g., airline, courier, mail).

On arrival in the laboratory, samples must be processed (including identification, if this has not already been conducted), excess tissues stored for future use, voucher specimens retained (or deposited in appropriate institutions) and collection data archived in a computerized database. A variety of approaches to long-term storage and archiving are available to the investigator, the choice among which depends both on the taxa and molecules of interest.

5.1   Transportation

Protocol 4 lists considerations relevant to the transportation of a variety of preserved specimens and tissues. Specific recommendations for packing and labelling are provided in Protocol 5.

Protocol 4   Common transportation methods

1. *Live specimens and cultures.* Invertebrate material is best shipped live from supply houses. However, transit times and export/import regulations may prevent this option when international shipping is involved. Small insects such as *Drosophila* or *Tenebrio* can be sent in vials containing limited amounts of food. Larger arthropods (e. g., scorpions) and many reptiles will tolerate several days without food or water. Most freshwater and marine organisms (adults or larvae) can be shipped live in sealed plastic bags.

Packages containing live specimens should be well padded, slightly humid and insulated with Styrofoam against external temperature fluctuations. Live field-collected invertebrates may also be transported as airline hold or carry-on baggage, again subject to the necessary export and import permits for live animals and the airline regulations.

Algal (and occasionally fungal) samples are usually provided from culture collections as liquid- or agar-grown cultures and may require specific light or temperature requirements during shipping. Although most phototrophic algae can survive several days without light, a prolonged delay in shipping time may be fatal, as may exposure to extreme temperatures.

Prompt treatment of organisms is required on receipt of the shipment. Specimens obtained for rearing should be placed on rearing media, whereas those obtained for DNA isolation should be immediately frozen at or below −20°C.

2. *Fresh plant samples.* Fresh plant tissue may be transported in two manners. For short transport duration (2–3 days, e. g., when acquiring samples from botanical gardens), leaf tissue or shoot cuttings in Ziploc bags can be carried at ambient temperature in insulated Styrofoam containers or mailed directly via overnight or express mail couriers, with or without Styrofoam containment.

A more reliable and expensive transport method involves placing the bags of tissue into an ice source inside a Styrofoam container. Wet ice may be used, but will melt and soak the package after a few days. Alternatively, a small amount of dry ice, wrapped in canvas, may be placed with the wet ice, to prevent the latter from melting. To avoid cold temperature burn, the Ziploc bags should be separated from the ice by means of newspaper.

Fresh tissue transport from remote locations requires phytosanitary, export and import permits. Tissue may be carried on wet ice in Styrofoam containers as airline hold or carry-on baggage [2]. Small packages require specific documentation to pass customs and agricultural inspection points in both shipping and receiving countries. Using airlines to ship packages is ill-advised, since many airlines refuse to handle a package originating abroad unless it has cleared customs, thus requiring a contact person. A copy of the USDA permit, phytosanitary certificate and letter identifying and describing the purpose of the plant tissue should be attached inside the package and outside with the shipping label (see below).

3. *Frozen samples.* Vertebrate tissue samples (e. g., blood and muscle) are routinely transported frozen. Transport of frozen tissue on dry ice or in liquid nitrogen is simple when air shipment is not required. However, authorisation for shipment of these chemicals on airlines is mandatory, since both are classified as "Dangerous Goods" [31]. Policies for packing, labelling and carrying such materials have been promulgated by the International Air Transport Association (IATA) in their *Dangerous Goods Regulations* manual, and general guidelines are provided on their website (Appendix 2). Several courier services have instituted similar policies for the transportation of dangerous goods, e. g. the FX 12 Operator Variation of Federal Express.

Airlines permit no more than 200 kg of dry ice in a single container [3, 4]. The "Shipper's Declaration of Dangerous Goods" is required and the package must be marked as Hazard Class 9 (solid carbon dioxide). Handwritten declarations are not accepted; forms must be typewritten or computer-generated (refer to Appendix 1 for the website of Saf-T-Pak, which provides free online software for computer-generated Shipper's Declaration forms). Since cargo facilities may require pound equivalents, both metric and English equivalents should be used for packages. The special black-and-white sticker is a prerequisite for acceptance as air cargo.

Airlines permit up to 50 kg of liquid nitrogen to be transported in a nonpressurised Dewar container, which must also be marked as hazardous (compressed gas). The Shipper's Declaration is again required and the package must be clearly labelled "NONPRESSURISED LIQUID NITROGEN", "THIS SIDE UP" and "DO NOT DROP – HANDLE WITH CARE" to prevent careless handling and spillage. It may also require a green label stating "NONFLAMMABLE GAS". Alternatively, the liquid nitrogen can be discarded before short flights and "Dry Shipper" (e. g., the Arctic Express Thermolyne CY 50915/50905 available from VWR Scientific Products, San Francisco, CA) used, thereby avoiding liquid nitrogen spillage. This is preferable for cryogenic collections because of the ease with which samples can be transported by air.

Sperm can be shipped on wet ice if it is collected undiluted in water and sodium azide has been added to a final concentration of 0.01–0.02%, in order to prevent bacterial growth [14]. For further guidelines on the transportation of frozen tissues, consult sections E1565 and E1566 in Volume 11.05 of the *Annual Book of ASTM Standards* of the American Society for Testing and Materials (West Conshohocken, PA) (Appendix 1).

4. *Desiccated samples.* Cultures of some fungi are stored lyophilised and can be shipped immediately on order [93]. Dried samples of skin, hair, feathers, muscle and blood from vertebrates may be placed in tubes or envelopes (in the case of blood spotted on paper) and mailed in regular mail [45]. Dried insects and dried plant tissues can be similarly transported: individual samples should be care-

fully wrapped in paper towel to prevent damage and placed inside vials or Ziploc bags. A small amount of desiccant (e. g., silica gel) should be included in the vials or bags. In the case of plant samples, this should be in contact with the leaf tissue. Sabrosky [152] and Piegler [153] provide guidelines for the packing and shipping of pinned insects. Desiccated samples must be shipped in airtight containers to prevent rehydration and indicator silica gel is recommended to verify that the samples have not rehydrated [2].

5. *Fluid-preserved samples.* The most convenient method for transporting tissue samples of most invertebrates, vertebrates and marine algae, is preserved in 70–100% ethanol. However, since ethanol is flammable, highly volatile and a strong solvent (especially at higher concentrations), the use of unbreakable, leak-proof and ethanol-resistant plastic vials is a prerequisite to avoid evaporation or spillage. Sealing the vials inside thick, ethanol-resistant plastic sleeves is advised as a further precaution. Similar precautions against leakage should be ensured when transporting samples stored in other fluids. For further details, refer to McCoy [154].

Protocol 5   Packing and labelling for transportation

1. *Packing.* Sturdy, internally padded packaging should be used when shipping specimens or tissue samples. All external labels should be clearly marked with permanent ink and covered with a plastic sheet. A list of materials included, with relevant names, addresses and telephone numbers should be enclosed for the benefit of customs officials and to help recipients determine if the package was tampered with.

2. *Labelling.* When pertinent, the outside of the package may be marked "BIOMEDICAL SAMPLES FOR SCIENTIFIC RESEARCH – PLEASE RUSH", though for most packages, the words "SCIENTIFIC SPECIMENS FOR RESEARCH" will suffice. Customs forms should clearly indicate that the contents have "NO COMMERCIAL VALUE".

3. *Shipping.* Before shipping, ensure that holidays (civic and religious) will not impede or delay the delivery of packages. It is also prudent to avoid shipping over the weekend. These considerations are especially critical for shipments of live or frozen samples. Parcels of live samples should not be sent during periods of extremely hot or cold weather. Recipients should be called, faxed or contacted via electronic mail on the date of dispatch and notified of the expected arrival of the package and the airbill or tracking number. Recipients should be informed as to whether they can expect delivery or should collect their samples, and are advised to bring multiple copies of any permits or documentation necessary for collection. If transporting across international boundaries, colleagues on both sides should be intimately involved in the process of shipment.

5.2 Long-term Storage

Frozen tissues
Long-term storage conditions should minimise variation in temperature, light and liquid volume. Therefore, most formal frozen tissue repositories maintain samples at –130–150°C in ultracold freezers or in the vapour phase of liquid nitrogen [155, 156]. Nonetheless, most plant, fungal and animal tissues will remain indefinitely stable for extraction of nucleic acids and proteins if maintained at –70–80°C. Furthermore, although long term storage at or below –80°C is the preferred method for animal samples (e. g., [65]), there are alternatives (Protocol 6).

For short-term storage, frozen algal samples can be kept in a standard freezer at –20°C, in a mixture of glycol and dry ice (~ –78°C) or in liquid nitrogen (–196°C). For long-term storage, algal viability is more likely to be maintained at or below –40°C. Routine, periodic monitoring of viability is essential for long-term storage of such samples.

Plant tissue samples have been stored at –20°C, after quick freezing at –70°C, with apparently no DNA degradation after 6–8 months [2]. However, marked degradation has also been observed when leaf tissue was stored at –20°C, compared with –70°C (no difference in DNA quality has been detected between tissue quick frozen in liquid nitrogen before being placed in the ultracold freezer relative to tissue simply placed in the freezer).

Storage at lower temperature is thus recommended and is essential if the tissue is to be used for isozyme analysis. Storage of tissues for RNA preparations, which require fresh or –70°C frozen tissues, similarly necessitates cryogenic conditions for successful nuclease inhibition. When tissues are stored for such purposes, –70°C freezers should be equipped with alarms and backup generators and temperatures should be checked on a regular basis, especially after storms or building maintenance [97]. Protocol 6 discusses additional considerations for the storage of frozen tissues.

If facilities for long-term storage of frozen samples are unavailable, investigators should consider depositing their samples in established frozen tissue collections upon completion of projects, for which guidelines for submission of samples are provided below. Further advice on cryopreservation, sources of the chemicals and the organisation of frozen tissue collections is provided by Dessauer et al. [3, 4], Simione [30], Baker and Hafner [97] and sections E1342, E1564, E1565 and E1566 in Volume 11.05 of the *Annual Book of ASTM Standards* (Appendix 1).

DNA
An alternative approach to tissue storage is to minimise or even eliminate storage in ultracold freezers and store only the DNA permanently (Protocol 6). Tissues are placed in the refrigerator immediately and DNA isolated as soon as possible. This is the preferred method in studies where fast DNA extraction

methods are available, tissue can be processed rapidly and no backup tissue source for additional extractions of DNA or proteins is required.

There are several benefits of this approach: 1) the need for ultracold storage space is diminished, thus lowering the costs of purchasing equipment, electrical power and maintenance 2) space is made available for more important tissues where a backup tissue source is required 3) DNA samples are quickly and efficiently obtained 4) breakdown of ultracold freezers is less serious since DNA samples (which are more stable than tissue samples) have already been obtained.

Protocol 6   Common long-term storage methods

1. *Frozen samples.* Frozen tissue may be stored in an ultracold freezer, on dry ice or in liquid nitrogen tanks. Chest-type freezers use interior space more efficiently and maintain colder temperatures than upright models, although the latter require less floor space [28]. Permanent marker eventually deteriorates at extreme cold temperature, as do many kinds of tape, so placing a pencil label inside the sample holder is advised. Plastic tubes are preferable to glass for frozen storage, since sudden temperature changes may crack glass when samples are removed from the freezer.

   The use of a cryoprotectant is advocated for frozen storage of many samples. For example, freezing and storage with a cryoprotectant has been advocated for preserving filamentous fungi in liquid nitrogen [157, 158], while algae can be stored as frozen samples with cryoprotectants such as glycerol, DMSO or dried milk solids [93].

2. *Desiccated samples.* Insect samples, plant material, feathers and hair may be stored dry at ambient temperature in the presence of a desiccant (e. g., silica gel). Spore-producing fungi have also been stored at ambient temperatures, usually for shorter periods of time, after drying in soil or silica gel. However, ambient temperatures cannot guarantee against DNA degradation in dry samples, especially if rehydration occurs [45]. In general, dry samples are a poor substitute for fresh, frozen or ethanol-preserved samples in molecular studies and should be placed in ultracold storage on receipt.

3. *Fluid-preserved samples.* DNA yield and quality from specimens preserved in 95–100% ethanol are usually almost as good as from fresh or frozen tissue. Samples can be stored at ambient temperature, but long-term storage of samples in ethanol must be conducted under fireproof conditions. Storage of ethanol samples at –20°C or colder is the best safeguard against long-term DNA degradation, since DNA in ethanol-preserved samples has been found to degrade after ca. six weeks at ambient temperature [65].

   Samples should be stored in sealed vials (to prevent evaporation) and the ethanol should be periodically replaced. Glass vials are recommended since ethanol may weaken, and ultimately crack, certain plastics. For detailed discussions of the containers required for long-term storage of fluid-preserved samples, consult Palmer [159], Legler [160] and De Moor [161]. Hoebeke [162] described a unit-storage system for fluid-preserved samples. Vertebrate tissue samples (e. g., blood) in 2% 2-phenoxyethanol (v/v), with glycerol or DMSO, are normally stored at –20°C [45, 73]. However, Asahida et al. [79] developed a high concentration TNES-urea buffer suitable for preserving fish muscle and liver samples at ambient temperature for up to three years. Sperm containing 0.01–0.02% sodium azide can be stored refrigerated (5°C) for 1–2 years, but must not be frozen [14].

4. *Fungal and algal cultures.* Fungal and algal samples may be stored as cultures, the long-term viability of which depends on species or strain, age of culture at time of storage, propagules present, culture medium (affecting nutritional state) and storage method. Extensive discussions of these techniques can be found in various reviews [105, 158, 163, 164]. Fungal stock cultures remain viable for at least 6 months on agar slants refrigerated at 5°C. However, cultures in vials (15–20 ml) remain viable for longer periods. Most fungi survive at ambient temperature for 2–4 weeks. Wrapping cultures in parafilm retards desiccation of the medium, thus increasing the length of viability; contamination is also lessened. Cultures on agar may be covered with sterile mineral oil, to lower oxygen levels, or kept in sterile distilled water at 4°C.

Some auxotrophic algae are quite similar to fungi in short-term storage requirements. However, most are phototrophic or photoauxotrophic and require adequate lighting. Temperatures slightly below ambient (~ 21°C) and dim lighting are best for culture storage in liquid medium or agar. Some algae require special conditions (e. g., bubbling with air or CO_2), even for short-term storage periods of 2–4 weeks. Detailed information on standard practices is contained in Stein [165].

Many fungal and algal cultures are viable after lyophilisation [93]. Lyophilisation of cultures has the advantage of preserving the genetic integrity of strains that might change during years of active growth.

5. *DNA.* Prepared DNA is best stored at 5°C or in ethanol at –20°C. High salt (> 1 M) in the presence of Na_2EDTA at pH 8.5 (Tris) buffer is recommended for long-term storage, along with storage in light-protected CsCl at 5°C. High molecular weight DNA is stable at 20°C for several days in buffer [1 mM Tris, 0.1 mM EDTA] and can be transported unrefrigerated.

5.3 Identification, Documentation and Storage of Voucher Specimens

Specimen identification

Organisms from which molecular data are obtained should be properly identified and documented. This is important for systematics and evolutionary studies because: 1) knowledge of the species involved may be crucial to interpretation

2) published molecular data may be utilised in ways other than for which they were initially obtained and 3) verifiability is one of the tenets of empirical science [14]. Given the effort required to obtain molecular data, it is the duty of the researcher to devote some care to proper identification and documentation.

If exemplar taxa are required as representatives of higher taxa, most groups contain common, well-known species that may be recognised by the nonspecialist with some confidence. However, the most reliable method of identification is to allow a specialist to provide identified material for molecular extraction, to identify specimens intended for use, or to identify voucher specimens from the same population. In addition to species identification, specialists can provide important information about taxon selection, unresolved relationships and model systems for studying comparative biological questions.

Publishing specialists can be located from the taxon indices of primary or reference journals like *Biological Abstracts*, specialist publications such as the *International Mycological Directory* [166] or *The Insect and Spider Collections of the World* [167], and the directories of societies, such as the *Directory and Guide of Resource Persons* of the American Society of Plant Taxonomists, *Resources in Entomology*, published by the Entomological Society of America, or the *Annuaire des Arachnologists Mondiaux* of the International Society of Arachnology. Several directories, e. g., *Index Herbariorum*, are now available online (Appendix 8). Unfortunately, although they are generally willing to help with identification, specialists are often backlogged with work and may be unable to assist in a timely fashion. The task of identification then rests with the investigator, who has an array of literature, varying from pictorial identification guides to primary taxonomic papers, from which to choose.

Pictorial guides (e. g., [168]), written for the layman, are easiest to use, but are least inclusive and reliable. Colour photographs or line drawings are accompanied by short descriptions and distributional range of the most common species occurring in the region. Where several closely related species occur in the region, usually only one is included. Moreover, these guides avoid using morphological characters necessary for correct identification, which is limiting given the range of ontogenetic and geographical variation found among conspecifics. Field guides are usually more detailed, providing keys in addition to descriptions and figures (e. g., [169, 170]), but only a few allow identification below the level of order or family (e. g., [171]). For such information, the investigator must consult monographic identification manuals, containing detailed keys written by specialists. These are superior for identifying the better known species in the region they cover, but require proficiency in morphological terminology. Monographic series, e. g., the *Synopses of the British Fauna* (*New Series*) published by the Linnean Society, containing keys and numerous illustrations, are also available for the identification of selected groups. Although regional in coverage, such monographs are often useful over a wider range than suggested by their titles (e. g., [172]).

Primary taxonomic literature, scattered in books, monographs and journals, forms the basis for all guides and manuals. If this is familiar to the investigator, then it is the best place to start. Monographs (e. g., taxonomic catalogues or regional floras and faunas) are an essential introduction if the investigator is unfamiliar with the literature. For example, Sims [173] referenced primary literature for invertebrates worldwide by taxonomic group and region. Additional access to primary literature on particular groups can be obtained from the bibliographies of identification manuals and catalogues.

Ease of identification varies with taxonomic group and region. Some groups (e. g., butterflies and birds) are better known, as are some regions (e. g., north temperate). For many groups and as many regions, comprehensive identification manuals do not exist at all. In these situations, museums and herbaria (which usually maintain extensive reference libraries and microscope facilities) should be accessed for identifying organisms.

Documentation and voucher specimens

Due to the dynamic nature of systematics and the nomenclatural changes which accompany the discovery of new species or the reanalysis of phylogenetic relationships, organisms from which molecular data are published should be documented in such a manner that their identity can be verified should question arise. An essential aspect of this documentation is the voucher specimen(s), which should originate from the same local population (deme) as the experimental animals, and be placed into a permanent depository (museum or herbarium). Huber [174] defined voucher specimens in the broad sense as "all biological specimens having the minimum information of collection locality (ideally specified by latitude, longitude, altitude) and date that are preserved to document biological research."

Molecular systematists should not be naïve about the importance of vouchers and the necessity for retaining information about the organisms from which they have sampled molecules. Voucher specimens "physically and permanently document data in an archival report by 1) verifying the identity of the organism(s) used in the study and 2) by so doing, ensure that a study which otherwise could not be repeated can be accurately reviewed or reassessed" [94]. Vouchers serve as a backup and act as important documentation for ongoing systematics investigations [45, 175].

Exactly what comprises a voucher specimen and how many such specimens should be collected depends upon the taxon in question, the characters required for its identification and the number of specimens available [174]. For many taxa, diagnostic features of the specimen used as a tissue sample may suffice and may be all that is available if only a single specimen could be obtained for molecular analysis. However, it is preferable to retain a second, intact specimen as the voucher (if there is a choice between two specimens, a sexually mature individual should always be retained in preference and the immature specimen used as the tissue sample). Additional specimens, representing sexual and ontogenetic variation, should always be acquired, if possible.

These may assist in the identification of sexually dimorphic taxa and may be invaluable in subsequent scoring of morphological data for a simultaneous cladistic analysis.

Photographs and sound recordings (e. g., of birds or anurans) should never be viewed as a replacement for vouchers, except in situations where endangered or threatened species are involved and populations have already been vouchered, e. g., Hawaiian Lobeliaceae [2]. However, photographic slide collections of organisms, from which tissue samples have been obtained, provide an excellent backup in documentation.

Although vouchers are usually obtained with field-collected material, they are often neglected when the tissue is obtained via an intermediate source (stock center, commercial supply company, botanical or zoological garden, colleague, etc.). In some cases (e. g., botanical and zoological gardens), collections are numbered by accession and the original voucher information (collector and number) can be traced in records maintained by the institution. Such information should be obtained and recorded. However, it is prudent to have a second voucher made at the same time as the tissue is collected, to clarify label switches or errors in collecting.

Publication of molecular data should include the locality where the organisms were collected, date of collection, name of collector, depository of the voucher specimens and their catalogue numbers (where available). Catalogue numbers allow access to further information (e. g., the name of the specialist who identified the specimens) and are invaluable for computer archiving as well as storage and retrieval of samples (see below). For further discussions on the importance of voucher specimens, the reader is referred to Lee et al. [94], Huber [174], Meester [175], Robinson [176] and Yates [177].

Storage of vouchers

In common with all specimens, long-term storage of vouchers requires proper labelling and preservation, including retention of the characters pertinent to identification [94]. Different taxonomic groups have different requirements for preservation as specimens. For example, insects, plants, fungi, corals, some sponges, echinoids, asteroids, the skins and bones of birds and mammals, and the shells of molluscs, are stored dry. Most soft-bodied arthropods, worms and marine invertebrates (including the soft parts of molluscs), reptiles and amphibians, marine algae, and fungi are stored in ethanol or formalin, though use of the latter is waning. Among fluid-preserved specimens, there are further group-specific requirements. For example organisms with calcareous tests, shells or bones, which dissolve in formalin, must be stored in buffered formalin or ethanol [178–180]. Soft-bodied aquatic invertebrates, which contract severely when placed directly in fixative, must be relaxed first [181]. Finally, many invertebrates and fungi require sectioning or removal of the genitalia (in arthropods) or radulae (in molluscs), which must be specially fixed and slide-mounted to be suitable as vouchers (e. g., [62, 182–185]).

It is beyond the scope of this chapter to elaborate on the myriad of storage techniques for museum and herbarium specimens (all of which apply to vouchers). Most taxonomic groups have literature specifically devoted to the topic, such as Mueller [186] or Lincoln and Sheals [187] for the preservation of marine invertebrates, Hall [188] or Wagstaffe and Fidler [189] for vertebrates and Savile [190], Ketchledge [191] or Smith [192] for botanical specimens. Harris [193] and Huber [174] provide general guidelines for preservation of the major animal phyla. The onus rests with the investigator to determine which methods are most appropriate for the group in question by consulting general texts on the preservation and curation of natural history collections (e. g., [94, 194–196]) and periodicals such as *Curator, Collection Forum* and the *Journal of Biological Curation*.

5.4 Archiving: Integrating Tissue Samples, Voucher Specimens and Collection Data into a Database

Practical considerations

Keeping track of tissue samples in the laboratory or in ultracold storage, of DNA (or protein) extracted from those samples, and of associated voucher specimens, is an important task when numerous concurrent molecular studies, involving many different investigators, are underway. Some laboratories maintain a decentralised system where each investigator keeps a separate record (e. g., in a spreadsheet) of their own samples. Others are fully centralised, all incoming tissues (and subsequently isolated DNAs) being assigned numbers upon arrival, from which all relevant information can be accessed in the database [156]. In either case, it is critical to include, or have referenced, all voucher information (collection locality, date of collection, collector, number of specimens, and depository). Additional information, e. g., quantity of tissue remaining, method of DNA extraction, date of extraction, quantity and quality of DNA, can also be recorded in the database [2].

Locating the frozen tissue or DNA sample listed in the database similarly requires appropriate organisation of freezers [97, 156]. Ultracold storage space is very expensive to purchase and maintain, so it is important that materials be stored in a space-efficient manner. It is also imperative that the access and inventory procedures for frozen tissue collections be extremely well organised. Separate boxes for holding frozen tissue samples can be maintained for each project or for related taxa, or can merely be assigned numbers according to date of acquisition. Cryoboxes (e. g., Nunc, Taylor-Wharton, Revco) are recommended for storing the samples in ultracold freezers or liquid nitrogen tanks (cryovats) and for storing DNAs in –20°C freezers. These boxes may be further partitioned internally, to hold up to 100 1.5–2.0 ml cryogenic microcentrifuge tubes, if required. Metal (Revco) racks, designed to hold several such boxes, are suitable for additional organisation.

The contents of both the ultracold storage boxes and the DNA storage boxes should be clearly marked to permit rapid entry and exit from the freezers or cryovats. For further discussion of the logistical aspects of long-term storage, consult Baker and Hafner [97].

Depositing samples into frozen tissue collections
As with the deposition of voucher specimens in a museum or herbarium collection, the deposition of tissue samples in a frozen tissue collection requires accurate documentation of the collection data pertinent to the sample and inclusion of those data with the sample [94, 97]. A record of what the sample was used for, and where the voucher specimens are deposited, should also be included. Specific institutions have their own requirements for deposition of samples in their collections and the investigator is advised to consult the latest information on their websites (Appendix 5) for further guidelines. If multiple samples are to be deposited, it is advisable to provide the data in machine-readable format or submit it electronically.

Computer databases
Irrespective of whether tissue samples and voucher specimens are deposited in the same laboratory, in separate collections within the same institution, or in separate institutions, a record of information pertinent to those samples and vouchers must be maintained in a computerized database. Numerous software packages are available to the investigator, varying from general database programs (e. g., Microsoft Access, Paradox) to programs designed specifically for the maintenance of biological collections (e. g., BIOTA, MUSE, Platypus, PRECIS). Woodward and Hlywka [197] developed a database strictly for managing frozen tissue collections and similar applications.

Ultracold freezers are very sensitive to even brief periods of temperature increase and every second that a freezer door is open while searching for a particular sample is energy-consuming and could eventually contribute to freezer failure [97]. Researchers must therefore know exactly where each sample is located before opening the freezer. Many of these problems can be minimised through the use of a computerized inventory system. Database programs specific to the organization of freezers include Freezerworks (Data-Works Development, Inc., Mountlake Terrace, WA), a commercial program which integrates a thermal transfer label printer and bar-code reader with the database software, and Frozen Cell Stock Monitor (FCSM), a virtual container program for individual workstations which is freely distributed by the authors [198].

Recently, there has been a drive to integrate collections maintenance software with software for other biological databases, such as cladistics, morphometrics, species description and virtual identification (e. g., DELTA, Specify, BIOLINK). Ultimately, the choice of software will depend upon the needs of the investigator, the flexibility of the software (including its stability, potential for addition of new data, platform independence, and generality of datafile formats), the hardware requirements and the cost (some software, e. g., DELTA, DELTA Access and Specify are free). Further discussion on the computerization of collections can be found in McAllister et al. [199], Arnold [200], Wingate [201] and Owen [202], while the reader is directed to the websites listed in Appendix 9 for information on database software.

6   Legal and Ethical Issues

Last, but by no means least, legal and ethical considerations are an integral component of any research program involving the acquisition of specimens or tissue samples. These issues concern not only the manner in which samples are acquired – be it from the field, from commercial suppliers or from collections-based institutions – but where they may be deposited, and the publication of results obtained from their analysis. Throughout the course of a research program, researchers are legally obligated to abide by collection, exportation and importation regulations, and by the regulations of institutions concerning the loan and deposition of samples. In addition, researchers are ethically obligated to conduct their investigations with due respect to the organisms and the country of origin.

6.1   Permits to Collect, Export and Import Specimens and Tissue Samples

International collecting regulations
Genetic resources were once treated as a common heritage, available without restriction for research and other usage, but this viewpoint was perceived as unfair to developing countries – the major source of genetic resources [203, 204]. Since the 1992 Convention on Biological Diversity declared that governments have the "sovereign right to exploit" the genetic resources under their domain, efforts to regulate access have commenced. Permits regulating where collecting is conducted, what is collected, and how specimens or tissue samples are transported, are now mandatory in many countries, and may present the greatest hurdle (besides financial support) to collecting in developing nations.

Regulations governing wildlife collection and transportation are complicated. Policy directives are not always clear and are generally covered by multiple administrative branches. Since these institutions are often autonomous and may be unaware of each other's requirements or enforcement efforts, there is no "one-stop shopping" for permits. For example, an expedition to the tepuis of Amazonas Territory, Venezuela, required seven different permits (from the Ministry of Environment, the Institute of National Parks, the Indian Affairs Bureau, the National Guard and the Governor of Amazonas Territory), more than one year's advance application, and a week of negotiations after arrival [2].

The profusion of regulations that apply at state or international level is beyond the scope of this chapter. However, some general recommendations are provided below.

Collecting permits typically require at least 6 months' advance application, but a year or more may be necessary for developing countries in Africa, Asia or Latin America. The regulations of countries may differ radically – e. g., Bolivia has no application process, whereas the processes in Mexico and Peru are lengthy and complex [2, 205] – and are subject to change without notice. Most permits require submission of a detailed proposal, an equally detailed report on completion of field work and reprints of publications ensuing from the work. In addition, permits usually require that some or all of the specimens collected (including any holotypes) be deposited in the national or local collection. Special permission may therefore be necessary to export unicate samples. Occasionally, a preliminary field trip report and/or a complete copy of the field notes must also be provided before the investigator is permitted to depart the country. Permits granted for an extended period of time (e. g., 6 or 12 months) may be renewable, but usually require periodic submission of progress reports or collaboration with local researchers [205].

Regulatory offices in many countries have neither time nor resources to respond to correspondence regarding permits from foreign nationals, hence it is advisable to first make contact with researchers in the host countries and with their consulates. Additional sources of information include societies such as the Association of Systematics Collections (Washington, D.C.), institutions that maintain staff or projects in the countries of interest (e. g., American Museum of Natural History, Smithsonian Institution, New York Botanical Garden) and fellow scientists who have recently travelled to such countries (Appendix 8). A professional venue for permits information is currently hosted by the Smithsonian Insitution (the listowner is Sally Shelton, Collections Officer of the National Museum of Natural History). This is a moderated cross-disciplinary listserver, intended to facilitate discussion and information flow on all issues related to the rapidly changing terrain of biological collecting, permits, access and import/export regulations. Refer to Appendix 2 for details on how to subscribe to the permits listserver as well as other relevant websites, e. g., the *Journal of International Wildlife Law & Policy* and the Wildlife Interest Group of the American Society of International Law, which include research bibliographies on legal issues and links to the full text of national and regional wildlife legislation in many countries.

Threatened or endangered species
Special restrictions apply to the collection and transportation of threatened or endangered species, for which the sampling of tissue and the preparation of DNA bear legal responsibilities akin to the collecting of whole specimens [45]. Researchers should familiarise themselves with the regulations before applying for permits. Detailed information concerning personnel who will handle endangered species must be provided in permit applications, because heavy fines

have been enforced for the violation of permit conditions [206]. Investigators are urged to include on their permit applications all students and technicians who perform field and laboratory protocols with materials from endangered species, and to promptly report any changes in protocols or personnel to the permit authority.

Researchers specialising in endangered taxa should also familiarise themselves with the Convention on International Trade in Endangered Species of Wild Fauna and Flora (CITES) permitting system. CITES is a comprehensive treaty signed by over 150 countries which regulates international trade to prevent the decline of species threatened (listed in Appendix I of CITES) or potentially threatened (listed in Appendix II) with extinction. "Trade" is defined as import, export or re-export of CITES-listed animal and plant specimens, regardless of whether or not commerce is involved. International shipment of endangered species listed in the CITES protocol is now strictly controlled in most countries and the importation into the U.S. (or any other participant nation) of any species (alive, dead or part) on the international endangered species list requires a permit. Trade is virtually prohibited in the case of Appendix I species and is restricted for Appendix II or III species. While there are many signatory nations to this treaty, each must pass its own enabling legislation, so that the implementation of CITES may vary from country to country. Because domestic authorities have no jurisdiction in foreign countries, they will seek to prosecute the importer of improperly shipped material. Accordingly, it is advisable to enlist the help of collaborators, fluent in the permitting process and in the languages on both sides of the border, when crossing international borders with CITES-regulated taxa. Similarly, researchers are advised to establish the reputation of individuals supplying tissues and to insist that they obtain the necessary permits to be included with the shipment of specimens. Otherwise, researchers may be guilty of receiving improperly packaged or transported materials.

Member countries and institutions of CITES can freely interchange materials with the proper permits. However, a CITES exemption or "Certificate of Scientific Exchange" possessed by many academic institutions does not replace the need for import, export or collecting permits. Such an exemption merely makes it possible for like institutions to conduct loan transactions without the need for management agency paperwork and there is no provision for individual collecting. For further information on endangered species, the researcher is referred to the websites of CITES and TRAFFIC, the wildlife trade monitoring program of the World Wildlife Fund (WWF) and the World Conservation Union (IUCN), provided in Appendix 2.

U.S. collecting regulations

State and federal agencies issue various permits regulating the collection, capture, holding and sampling procedures that pertain to plants and most wild or feral animals in the U.S.A. Where vertebrates are concerned, regulations may dictate the numbers and circumstances under which investigators may collect blood, hair, feathers, urine, saliva, semen, milk, eggs, venom, body tissues and whole carcasses, including road kills.

In contrast, few restrictions apply to invertebrates, unless they occur specifically on protected lands. However, the collection of certain freshwater or nearshore "shellfish" (edible molluscs, crustaceans and echinoderms) is controlled by state government and may require a collecting permit or fishing license. Investigators should determine beforehand whether the taxa of interest have any special restrictions or are unregulated, as in the case of designated pest species. The U.S. and international endangered species lists are published and updated annually in the U.S. Code of Federal Regulations, Title 50, parts 17 and 23 (see below). Copies of the lists and permit information may be obtained from the U.S. Fish and Wildlife Service (Appendix 2). Many states in the U.S. also have separate lists of endangered and threatened species for which permission to collect must be obtained from the state natural resource agency.

In general, few regulations pertain to collections on privately-owned land, unless threatened or endangered species are involved. However, permits are compulsory for collections made on protected federal lands, such as national parks, wildlife refuges, wilderness areas, conservancies, marine sanctuaries and monuments. Such permits are usually specific to species and quantity of material allowed for collection and expire within 6 or 12 months.

Similar regulations also apply to state parks and streams in most states. However, state regulations vary, depending on jurisdiction (fisheries *versus* land and natural resources offices). For example, all intertidal and marine organisms are protected in California and any collecting conducted in the intertidal zone, or by scuba close to shore, requires a permit. General information regarding federal regulations can be obtained from the U.S. Fish and Wildlife Service, whereas information concerning regulations in a particular state must be acquired from the pertinent natural resource agency of the state. Inquiries concerning collecting in specific reserves should be addressed to the manager, superintendent or ranger in charge.

Researchers are urged to contact relevant agencies (Appendix 2) before planning experiments, especially if the investigation requires unstable or easily degraded materials, e. g., mRNAs of certain age or tissue specificity. When annual restrictions on take (e. g., sex, breeding *versus* nonbreeding season) apply to an organism, samples may only be obtainable at certain times, thereby influencing academic schedules and the starting dates of grants.

International transportation regulations
Many countries enforce strict regulations regarding the export of biological material, which may be more stringently controlled than regulations governing the collection of samples. For example, permission to collect specimens and tissue samples may only be granted on condition that some or all of the samples remain in the country of origin (as discussed above). The onus rests with the investigator to apply to the appropriate authorities in each country.

Import permits are similarly required when material reaches its destination. There are no prohibitions *per se* against the importation into the U.S. (or most other countries) of specimens or samples preserved in ethanol or formalin, or as extracted nucleic acids (insofar as these are all assumed to be sterile), unless they originate from an endangered species. However, customs officials may be unable to judge whether the samples are legal and may confiscate questionable material until that has been verified. In order to avoid such situations, the researcher should contact the appropriate agency beforehand. If a permit is not required, a letter to that effect should be obtained from the agency.

For some studies, it may be necessary to import fresh or live material (e. g., where organisms are small and must be cultured to obtain sufficient DNA, or if mechanistic or developmental studies are intended). Live plants and animals fall under the regulations of various agencies. For example, in Canada, live animal and animal product importation permits are issued by the Canadian Food Inspection Agency (CFIA), which maintains an Automated Import Reference System (AIRS) for internal use in monitoring permit requirements. In order to be issued with a permit by the CFIA, researchers must indicate the animal species, its country of origin and the reason for its importation (Appendix 2).

Procedures may stipulate how animals must be inspected and transported, in addition to the regular permits required for collecting, exporting and importing. For example, EU countries have strict regulations governing the humane transport of live animals, which include regulations stipulating the amount of time they are allowed to spend in transit (Appendix 2). These rules are based on guidelines published by the International Air Transport Association (IATA) in the IATA publication *Live Animals Regulations*.

Live animal importation permits usually require some testing certification or affidavit of disease-free status from the country of origin, in addition to their export permit. Similarly, a phytosanitary certificate is routinely required to demonstrate that plant material (fresh or desiccated) is pest-free before it can be exported. In general, African and Asian countries cannot certify their material, which therefore cannot be imported into EU countries or the U.S. Import permits may also require that the receiving laboratory and/or quarantine facility be government-inspected. Material of economically important plant groups (e. g., Rutaceae, Poaceae, Orchidaceae) are subject to especially stringent inspection for arthropods, viruses and fungal pathogens [2], while live animals of unknown status may require that a risk assessment be conducted prior to issuance of an importation permit. Finally, researchers are advised to ascertain whether state or provincial permits are required in addition to the federal or national regulations.

U.S. transportation regulations
Many different regulations govern the transportation of animals, plants or their parts into and out of the U.S.A. Lists of wildlife and plant species that specifically require a federal permit for importation include species that are endangered or threatened, protected by CITES, or deemed injurious, and include all migratory

birds and marine mammals. Other restricted articles include, but are not limited to, the following: all sea turtle products; many reptile skins and leathers, especially those originating from South American countries; most wild bird feathers, mounted birds or bird skins; ivory from elephants, whales, walruses and narwhals; furs from most spotted cats and all marine mammals; corals; and many plant species including orchids, cacti and cycads. Federal regulations further prevent the importation of fish or wildlife into any state if that state's laws or regulations are more restrictive than any applicable federal treatment. Wild animals taken, killed, sold, possessed or exported to the U.S. in violation of any foreign laws are also denied entry. Applicable U.S. legislation is as follows:

1. *Endangered Species Act*: prohibits the import and export of species listed as endangered and most species listed as threatened. More than 1 000 species of animals and plants are officially listed under U.S. law as endangered or threatened. Refer to *Endangered and Threatened Wildlife and Plants* (50 CFR 17.11 and 17.12) for annually updated lists of these taxa.

2. *Lacey Act*: prohibits the import, export, transport, sale, receipt, procurement or purchase in interstate or foreign commerce of animal species that have been taken, possessed, transported or sold in violation of any state or foreign law or taken or possessed in violation of other federal law or Indian tribal law.

3. *Marine Mammal Protection Act*: prohibits the import of marine mammals and their parts and products.

4. *Wild Bird Conservation Act*: regulates or prohibits the import of many exotic bird species.

The U.S. Department of Agriculture (USDA) permit is a prerequisite for entry of fresh plant or animal material into the U.S. from other countries, including Canada. Although the USDA permit can be obtained for specific taxa and investigators, it is prudent to include several investigators and an array of taxa in the application. Copies of the USDA import permit should be carried at all times and forwarded to colleagues who will be sending material from other countries, so that copies may be included inside the parcel and outside with the shipping label.

The USDA lists animals for which import permits are "not required." However, these may still require inspection upon arrival. Plants, cuttings, seeds, unprocessed plant parts and certain endangered species either require an import permit or are prohibited from entering the U.S. Endangered or threatened species of plants and plant products, if importation is not prohibited, will require an export permit from the country of origin. Every plant or plant part has to be declared to Customs and must be presented for inspection. Researchers planning to import fish, wildlife or any product or part thereof, are advised to check with the Customs or Fish and Wildlife Service in advance (Appendix 2).

If animals are to be shipped alive, state and federal regulations usually require quarantine, including agricultural inspections of cage litter for noxious weeds or invertebrates. Lengthy quarantine is also required when shipping animals between the continental U.S. and Hawaii or Alaska. Severe delays may likewise be expected when importing samples of public health, agricultural or veterinary importance (e. g., disease organisms, plant pathogenic fungi or insect pests) into the U.S., as the permit process requires official state and federal agriculture approval, which can take several months. Researchers should contact the Foreign Quarantine Program of the U.S. Public Health Service and the Animal and Plant Health Inspection Service (APHIS) of the USDA (Appendix 2). APHIS performs inspections, offers guidelines and handles import permits. All wildlife and wildlife products must enter or exit the U.S. at one of the following designated USDA inspection ports: Baltimore, MD; Boston, MA; Chicago, IL; Dallas/Ft. Worth, TX; Honolulu, HI; Los Angeles, CA; Miami, FL; New Orleans, LA; New York, NY; Portland, OR; San Francisco, CA; Seattle, WA.

For a more detailed discussion of the regulations governing transport of specimens and tissue samples into the U.S., including examples of the official application forms, refer to Dessauer and Hafner [31]. Sheldon and Dittman [122] provide a discussion of the permitting procedures required for import and export of samples from U.S. frozen tissue collections.

6.2   Legal Issues Concerning Specimens and Tissue Samples in Collections

Depositing samples
Investigators wishing to deposit tissue samples in biorepositories must be prepared to sign an affidavit stating that the sample was collected in accordance with all applicable laws and regulations [207–209]. Collection files should contain copies of collecting permits issued to the original collector of the specimens and, if the material is imported into the U.S., should contain copies of the requisite USDA importation form [97, 205].

Loaning samples
Collections-based institutions vary in their policies regarding access to materials [210–212], especially when destructive sampling is involved [123]. Some are more flexible than others, depending on their experience with individual investigators and their familiarity with the proposed techniques. Researchers should understand that the dictates of the curator, who intends to conserve materials, are intellectually opposed to the ideals of the experimentalist, who intends to destroy them. Accordingly, investigators must realise that destructive sampling of specimens from some of these institutions is simply not feasible or, if it is, that rare specimens or taxa which do not have a plentiful representation in collections are unlikely to be made available. Rather than view the curator as a block to progress, the experimentalist should design laboratory protocols that maximise use of existing materials or sample precious materials in nondestructive ways, and avoid the use of exhibition-quality specimens when partial, but well-documented items will suffice.

Each institution has its own policies regarding access to, and amounts available from, any given specimen and these guidelines should be consulted, or the curators contacted, before loans are requested. Critical information on which loans are made may include the potential for significant new knowledge, experimental design, skill of the researchers, site of proposed research, proposed quantity of tissue to be consumed, quantity of tissue available for loan, past or planned contribution to the collection, collaboration with contributors to the collection, etc. [97, 123].

Most institutions require that a researcher requesting tissues from a collection submit a proposal to the curator or collections manager, wherein the importance of the proposed work is outlined and the capability of the researcher to successfully use the loaned tissues is demonstrated. Graduate students may be required to submit supplementary materials from their thesis advisors, which include evidence that their advisor is conversant with the techniques involved (i. e., peer-reviewed publications using such methods). Almost all collections-based institutions also require that unused tissues and nucleic acids are returned to the source collection for storage and archiving after the research has been completed, and that molecular data are deposited in GenBank, EMBL or a comparable database. For a primer on policies concerning the acquisition and deacquisition of specimens and samples in natural history collections, refer to Hoagland [212].

6.3   Ethical Issues

Illegal samples
Legal regulations are intended to prohibit the wanton destruction of biodiversity for commercial profit. Unfortunately, these regulations are often difficult to enforce, especially where the commercial trade in plants and animals (or their products) is concerned [213]. Illegally collected plants and animals are constantly smuggled across international borders and may find their way into legally imported consignments from neighbouring countries. The researcher may be confronted with a rare sample, the source of which is dubious (e. g., alleged country of origin lies outside the species' known distributional range) and which may therefore be illegal. Although this may be impossible to determine (e. g., the dealer may supply false paperwork), the researcher must make a moral decision to reject the sample. It is unethical to deal with any institution or individual that does not abide by the regulations of state, national and international organisations.

Handling animals
Ethical issues also concern the treatment of vertebrates (and occasionally invertebrates) to be sampled for molecular studies. Most scientific institutions receiving federal grants carry internally constituted animal ethics advisory

committees, which evaluate research proposals that include use of vertebrates. Permission to take small quantities of blood, urine, feathers, feces or fur from healthy animals is usually granted routinely, except in cases where investigators plan experiments that require animals to be housed on-site or that utilise protocols which can be expected to produce significant but unavoidable stress.

Besides specifying numbers to be sampled and times at which animals can be collected, regulatory agencies may dictate the preferred or permitted method of collecting for ethical reasons. In the case of passerine birds, the use of anesthesia for bleeding or even minor abdominal surgery, such as laparotomies, is not recommended [214]. The time permitted to hold animals for processing, while collecting bodily fluids and making morphological measurements, may be limited to one hour to minimise stress. Handling of birds, bats and animals with chemical defences requires special skills, hence close collaboration with a skilled field biologist, certified to handle such species, is advised in these cases. Further guidelines on methods of anesthesia are provided by Dessauer et al. [3, 4].

Research attitudes

A further ethical consideration when obtaining samples concerns the attitudes of researchers to regulations imposed by foreign countries. Many developing countries prohibit the export of any part (including isolated DNA and proteins or even PCR products that still contain some of the template DNA) of endangered or economically important species, which may encompass all biodiversity in some cases [205]. Although these regulations may be viewed as an obstacle to scientific progress and there may be technical ways of circumventing them (e. g., [215]), it is important to realise that such organisms are resources for the countries in question and should be respected as such [216].

Similarly, researchers and scientific administrators should become more aware of the ethical responsibility of providing credit to colleagues in developing countries for assistance rendered, which may include co-authorship of publications, acknowledgements and inclusion of foreign researchers on proposals to granting agencies [2, 97]. Many developing countries now encourage collaboration with local researchers for the promotion of permit applications [203]. For a primer on genetic resources policy, including discussions on national sovereignty, access to and ownership of intellectual property rights, and the "common heritage of mankind", refer to Stenson and Gray [204] and Hoagland and Rossman [217].

Acknowledgements

We thank Elizabeth Bonwich, Darrel Frost, Gonzalo Giribet, Lisa Kronthal, Mark Siddall and Ward Wheeler for discussions on the above topics and assistance with the voluminous literature. During the writing of this paper, LP was supported by the Foundation for Research Development (Prestigious Scholarship), the American Museum of Natural History (Collection Study Grant) and the Skye Foundation and Charitable Trust.

Appendix

1 Molecular Biology Websites, Products and Services

- Molecular Biology Techniques Forum and Molecular Biology Resources, hosted by the Northwest Fisheries Science Center (NWFSC), National Marine Fisheries Service (NMFS), National Oceanographic and Atmospheric Administration (NOAA), US Department of Commerce, Seattle, WA, USA: http://www.nwfsc.noaa.gov/protocols/methods/methods.html; http://www.nwfsc.noaa.gov/protocols/resources.html
- Molecular Biology Databases on the Internet, hosted by the Weizmann Institute of Science, Rehovot, Israel: http://bioinfo.weizmann.ac.il/mb/db/species_specific_databases.html
- BioLinks, hosted by the Department of Molecular and Cell Biology (MCB), Harvard University, Cambridge, MA, USA: http://mcb.harvard.edu/BioLinks.html
- *Ancient DNA Newsletter*, The Zoological Society of London, Institute of Zoology, London, UK: http://www.londonzoo.co.uk/ioz
- *BioTechniques*, Eaton Publishing Co., Natick, MA, USA: http://www.biotechniques.com
- BioSupplyNet, Cold Spring Harbor Laboratory Press, Cold Spring Harbor, NY, USA: http://www.biosupplynet.com/index.cfm
- Perkin-Elmer, Inc., Wellesley, MA, USA: http://www.perkinelmer.com
- Kimberly-Clark Corporation, Dallas, TX, USA: http://www.kimberly-clark.com
- VWR Scientific Products, San Francisco, CA, USA: http://vwrsp.com
- Nalge-Nunc International Corporation, Rochester, NY, USA: http://www.nalgenunc.com
- Revco Technologies, Inc., Asheville, NC, USA: http://www.revcotech.com
- Taylor-Wharton International, Theodore, AL, USA: http://www.taylor-wharton.com
- For shipping products and services: Saf-T-Pak Inc., Edmonton, AB, Canada: http://www.saftpak.com

- To order sections E1342, E1564, E1565, E1566 of the *Annual Book of ASTM Standards* (Vol. 11.05), American Society for Testing and Materials (ASTM), West Conshohocken, PA, USA: E1342 (Practice for Preservation by Freezing, Freeze-Drying, and Low Temperature Maintenance of Bacteria, Fungi, Protista, Viruses, Genetic Elements, and Animal and Plant Tissues); E1564 (Guide for Design and Maintenance of Low-Temperature Storage Facilities for Maintaining Cryopreserved Biological Materials); E1565 (Guide for Inventory Control and Handling of Biological Material Maintained at Low Temperatures); E1566 (Guide for Handling Hazardous Biological Materials in Liquid Nitrogen): http://www.astm.org

2 Permits and Regulations

- The International Air Transport Association (IATA), Dangerous Goods Online: http://www.iata.org/cargo/dg
- To order IATA publications: *Dangerous Goods Regulations*: http://www. iata.org/ads/dgr.htm; *Live Animals Regulations*: http://www.iata.org/ads/ lar.htm
- Import-export regulations for selected countries, hosted by the Biodiversity and Biological Collections Web Server: gopher://biodiversity.bio.uno.edu/11/ curation/permits
- Treaty Database, American Society of International Law (ASIL), Wildlife Interest Group, Burlingame, CA, USA: http://www.eelink.net/~asilwildlife/ docs.html
- *Journal of International Wildlife Law & Policy*, Kluwer Academic Publishers, Norwell, MA, USA: http://ww.jiwlp.com
- Permits for collecting and exporting samples from Australia: http://pioneer. mov.vic.gov.au/chaec/open.html
- Instituto Nacional de Biodiversidad (INBio), collecting permits for Costa Rica: http://www.inbio.ac.cr/en/default.html
- European Union environmental legislation: http://europa.eu.int/comm/ environment/legis_en.htm
- Comisión Nacional para el Conocimiento y Uso de la Biodiversidad (CONABIO), collecting permits for Mexico: http://www.conabio.gob.mx
- To obtain current listings of endangered and threatened species, and federal permits for the US: US Fish and Wildlife Service (FWS), International Affairs, Arlington, VA: http://international.fws.gov
- To find out about quarantine requirements in shipping to the US: US Department of Agriculture (USDA), Animal and Plant Health Inspection Service (APHIS), Hyattsville, MD: http://www.aphis.usda.gov; USDA APHIS import-export directory: http://www.aphis.usda.gov/oa/imexdir.html; USDA APHIS, Veterinary Services, National Center for Import and Export (NCIE): http://www.aphis.usda.gov/ncie; USDA, APHIS, Plant Protection and Quarantine (PPQ): http://www.aphis.usda.gov/ppq

- US Customs Service, Washington, DC, Importing and Exporting: http://www.customs.gov/impoexpo/impoexpo.htm
- To determine if the sample from a vertebrate is disease-free: National Wildlife Health Center, Madison, WI, USA: http://www.nwhc.usgs.gov
- To determine whether importation to the US requires special permits because of disease risks: Centers for Disease Control and Prevention (CDC), US Department of Health and Human Services, Atlanta, GA: http://www.cdc.gov
- Canadian Food Inspection Agency (CFIA), Nepean, ON, Canada: http://www.cfia-acia.agr.ca
- Convention on International Trade in Endangered Species of Wild Fauna and Flora (CITES): http://www.unep.ch/cites
- TRAFFIC, Cambridge, UK: http://www.traffic.org
- World Wildlife Fund (WWF), Washington, DC, USA: http://www.worldwildlife.org
- The World Conservation Union (IUCN), Gland, Switzerland: http://www.iucn.org
- To join the permits listserver: Email listserv@sivm.si.edu. No subject is required. In the body of the message, issue the command: Subscribe PERMIT-L Firstname Surname
- Examples of frozen tissue collections policy: http://abscweb.wr.usgs.gov/research/ammtap/request.htm; http://www.uaf.edu/museum/af/using.html; http://www.washington.edu/burkemuseum/tissuepolicy.html; http://www.lms.si.edu/tissuepolicy.html

3   Live Stock Centers and Seed Banks

- Database of Laboratory Animals in Japan, University of Tokushima, Tokushima, Japan: http://www.anex.med.tokushima-u.ac.jp/index.html
- *E. coli* Collection, National Institute of Genetics, Mishima, Japan: http://shigen.lab.nig.ac.jp/ecoli/strain
- *E. coli* Reference Collection (ECOR) and DEC *E. coli* Collection, Penn State University, University Park, PA, USA: http://www.bio.psu.edu/People/Faculty/Whittam/Lab/ecor; http://www.bio.psu.edu/People/Faculty/Whittam/Lab/deca
- *E. coli* Genetic Stock Center (CGSC), Yale University, New Haven, CT, USA: http://cgsc.biology.yale.edu
- *Salmonella* Genetic Stock Center (SGSC), University of Calgary, AB, Canada: http://www.ucalgary.ca/~kesander
- Soil Microbiology Division, International Rice Research Institute, Manila, Philippines: http://www.irri.org
- *Chlamydomonas* Genetics Center, Duke University, Durham, NC, USA: http://www.biology.duke.edu/chlamy

- *Pseudomonas* Genetics Stock Center (PGSC), University of East Carolina, Greenville, NC, USA: http://www.pseudomonas.med.ecu.edu/Pseudomonas/index1.html
- Microalgal Biotechnology Labs and Biomass Conversion/Organic Synthesis Labs, Center for Basic Sciences, National Renewable Energy Laboratory (NREL), Golden, CO, USA: http://www.nrel.gov/basic_sciences/biosci.html
- Fungal Genetics Stock Center (FGSC), University of Kansas Medical Center, Kansas City, KS, USA: http://www.fgsc.net
- US Department of Agriculture Agricultural Research Service (USDA-ARS), US Plant, Soil and Nutrition Laboratory, Ithaca, NY, USA: http://www.arserrc.gov/naa/home/fedpsnl.htm
- USDA Forest Service, Forest Products Laboratory (Center for Forest Mycology), Madison, WI, USA: http://www.fpl.fs.fed.us
- National Germplasm Resources Laboratory (NGRL), Beltsville, MD, USA, and National Germplasm Repositories of the USDA National Plant Germplasm System (NPGS): http://www.barc.usda.gov/psi/ngrl/ngrl.html; http://www.ars-grin.gov/npgs/rephomepgs.html; http://www.ars-grin.gov/npgs/holdings.html
- National Seed Storage Laboratory (NSSL), Fort Collins, CO, USA: http://www.ars-grin.gov/ars/NoPlains/FtCollins
- The *Arabidopsis* Information Resource (TAIR): http://www.arabidopsis.org
- SENDAI *Arabidopsis* Seed Stock Center (SASSC), Department of Biology, Miyagi University of Education, Miyagi, Japan: http://shigen.lab.nig.ac.jp/arabidopsis
- Nottingham *Arabidopsis* Stock Center (NASC), The University of Nottingham, Nottingham, UK: http://nasc.nott.ac.uk/home.html
- *Arabidopsis* Information Management System (AIMS), *Arabidopsis* Biological Resource Center, Columbus, OH, USA: http://aims.cse.msu.edu/aims
- Barley Germplasm Database, National Institute of Genetics, Mishima, Japan: http://www.shigen.nig.ac.jp/barley/Barley.html
- Maize Genetics Cooperation Stock Center, University of Illinois at Urbana, Champaign, IL, USA: http://w3.aces.uiuc.edu/maize-coop
- Maize Germplasm Collection, North Central Regional Plant Introduction Station, Iowa State University, Ames, IA, USA: http://w3.ag.uiuc.edu/maize-coop/ncrpis.html
- Oryzabase, Rice Genetic Resources Center, National Institute of Genetics, Mishima, Japan: http://shigen.lab.nig.ac.jp/rice/oryzabase
- US Potato Gene Bank, Sturgeon Bay, WI, USA: http://www.ars-grin.gov/ars/MidWest/NR6
- Tomato Genetic Resources Center (TGRC), University of California, Davis, CA, USA: http://tgrc.ucdavis.edu
- US National Arboretum, Woody Landscape Plant Germplasm Repository, Glenn Dale, MD, USA: http://www.ars-grin.gov/cgi-bin/npgs/html/site_holding.pl?NA

- The Millenium Seed Bank Project, Science and Horticulture, Royal Botanic Gardens Kew, Richmond, UK: http://www.rbgkew.org.uk/seedbank
- *Index Seminum*: Seed Exchange (selected websites): http://www.unimo.it/ortobot/seedexc/seedexc.htm;   http://www.helsinki.fi/ml/botgard/indesem.html; http://www.toyen.uio.no/botanisk/bothage/catalog
- American Fern Society (AFS) Spore Exchange, Hacienda Hts., CA, USA: http://amerfernsoc.org/sporexyy.html
- Nematode Collection, University of California, Davis, CA, USA: http://ucdnema.ucdavis.edu/imagemap/nemmap/museum.htm
- *Caenorhabditis* Genetics Center (CGC), University of Minnesota, St. Paul, MN, USA: http://biosci.umn.edu/CGC/CGChomepage.htm
- *Drosophila* Genetic Resources, National Institute of Genetics, Mishima, Japan: http://www.shigen.nig.ac.jp/fly/nighayashi.html
- *Drosophila* Stock Center, Facultad de Ciencias, Universidad Nacional Autónoma (UNAM), México: http://hp.fciencias.unam.mx/Drosophila/LOSHTML/portada.html
- The *Drosophila* Species Stock Center, Tucson, AZ, USA: http://stockcenter.arl.arizona.edu
- *Drosophila melanogaster* Stock Center, Indiana University, Bloomington, IN, USA: http://flybase.bio.indiana.edu/stocks
- Szeged *Drosophila melanogaster* P Insertion Mutant Stock Center, Department of Genetics, Jószef Attila University, Szeged, Hungary: http://gen.bio.u-szeged.hu/stock
- Exelixis Inc., EP Flystation, San Francisco, CA, USA: http://cdigraphics.com/flystation2
- Mosquito Colonies, Department of Medical Entomology, South African Institute for Medical Research (SAIMR), Johannesburg, South Africa: http://www.wits.ac.za/fac/med/entomology/resource.htm
- Instituto Butantan, São Paolo, Brazil: http://bernard.pitzer.edu/~lyamane/butantan.htm
- *Tribolium* Stock Center, Kansas State University, Manhattan, KS, USA: http://bru.usgmrl.ksu.edu/beeman%5Ctribolium
- Zebrafish Information Network (ZFIN): http://zfin.org
- Zebrafish International Resource Center (ZFIN), University of Oregon, Eugene, OR, USA: http://zfin.org/zf_info/stckctr/stckctr.html
- The Indiana University Axolotl Colony, Indiana University, Bloomington, IN, USA: http://www.indiana.edu/~axolotl
- Jax® Mice, The Jackson Laboratory, Bay Harbor, ME, USA: http://jaxmice.jax.org/index.shtml
- Mutant Mouse Database, Oak Ridge National Laboratory, Oak Ridge, TN, USA: http://lsd.ornl.gov/htmouse
- *Peromyscus* Genetic Stock Center, University of South Carolina, Columbia, SC, USA: http://stkctr.biol.sc.edu

4 Cell Lines and Culture Collections

- Common Access to Biological Resources and Information (CABRI) Consortium: http://www.cabri.org
- Shared Information of Genetic Resources (SHIGEN), National Institute of Genetics, Mishima, Japan: http://shigen.lab.nig.ac.jp
- Genetic Resource Databank (GRD), National Institute of Genetics, Mishima, Japan: http://www.shigen.nig.ac.jp/grd
- Cell Line Database (CLDB) and Interlab Project, Genova, Italy: http://www.biotech.ist.unige.it/interlab/cldb.html
- Interlab Cell Line Collection (ICLC), Genova, Italy: http://www.iclc.it
- Coriell Cell Repositories, National Institute of General Medical Sciences (NIGMS), National Institutes of Health, Bethesda, MD, USA: http://locus.umdnj.edu
- The University of Michigan Breast Cell/Tissue Bank and Data Base, University of Michigan, Ann Arbor, MI, USA: http://www.cancer.med.umich.edu/umbnkdb.html
- Japanese Collection of Research Bioresources (JCRB) Cell Bank and Human Science Research Resources Bank (HSRRB), National Institute of Health Sciences (NIHS), Tokyo, Japan: http://cellbank.nihs.go.jp
- The National Laboratory for the Genetics of Israeli Populations (NLGIP), Tel Aviv University, Tel Aviv, Israel: http://www.tau.ac.il/medicine/NLGIP/nlgip.htm
- American Type Culture Collection (ATCC), Manassas, VA, USA: http://www.atcc.org
- European Collection of Cell Cultures (ECACC), Center for Applied Microbiology and Research (CAMR), Salisbury, UK: http://www.biotech.ist.unige.it/cldb/descat5.html
- World Federation for Culture Collections-Microbial Resources Centers (WFCC-MIRCEN), World Data Centre for Microorganisms (WDCM), Wako, Saitama, Japan: http://wdcm.nig.ac.jp
- Home Pages of Culture Collections in the World, hosted by WFCC-MIRCEN: http://wdcm.nig.ac.jp/hpcc.html
- Microbial Strain Data Network (MSDN), Sheffield, UK: http://panizzi.shef.ac.uk/msdn
- UK National Culture Collections (UKNCC): http://www.ukncc.co.uk
- US Federation for Culture Collections (USFCC): http://methanogens.pdx.edu/usfcc
- Deutsche Sammlung von Mikroorganismen und Zellkulturen GmbH (DSMZ), The German Collection of Microorganisms and Cell Cultures, Braunschweig, Germany: http://www.dsmz.de/index.html
- The German Collection of Human and Animal Cell Cultures, The German Collection of Microorganisms and Cell Cultures, Braunschweig, Germany: http://www.biotech.ist.unige.it/cldb/coll121.html

- The Netherlands Culture Collection of Bacteria (NCCB), formerly LMD and Phabagen, Utrecht University, Utrecht, The Netherlands: http://www.cbs.knaw.nl/nccb/database.htm
- The Culture Collection of Algae and Protozoa, Institute of Freshwater Ecology (IFE) Windermere Laboratory at Ambleside and Dunstaffnage Marine Laboratory (DML) at Oban, UK: http://www.ife.ac.uk/ccap
- The Culture Collection of Algae at The University of Texas Austin (UTEX), Austin, TX, USA: http://www.bio.utexas.edu/research/utex
- Culture Collection of Algae, Department of Cell Biology and Applied Botany, Philipps-Universität, Marburg, Germany: http://staff-www.uni-marburg.de/~cellbio/welcomeframe.html
- National Collection of Type Cultures and Pathogenic Fungi (NCTC), Public Health Laboratory Service (PHLS), UK: http://www.phls.co.uk/services/nctc/index.htm
- University of Alberta Microfungus Collection and Herbarium (UAMH), University of Alberta, Edmonton, AB, Canada: http://www.devonian.ualberta.ca/uamh
- Culture Collection of Microorganisms, Institute for Fermentation, Osaka (IFO), Japan: http://wwwsoc.nacsis.ac.jp/ifo/microorg/microorg.htm
- USDA Agricultural Research Service (ARS) Microbial Culture Collection (NRRL), Peoria, IL, USA: http://nrrl.ncaur.usda.gov
- Molecular Pathology Shared Resource, Cancer Center, University of California San Diego (UCSD), CA, USA: http://cancer.ucsd.edu/molpath/examples.htm
- Molecular Pathology Collection (frozen and archival human pancreatic ductal carcinomas, metastatic lesions, preneoplastic lesions, and tumor-derived cell lines), Allegheny University, Philadelphia, PA, USA: http://www.mcphu.edu/research/rspt/resdb/frozenandarchival.html
- The Cloning Vector Collection, National Institute of Genetics, Mishima, Japan: http://www.shigen.nig.ac.jp/cvector/cvector.html
- cDNA libraries available commercially: CLONTECH Laboratories, Inc., Palo Alto, CA, USA: http://www.clontech.com/index.shtml; OriGene Technologies, Inc., Rockville, MD, USA: http://www.origene.com/cdnalib.htm; Stratagene, La Jolla, CA, USA: http://www.stratagene.com/home.htm

5   Frozen Tissue Collections

- Departamento de Ciencias Biológias, Facultad de Ciencias Agrarias, Universidad Nacional de Cuyo, Mendoza, Argentina: http://www.uncu.edu.ar
- Department of Biological Sciences, Macquarie University, Sydney, NSW, Australia: http://www.bio.mq.edu.au
- School of Biological Science, University of New South Wales, Sydney, NSW, Australia: http://www.bioscience.unsw.edu.au
- School of Biological Sciences, University of New England, Armidale, NSW, Australia: http://www.une.edu.au/sciences/schoolbiosci/index.html

- CSIRO Marine Laboratories, Cleveland, QLD, Australia: http://www.bne. marine.csiro.au
- The Australian Biological Tissue Collection, Evolutionary Biology Unit, South Australia Museum, Adelaide, SA, Australia: http://www.samuseum.sa. gov.au/ebu.htm
- Institut für Medizinische Chemie, Veterinärmedizinische Universität Wien, Wien, Austria: http://www-med-chemie.vu-wien.ac.at
- Royal Ontario Museum and Department of Zoology, University of Toronto, ON, Canada: http://www.rom.on.ca/index.html; http://www.zoo.utoronto.ca
- Redpath Museum, McGill University, Montreal, QC, Canada: http://www. redpath-museum.mcgill.ca
- Finnish Game and Fisheries Research Institute, Ahvenjarvi Game Research Station, Ilomantsi, Finland: http://www.rktl.fi/english
- Section de Biologie, Institut Curie, Paris, France: http://www.curie.fr
- Institut des Sciences de l'Evolution, Université Montpellier II, Montpellier, France: http://www.isem.univ-montp2.fr
- Departement de Biologie-Ecologie, Université Paul Valéry, Université Montpellier III, Montpellier, France: http://alor.univ-montp3.fr/epe/Index.html
- Department of Zoology, Hessisches Landesmuseum Darmstadt (HLMD), Darmstadt, Germany: http://www.darmstadt.gmd.de/Museum/HLMD
- Biology Department, Philipps-Universität, Marburg, Germany: http:// www.uni-marburg.de/biologie/welcomeengl.html
- Institut für Physiologie, Universität Regensburg, Regensburg, Germany: http://www.biologie.uni-regensburg.de/InstituteNWFIII/physiologie.html
- Max Planck Institute of Biochemistry, Martinsried, Germany: http://www. biochem.mpg.de/home_en.html
- The Institute of Evolution, University of Haifa, Haifa, Israel: http://research. haifa.ac.il/~evolut/~evolut.html
- Department of Chemistry, Faculty of Science and Graduate School of Science, Tohoku University, Aobayama, Sendai, Japan: http://www.chem. tohoku.ac.jp/index-e.html
- Department of Biology, Universiti Putra Malaysia (UPM), Serdang, Selangor, Malaysia: http://fsas.upm.edu.my/~arahim/jabbio/index.html
- Institute of Biological Sciences, University of Malaya, Kuala Lumpur, Malaysia: http://biology.um.edu.my/home.html
- Department of Biochemistry, University of Nijmegen, Nijmegen, The Netherlands: http://www-bioch.sci.kun.nl/bioch/ger.html
- Department of Clinical and Human Genetics, Faculty of Medicine, Free University, Amsterdam, The Netherlands: http://www.med.vu.nl/org/afd/kga
- Department of Human and Clinical Genetics, University of Leiden, Leiden, The Netherlands: http://www.medfac.leidenuniv.nl/humangenetics/ stages_nl.html
- Department of Biology and Nature Conservation, Zoological Institute, Agricultural University of Norway, Ås, Norway: http://wwwnlh.nlh.no/institutt/ ibn/default_english.htm

- Department of Anatomy, Faculty of Medicine, Jagiellonian University, Krakow, Poland: http://www.cm-uj.krakow.pl/guide/cm4133.html
- Institute of Zoology, Academia Sinica, Taipei, Taiwan, Republic of China: http://www.sinica.edu.tw/zool/english/eindex.htm
- Department of Anatomy and Biology, National Defense Medical Center, Taipei, Taiwan, Republic of China: http://www.ndmctsgh.edu.tw
- Laboratory of Isotopic Investigations, Engelhardt Institute of Molecular Biology, Moscow, Russia: http://www.imb.ac.ru
- Department of Genetics and Pathology, Uppsala University, Uppsala, Sweden: http://www.genpat.uu.se
- Zoological Institute, Department of Biology, University of Bern, Bern, Switzerland: http://www.cx.unibe.ch/zos/zoologie
- Institute of Ecology – Laboratory for Zoology, Université de Lausanne, Lausanne-Dorigny, Switzerland: http://www.unil.ch/izea
- Laboratory of Molecular Biology, Institute of Clinical Pathology, University Hospital, Zürich, Switzerland: http://www.unizh.ch/pathol
- Department of Biology, School of Environment and Life Sciences, University of Salford, Salford, UK: http://www.els.salford.ac.uk/proper/biology.htm
- Department of Human Sciences, Faculty of Science, Loughborough University, Leicestershire, UK: http://www.lboro.ac.uk/departments/hu/index.html
- School of Biology and Biochemistry, Queen's University, Belfast, Northern Ireland, UK: http://www.qub.ac.uk/bb
- National Marine Fisheries Service (NMFS) Laboratories, National Oceanographic and Atmospheric Administration (NOAA), US Department of Commerce, USA: http://www.websites.noaa.gov/guide/government/nmfs.html; http://www.nmfs.noaa.gov/regional.htm
- Department of Biological Sciences, University of Alabama, Tuscaloosa, AL, USA: http://www.as.ua.edu/biology/scf
- Alaska Frozen Tissue Collection, University of Alaska Museum, University of Alaska Fairbanks, Fairbanks, AK, USA: http://zorba.uafadm.alaska.edu/museum/af/index.html
- Alaska Department of Fish and Game, Gene Conservation Laboratory, Anchorage, AK, USA: http://www.cf.adfg.state.ak.us/geninfo/research/genetics/genetics.htm
- Auke Bay National Marine Fisheries Laboratory, National Marine Fisheries Service (NMFS), National Oceanographic and Atmospheric Administration (NOAA), US Department of Commerce, Juneau, AK, USA: http://www.afsc.noaa.gov/abl
- Alaska Marine Mammal Tissue Archival Project (AMMTAP), Anchorage, AK, USA: http://abscweb.wr.usgs.gov/research/ammtap/index.htm
- Department of Pharmacology, University of Arizona College of Medicine, Tucson, AZ, USA: http://www.pharmacology.arizona.edu/Pharm_Tox
- Laboratory of Molecular Systematics, California Academy of Sciences, San Francisco, CA, USA: http://www.calacademy.org/research

- Museum of Vertebrate Zoology (MVZ), University of California, Berkeley, CA, USA: http://www.mip.berkeley.edu/mvz/collections/TissueCollection.htm
- Center for Reproduction of Endangered Species (CRES), Zoological Society of San Diego, San Diego, CA, USA: http://sandiegozoo.org/cres/frozen_resourcelist.html
- Division of Biological Sciences, Section of Ecology and Evolution, University of California, Davis, CA, USA: http://www.eve.ucdavis.edu
- Beckman Research Institute, Duarte, CA, USA: http://bricoh.coh.org/docs
- Department of Organismic Biology, Ecology and Evolution, University of California, Los Angeles, CA, USA: http://www.obee.ucla.edu
- Department of Biochemistry and Molecular Biology, Colorado State University, Fort Collins, CO, USA: http://www.bmb.colostate.edu
- Department of Biological Sciences, University of Connecticut, Storrs, CT, USA: http://www.biology.uconn.edu
- Peabody Museum of Natural History, Yale University, New Haven, CT, USA: http://www.peabody.yale.edu/collections
- Department of Biology, University of Miami, Coral Gables, FL, USA: http://fig.cox.miami.edu
- Department of Chemistry, Florida State University, Tallahassee, FL, USA: http://www.chem.fsu.edu
- Conservation Genetics Laboratory, Nova Southeastern University Oceanographic Center, Dania Beach, FL, USA: http://www.nova.edu/ocean
- Department of Biochemistry and Molecular Biology, University of Florida Medical School, Gainesville, FL, USA: http://www.med.ufl.edu/biochem
- Department of Small Animal Clinical Sciences, University of Florida College of Veterinary Medicine, Gainesville, FL, USA: http://www.vetmed.ufl.edu/sacs/index.htm
- Department of Zoology, University of Florida, Gainesville, FL, USA: http://www.zoo.ufl.edu
- Savannah River Ecology Laboratory, Institute of Ecology, Aiken, SC, USA: http://www.uga.edu/srel/DNA_Lab/dna_lab.htm
- School of Integrative Biology, University of Illinois, Urbana, IL, USA: http://www.life.uiuc.edu/sib
- Field Museum of Natural History, Chicago, IL, USA: http://www.fmnh.org/research_collections
- Organismal Biology and Anatomy, Division of Biological Sciences, University of Chicago, Chicago, IL, USA: http://pondside.uchicago.edu/oba
- Department of Biology, Indiana University, Bloomington, IN, USA: http://www.bio.indiana.edu
- Department of Microbiology and Immunology, Indiana University School of Medicine, Indianapolis, IN, USA: http://www.iupui.edu/~micro
- Museum of Natural History, University of Kansas, Lawrence, KS, USA: http://www.nhm.ku.edu/research.html

- Collection of Genetic Resources, Museum of Natural Science, Louisiana State University, Baton Rouge, LA, USA: http://www.museum.lsu.edu/LSUMNS/Museum/NatSci/tissues.html
- Division of Biochemistry and Molecular Biology, Department of Biological Sciences, Louisiana State University, New Orleans, LA, USA: http://www.biology.lsu.edu/bmb
- Department of Biology, College of Life Sciences, University of Maryland, College Park, MD, USA: http://www.life.umd.edu/biology/index.html
- Laboratory of Genomic Diversity, Frederick Cancer Research and Development Center, National Cancer Institute, National Institutes of Health, Frederick, MD, USA: http://rex.nci.nih.gov/lgd/front_page.htm
- Captive Propagation Research Group, Patuxent Wildlife Research Center, U.S. Geological Survey (USGS) and U.S. Fish and Wildlife Service (FWS), Laurel, MD, USA: http://www.pwrc.usgs.gov; http://patuxent.fws.gov
- Department of Entomology, University of Massachusetts, Amherst, MA, USA: http://www.umass.edu/ent/index.html
- Department of Anatomy and Cell Biology, Wayne State University School of Medicine, Detroit, MI, USA: http://www.med.wayne.edu/anatomy
- Department of Biology, University of Michigan, Ann Arbor, MI, USA: http://www.biology.lsa.umich.edu
- Bell Museum of Natural History, University of Minnesota, Minneapolis, MN, USA: http://www1.umn.edu/bellmuse/collections.html
- Division of Biological Sciences, University of Montana, Missoula, MT, USA: http://biology.umt.edu
- Center of Theoretical and Applied Genetics, Department of Ecology, Evolution and Natural Resources, Rutgers University, New Brunswick, NJ, USA: http://www.rci.rutgers.edu/%7Edeenr/index.html
- Rutgers University Cell and DNA Repository, Nelson Biological Laboratories, Rutgers University, Piscataway, NJ, USA: http://lifesci.rutgers.edu/~genetics
- Division of Biological Materials, Museum of Southwestern Biology, University of New Mexico, Albuquerque, NM, USA: http://www.unm.edu/~museum
- Ambrose Monell Collection for Molecular and Microbial Research, American Museum of Natural History, New York, NY, USA: http://www.amnh.org/science/genomics/research/frozen.html
- Bassett Healthcare Research Institute, Cooperstown, NY, USA: http://www.bassetthealthcare.org/research/programs.html
- Department of Biological Sciences, Fordham University, Bronx, NY, USA: http://www.fordham.edu/biology
- Department of Biology, University of Rochester, Rochester, NY 14267, USA: http://www.rochester.edu/College/BIO
- Department of Ecology and Evolutionary Biology, Cornell University, Ithaca, NY, USA: http://www.es.cornell.edu
- Bronx Zoo and Wildlife Conservation Society, Bronx, NY, USA: http://wcs.org/home/science/genetics

- Department of Biology, King's College, Briarcliff Manor, NY, USA: http://www.kings.edu/biology/bioweb.html
- Roswell Park Memorial Institute, Buffalo, NY, USA: http://www.cog.ufl.edu/publ/inst/inpg4920.htm
- Biology Department, Belmont Abbey College, Belmont, NC, USA: http://www.belmontabbeycollege.edu
- Cincinnati Museum of Natural History, University of Cincinnati, Cincinnati, OH, USA: http://www.cincymuseum.org
- Forensic Laboratory, U.S. Fish and Wildlife Service, Ashland, OR, USA: http://www.lab.fws.gov
- Academy of Natural Sciences, Philadelphia, PA, USA: http://www.acnatsci.org
- Department of Psychiatry, University of Pittsburgh School of Medical, Pittsburgh, PA, USA: http://www.wpic.pitt.edu
- Carnegie Museum of Natural History, Pittsburgh, PA, USA: http://www.clpgh.org/cmnh
- Veterinary Department, Riverbanks Zoo and Botanical Garden, Columbia, SC, USA: http://www.riverbanks.org
- National Biomonitoring Specimen Bank (NBSB), National Institute of Standards and Technology (NIST), Chemical Science and Technology Laboratory, Charleston, SC, USA: http://www.nist.gov/public_affairs/gallery/specimen.htm
- National Marine Mammal Tissue Bank, National Marine Fisheries Service (NMFS) Marine Mammal Health and Stranding Response Program, Ft. Johnson National Oceanographic and Atmospheric Administration (NOAA) Facility, Charleston, SC, USA: http://www.nmfs.noaa.gov/prot_res/PR2/Health_and_Stranding_Response_Program/mmhsrp.html
- Division of Biology, University of Tennessee, Knoxville, TN, USA: http://www.bio.utk.edu
- Department of Biology, Natural History Collection, Angelo State University, San Angelo, TX, USA: http://www.angelo.edu/dept/biology/info/asnhc.htm
- Department of Biological Sciences, University of North Texas, Denton, TX, USA: http://www.ias.unt.edu
- Natural Science Research Laboratory, Museum of Texas Tech University, Lubbock, TX, USA: http://www.nsrl.ttu.edu/collecti.htm
- Texas Cooperative Wildlife Collection, Texas A&M University, College Station, TX, USA: http://wfscnet.tamu.edu/tcwc/tissues.htm
- Texas Memorial Museum and Section of Integrative Biology, University of Texas, Austin, TX, USA: http://www.tmm.utexas.edu/research/index.html; http://www.biosci.utexas.edu/IB
- Department of Biology and Biochemistry, University of Houston, Houston, TX, USA: http://www.bchs.uh.edu
- Department of Zoology, College of Biology and Agriculture, Brigham Young University, Provo, UT, USA: http://bioag.byu.edu/zoology

- Department of Biology, College of Agriculture and Life Sciences, University of Vermont, Burlington, VT, USA: http://www.uvm.edu/~biology
- Burke Museum, University of Washington, Seattle, WA, USA: http://www.washington.edu/burkemuseum/tissuepolicy.html
- Northwest Fisheries Science Center (NWFSC), National Marine Fisheries Service (NMFS), National Oceanographic and Atmospheric Administration (NOAA), US Department of Commerce, Seattle, WA, USA: http://research.nwfsc.noaa.gov/nwfsc-homepage.html
- Washington Department of Fish and Wildlife, Olympia, WA, USA: http://www.wa.gov/wdfw
- Laboratory of Molecular Systematics, National Museum of Natural History, Smithsonian Institution, Washington, DC, USA: http://www.lms.si.edu/frozencollections.html
- National Zoological Park, Smithsonian Institution, Washington, DC, USA: http://natzoo.si.edu
- Department of Biological Sciences, Marshall University, Huntington, WV, USA: http://www.marshall.edu/biology
- Department of Physiology, University of Wisconsin Medical School, Madison, WI, USA: http://www.physiology.wisc.edu
- International Crane Foundation, Baraboo, WI, USA: http://www.savingcranes.org
- Molecular Systematics Laboratory, Zoological Museum, University of Wisconsin, Madison, WI, USA: http://www.wisc.edu/zoology/museum/museum.html
- Department of Biological Sciences, University of Wisconsin-Parkside, Kenosha, WI, USA: http://www.uwp.edu/academic/biology/biologymenu/Index.htm

6 Commercial Supply Companies

- Berkshire Biological Supply Co., Westhampton, MA, USA: http://www.crocker.com/~berkbio
- Carolina Biological Supply Co., Burlington, NC, USA: http://www.carolina.com
- Fisher Scientific, Educational Materials Division, Chicago, IL, USA: https://www1.fishersci.com/main.jsp
- Fluker Farms, Inc., Port Allen, LA, USA: http://www.flukerfarms.com
- Glades Herp., Inc., Ft. Myers, FL, USA: http://www.gherp.com
- Gulf Specimen Marine Laboratories, Inc., Panacea, FL, USA: http://www.gulfspecimen.org
- Marine Biological Laboratory, Marine Resources Center, Woods Hole, MA, USA: http://zeus.mbl.edu/public/organisms/catalog.php3
- Pacific Biological Supply, Sherman Oaks, CA, USA: http://www.pacificbio.com

- Ward's Natural Science Establishment, Inc., Rochester, NY and Santa Fe Springs, CA, USA: http://www.wardsci.com

7 Botanical and Zoological Gardens

- Botanic Gardens Conservation International (BCGI), an International Clearing House Mechanism for Botanic Gardens of the World, Royal Botanic Gardens Kew, Richmond, UK: http://www.bgci.org.uk
- Internet Directory for Botany (IDB), University of Alberta, Edmonton, AB, Canada: http://www.botany.net/IDB
- Lists of Botanical Gardens and Arboreta: http://dir.yahoo.com/Science/ Biology/Botany/Botanical_Gardens; http://dir.yahoo.com/Science/Biology/ Botany/Botanical_Gardens/Arboretums; http://www.botany.net/IDB/subject/ botgard.html
- CAUZ Directory, The Consortium of Aquariums, Universities and Zoos (CAUZ) for Worldwide Conservation: http://www.selu.com/bio/cauz
- American Zoo and Aquarium Association (AZA) Zoo and Aquarium Directory: http://www.aza.org/members/zoo
- WWW Virtual Library: Zoos, hosted by ZooNet: http://zoonet.home. mindspring.com/www_virtual_lib/zoos.html
- The Cyber ZooMobile, Zoo Links: http://www.primenet.com/~brendel/ zoo.html
- The Good Zoo Guide Online: http://www.goodzoos.com
- ZooNet: http://www.zoonet.org
- ZooWeb: World Wide Link to Zoos and Aquariums: http://www.zooweb.net
- Lists of Aquariums and Aviaries: http://dir.yahoo.com/Science/Biology/ Zoology/Zoos/Aquariums; http://dir.yahoo.com/Science/Biology/Zoology/ Zoos/Aviaries
- Public butterfly gardens and zoos, hosted by The Butterfly Website: http:// www.mgfx.com/butterfly/gardens/index.cfm
- Insect Zoos, Butterfly Gardens and Museums in North America, hosted by the Entomological Society of America: http://www.entsoc.org/education/ insect_zoos.htm

8 Natural History Museums, Herbaria and Directories of Specialists

- Association of Systematics Collections (ASC), Washington, DC, USA: http:// www.ascoll.org
- Society for the Preservation of Natural History Collections (SPNHC): http:// www.spnhc.org
- The Systematics Association, The Natural History Museum, London, UK: http://www.systass.org
- The Biodiversity and Biological Collections Web Server, University of New Orleans, New Orleans, LA, USA: http://biodiversity.uno.edu

- General Systematic Research Tools and Resources for Systematics Research, University of Michigan Herbarium, Ann Arbor, MI, USA: http://www.herb.lsa.umich.edu/tool_dir.htm
- Center for Biosystematics, University of California, Davis, CA, USA: http://cbshome.ucdavis.edu/index.html
- Virtual Library Museums Pages: Directory of Online Museums: http://www.icom.org/vlmp
- UCMP lists of Natural History Collections, Societies and Organisations, hosted by the Museum of Paleontology, University of California (UCMP), Berkeley, CA, USA: http://www.ucmp.berkeley.edu/collections/other.html; http://www.ucmp.berkeley.edu/subway/nathistmus.html; http://www.ucmp.berkeley.edu/subway/nathistorg.html
- Directory of Research Systematics Collections (DRSC) of the National Biological Information Infrastructure (NBII), ASC and USGS, Washington, DC, USA: http://bp.cr.usgs.gov/drsc/drsc.cfm
- Natural History Museums and Collections: http://www.lib.washington.edu/sla/natmus.html; http://fsas.upm.edu.my/~arahim/jabbio/museum.html
- Biological Museums on the Web, hosted by The Biodiversity and Biological Collections Web Server, University of New Orleans, New Orleans, LA, USA: http://biodiversity.uno.edu/cgi-bin/hl?museum
- The Insect and Spider Collections of the World, hosted by the Bernice Pauahi Bishop Museum, Honolulu, HI, USA: http://www.bishopmuseum.org/bishop/ento/codens-r-us.html
- Insects and Entomological Resources, hosted by Saint Anselm College, Manchester, NH, USA: http://www.anselm.edu/homepage/chieber/insects.html
- Iowa State Entomology Index: Insect Collections, Iowa State University, IA, USA: http://www.ent.iastate.edu/List/insect_collections.html
- Council of Heads of Australian Entomological Collections (CHAEC), hosted by Museum Victoria, Melborne, VIC, Australia: http://pioneer.mov.vic.gov.au/chaec
- International Society of Arachnology, National Museum of Natural History, Washington, DC, USA: http://www.ufsia.ac.be/Arachnology/Pages/ISA.html
- Polar Museums Directory, The Scott Polar Research Institute (SPRI), University of Cambridge, UK: http://www.spri.cam.ac.uk/lib/museums.htm
- Botanical Museums, Herbaria, Natural History Museums on the Internet Directory for Botany (IDB), University of Alberta, Edmonton, AB, Canada: http://www.botany.net/IDB/subject/botmus.html
- A list of herbaria, hosted by Tel Aviv Herbarium (TELA), Tel Aviv, Israel: http://www.tau.ac.il/lifesci/botany/herbaria.htm
- *Index Herbariorum* and *Plant Specialists Index*, Online Search Index for Institutions and People, hosted by The New York Botanical Garden, Bronx, NY, USA: http://www.nybg.org/bsci/ih/ih.html
- American Society of Plant Taxonomists (ASPT), Membership Directory: http://www.sysbot.org/members.htm

- Worldwide List of Internet Accessible Herbaria: http://www.ibiblio.org/botnet/flora/wwwlist2.html
- Further listings of herbaria: http://dir.yahoo.com/Science/Biology/Botany/Herbaria
- Mycologists Online, Worldwide Directory for Mycology and Lichenology, hosted by the Institute of Botany, Bratislava, Slovakia: http://web.savba.sk/botu/myco

9 Biodiversity and Biological Collections Database Software

- The Biodiversity and Biological Collections Web Server, University of New Orleans, New Orleans, LA, USA: http://biodiversity.uno.edu
- General Systematic Research Tools and Resources for Systematics Research, University of Michigan Herbarium, Ann Arbor, MI, USA: http://www.herb.lsa.umich.edu/tool_dir.htm
- Digital Taxonomy – An Information Resource on Biodiversity Data Management: http://www.geocities.com/RainForest/Vines/8695
- BIOLINK, CSIRO Publishing, Melbourne, VIC, Australia: http://www.biolink.csiro.au
- BIOTA: The Biodiversity Database Manager, Sinauer Associates, Inc., Sunderland, MA, USA: http://viceroy.eeb.uconn.edu/biota
- DELTA: http://biodiversity.bio.uno.edu/delta
- MUSE and Specify, Biodiversity Research Center, Lawrence, KS, USA: http://biodiversity.bio.uno.edu/muse; http://usobi.org/specify/Muse.html; http://www.usobi.org/specify
- Platypus, CSIRO Publishing, Melbourne, VIC, Australia: http://www.environment.gov.au/abrs/platypus.html
- Freezerworks, Dataworks Development, Inc., Mountlake Terrace, WA, USA: http://www.freezerworks.com

References

1   DeSalle R, Williams AK, George M (1994) Isolation and characterisation of animal mitochondrial DNA. In: EA Zimmer, TJ White, RL Cann, AC Wilson (eds): *Methods in Enzymology Vol. 224, Molecular Evolution: Producing the Biochemical Data*. Academic Press, San Diego, CA, 176–204

2   Sytsma KJ, Givnish TJ, Smith JF, Hahn WJ (1994) Collection and storage of land plant samples for macromolecular studies. In: EA Zimmer, TJ White, RL Cann, AC Wilson (eds): *Methods in Enzymology Vol. 224, Molecular Evolution: Producing the Biochemical Data*. Academic Press, San Diego, CA, 23–37

3   Dessauer HC, Cole CJ, Hafner MS (1990) Collection and storage of tissues. In: DM Hillis, C Moritz (eds): *Molecular Systematics*. Sinauer, Sunderland, MA, 25–41

4   Dessauer HC, Cole CJ, Hafner MS (1996) Collection and storage of tissues. In: DM Hillis, C Moritz, BK Mable (eds): *Mo-*

lecular Systematics. 2nd Edition. Sinauer, Sunderland, MA, 29–47

5   Brown TA (1999) How ancient DNA may help in understanding the origin and spread of agriculture. *Philosophical Transactions of the Royal Society of London, Series B – Biological Sciences* 354: 89–97

6   Arctander P (1988) Comparative studies of avian DNA by restriction fragment length polymorphism analysis: Convenient procedures based on blood samples from live birds. *Journal of Ornithology* 129: 205–216

7   Lipkin E, Shalom A, Khatib H, Soller M et al. (1993) Milk as a source of deoxyribonucleic acid and as a substrate for the polymerase chain reaction. *Journal of Dairy Science* 76: 2025–2032

8   Gonser RA, Collura RV (1996) Waste not, want not: toe-clips as a source of DNA. *Journal of Herpetology* 30: 445–447

9   Mockford SW, Wright JM, Snyder M, Herman TB (1999) A non-destructive source of DNA from hatchling freshwater turtles for use in PCR base assays. *Herpetological Review* 30: 148–149

10   Goossens B, Waits LP, Taberlet P (1998) Plucked hair samples as a source of DNA – Reliability of dinucleotide microsatellite genotyping. *Molecular Ecology* 7: 1237–1241

11   Khoshoo TN (1998) Assessing genetic diversity in wild mega animals using noninvasive methods. *Current Science* 74: 13–14

12   Taberlet P, Waits LP (1998) Noninvasive genetic sampling. *Trends in Ecology & Evolution* 13: 26–27

13   Strathmann MF (1987) *Reproduction and Development of Marine Invertebrates of the Northern Pacific Coast.* University of Washington Press, Seattle, WA

14   Dick M, Bridge DM , Wheeler WC, DeSalle R (1994) Collection and storage of invertebrate samples. In: EA Zimmer, TJ White, RL Cann, AC Wilson (eds): *Methods in Enzymology Vol. 224, Molecular Evolution: Producing the Biochemical Data.* Academic Press, San Diego, CA, 51–65

15   Koch DA, Duncan GA, Parsons TJ, Pruess KP et al. (1998) Effects of preservation methods, parasites, and gut contents of black flies (Diptera, Simuliidae) on polymerase chain reaction products. *Journal of Medical Entomology* 35: 314–318

16   Dorn PL, Selgean S, Guillot M (1997) Simplified method for preservation and polymerase chain reaction amplification of *Trypanosoma cruzi* DNA in human blood. *Memorias do Instituto Oswaldo Cruz* 92: 253–255

17   Mori SA, Holm-Nielsen LB (1981) Recommendations for botanists visiting neotropical countries. *Taxon* 30: 87–91

18   Davis SD, Droop SJM, Gregerson P, Henson L et al. (1986) *Plants in Danger. What do we know?* International Union for the Conservation of Nature and Natural Resources, Cambridge, UK

19   Soltis PS, Soltis DE (1993) Ancient DNA. Prospects and limitations. *New Zealand Journal of Botany* 31: 203–209

20   Handt O, Hoss M, Krings M, Pääbo S (1994) Ancient DNA – methodological challenges. *Experientia* 50: 524–529

21   Thomas WK, Pääbo S (1994) DNA sequences from old tissue remains. In: EA Zimmer, TJ White, RL Cann, AC Wilson (eds): *Methods in Enzymology Vol. 224, Molecular Evolution: Producing the Biochemical Data.* Academic Press, San Diego, CA, 406–419

22   Pääbo S (1989) Ancient DNA: Extraction, characterization, molecular cloning, and enzymatic amplification. *Proceedings of the National Academy of Sciences U.S.A.* 86: 1939–1943

23   Yagi N, Satonaka K, Horio M, Shimogaki H et al. (1996) The role of DNAse and EDTA on DNA degradation in formaldehyde fixed tissues. *Biotechnic & Histochemistry* 71: 123–129

24   Adams RP, Zhong M, Fei Y (1999) Preservation of DNA in plant specimens – inactivation and reactivation of DNAses in field specimens. *Molecular Ecology* 8: 681–683

25   Briggs DEG (1999) Molecular taphonomy of animal and plant cuticles – selective preservation and diagenesis. *Philo-*

sophical *Transactions of the Royal Society of London, Series B – Biological Sciences* 354: 7–16

26 Bada JL, Wang XYS, Hamilton H (1999) Preservation of key biomolecules in the fossil record – current knowledge and future challenges. *Philosophical Transactions of the Royal Society of London, Series B – Biological Sciences* 354: 77–86

27 Sytsma KJ, Schaal BA (1985) Genetic variation, differentiation, and evolution in a species complex of tropical shrubs based on isozymic data. *Evolution* 39: 582–593

28 Dessauer HC, Menzies RA, Fairbrothers DE (1984) Procedures for collecting and preserving tissues for molecular studies. In: HC Dessauer, MS Hafner (eds): *Collections of Frozen Tissues: Value, Management, Field and Laboratory Procedures, and Directory of Existing Collections*. Association of Systematics Collections, University of Kansas Press, Lawrence, KS, 21–24

29 Simione FP, Brown EM (1991) *ATCC Preservation Methods: Freezing and Freeze-Drying*. 2nd Edition. American Type Culture Collection, Rockville, MD

30 Simione FP (1992) *Cryopreservation Manual*. Nalge Company, Rochester, NY

31 Dessauer HC, Hafner MS (1984) *Collections of Frozen Tissues: Value, Management, Field and Laboratory Procedures, and Directory of Existing Collections*. Association of Systematics Collections, University of Kansas Press, Lawrence, KS

32 Dessauer HC, Braun MJ, Neville S (1983) A simple hand centrifuge for field use. *Isozyme Bulletin* 16: 91.

33 Hay RJ, Gee GF (1984) Procedures for collecting cell lines under field conditions. In: HC Dessauer, MS Hafner (eds): *Collections of Frozen Tissues: Value, Management, Field and Laboratory Procedures, and Directory of Existing Collections*. Association of Systematics Collections, University of Kansas Press, Lawrence, KS, 25–26

34 Mazur P (1980) Fundamental aspects of the freezing of cells with emphasis on mammalian ova and embryos. In: *Proceedings of the Ninth International Congress on Animal Reproduction and Artificial Insemination*, 99–114

35 Gordh G, Hall JC (1979) A critical point drier used as a method of mounting insects from alcohol. *Entomological News* 90: 57–59

36 Austin AD, Dillon N (1997) Extraction and PCR of DNA from parasitoid wasps that have been chemically dried. *Australian Journal of Entomology* 36: 241–244

37 Logan JA (1999) Extraction, polymerase chain reaction, and sequencing of a 440 base-pair region of the mitochondrial cytochrome oxidase I gene from 2 species of acetone-preserved damselflies (Odonata, Coenagrionidae, Agrionidae) *Environmental Entomology* 28: 143–147

38 Post RJ, Flook PK, Millest AL (1993) Methods for the preservation of insects for DNA studies. *Biochemical Systematics and Ecology* 21: 85–90

39 Phillips AJ, Simon C (1995) Simple, efficient, and nondestructive DNA extraction protocol for arthropods. *Annals of the Entomological Society of America* 88: 281–283

40 Toe L, Back C, Adjami AG, Tang JM et al. (1997) *Onchocerca volvulus* – Comparison of field collection methods for the preservation of parasite and vector samples for PCR analysis. *Bulletin of the World Health Organization* 75: 443–447

41 Makowski GS, Davis EL, Hopfer SM (1997) Amplification of Guthrie card DNA – Effect of guanidine thiocyanate on binding of natural whole-blood PCR inhibitors. *Journal of Clinical Laboratory Analysis* 11: 87–93

42 Ostrander EA, Maslen CL, Hallick LM (1987) Recovery of critical DNA samples from laboratory bench paper. Focus 9: 9

43 McCabe ERB, Huang S-Z, Seltzer WK, Law ML (1987) DNA microextraction from dried blood spots on filter paper blotters: potential application to newborn screening. *Human Genetics* 75: 213–217

44 Weisberg EP, Giorda R, Trucco M, Lampasona V (1993) Lyophilization as a method to store samples of whole-blood. *BioTechniques* 15: 64

45 Cann RL, Feldman RA, Freed LA, Lum JK et al. (1994) Collection and storage of vertebrate samples. In: EA Zimmer, TJ White, RL Cann, AC Wilson (eds): *Methods in Enzymology Vol. 224, Molecular Evolution: Producing the Biochemical Data*. Academic Press, San Diego, CA, 38–51

46 Croat TB (1979) Use of a portable propane gas oven for field drying plants. *Taxon* 28: 573–580

47 Hower RO (1979) *Freeze-drying Biological Specimens*. Smithsonian Institution Press, Washington, DC

48 Rogers SO, Bendich AJ (1985) Extraction of DNA from milligram amounts of fresh, herbarium and mummified plant tissues. *Plant Molecular Biology* 5: 69–76

49 Hamrick JL, Loveless MD (1986) Isozyme variation in tropical trees: Procedures and preliminary results. *Biotropica* 18: 201–207

50 Doyle JJ, Dickson EE (1987) Preservation of plant samples for DNA restriction endonuclease analysis. *Taxon* 36: 715–722

51 Pyle MM, Adams RP (1989) *In situ* preservation of DNA in plant specimens. *Taxon* 38: 576–581

52 Tai TH, Tanksley SD (1990) A rapid and inexpensive method for isolation of total DNA from dehydrated plant tissue. *Plant Molecular Biology Reporter* 8: 297–303

53 Harris SA (1993) DNA analysis of tropical plant species. An assessment of different drying methods. *Plant Systematics and Evolution* 188: 57–64

54 Khadhair AH, Momani F, Hiruki C (1995) Molecular stability of clover proliferation phytoplasma DNA in periwinkle plant tissues after thermal treatment under microwave conditions. *Proceedings of The Japan Academy Series B – Physical and Biological Sciences* 71: 265–268

55 Chase M, Hills J (1991) Silica gel: an ideal material for field preservation of leaf samples for DNA studies. *Taxon* 40: 215–220

56 Hanlin RT (1972) Preservation of fungi by freeze-drying. *Bulletin of the Torrey Botanical Club* 99: 23–27

57 Bolster P (1978) *Preserving Flowers with Silica Gel*. Canadian Department of Agriculture Publication 1649, Ottawa

58 Liston A, Rieseberg LH, Adams RP, Do N et al. (1990) A method for collecting dried plant specimen for DNA and isozymes analysis, and the results of a field test in Xinjiang, China. *Annals of the Missouri Botanical Garden* 77: 859–864

59 Mellor JD (1978) *Fundamentals of Freeze-Drying*. Academic Press, New York

60 Schierenbeck KA (1994) Modified polyethylene-glycol DNA extraction procedure for silica gel-dried tropical woody plants. *BioTechniques* 16: 392–394

61 Holzmann M, Pawlowski J (1996) Preservation of Foraminifera for DNA extraction and PCR amplification. *Journal of Foraminiferal Research* 26: 264–267

62 Winsor L (1998) Collection, handling, fixation, histological and storage procedures for taxonomic studies of terrestrial flatworms (Tricladida, Terricola). *Pedobiologia* 42: 405–411

63 Noguchi M, Furuya S, Takeuchi T, Hirohashi S (1997) Modified formalin and methanol fixation methods for molecular biological and morphological analyses. *Pathology International* 47: 685–691

64 Greer CE, Lund JK, Manos MM (1991) PCR amplification from paraffin embedded tissues: recommendations on fixatives for long-term storage and prospective studies. *PCR Methods and Applications* 1: 46–50

65 Reiss RA, Schwert DP, Ashworth AC (1995) Field preservation of Coleoptera for molecular genetic analyses. *Environmental Entomology* 24: 716–719

66 Taylor WR (1981) Preservation practices: water in tissues, specimen volume, and alcohol concentration. *Curation Newsletter* 2: 1–3

67 Masner L (1994) Effect of low temperature on preservation and quality of insect specimens stored in alcohol. *Insect Collection News* 9: 14–15

68 Flournoy LE, Adams RP, Pandy RN (1996) Interim and archival preservation of plant specimens in alcohols for

DNA Studies. *BioTechniques* 20: 657–660

69 Bahl A, Pfenninger M (1996) A rapid method of DNA isolation using laundry detergent. *Nucleic Acids Research* 24: 1587–1588

70 Kuch U, Pfenninger M, Bahl A (1999) Laundry detergent effectively preserves amphibian and reptile blood and tissues for DNA isolation. *Herpetological Review* 30: 80–82

71 Sansinforiano ME, Padilla JA, Demendoza JH, Demendoza MH et al. (1998) Rapid and easy method to extract and preserve DNA from *Cryptococcus neoformans* and other pathogenic yeasts. *Mycoses* 41: 195–198

72 Nakanishi M, Wilson AC, Nolan RA, Gorman GC et al. (1969) Phenoxyethanol: Protein preservative for taxonomists. *Science* 163: 681–683

73 Birungi J, Roy MS, Arctander P (1998) DMSO-preserved samples as a source of messenger RNA for RT-PCR. *Molecular Ecology* 7: 1429–1430

74 Quinn T, White, B (1987) Identification of restriction-fragment-length polymorphisms in genomic DNA of the lesser snow goose (*Anser calrulescens calrulescens*) *Molecular Biology and Evolution* 4: 126–143

75 Lansman RA, Shade RO, Shapira JF, Avise JC (1981) The use of restriction endonucleases to measure mitochondrial DNA sequence relatedness in natural populations. *Journal of Molecular Evolution* 17: 214–226

76 Gelhaus A, Urban B, Pirmez C (1995) DNA extraction from urea-preserved blood or blood clots for use in PCR. *Trends in Genetics* 11: 129–129

77 Laulier M, Pradier E, Bigot Y, Periquet G (1995) An easy method for preserving nucleic acids in field samples for later molecular and genetic studies without refrigerating. *Journal of Evolutionary Biology* 8: 657–663

78 Tuttle RM, Waselenko JK, Yosseffi P, Weigand, N et al. (1998) Preservation of nucleic acids for polymerase chain reaction after prolonged storage at room temperature. *Diagnostic Molecular Pathology* 7: 302–309

79 Asahida T, Kobayashi T, Saitoh K, Nakayama I (1997) Tissue preservation and total DNA extraction from fish stored at ambient temperature using buffers containing high concentration of urea. *Fisheries Science* 62: 727–730

80 Dawson MN, Raskoff KA, Jacobs DK (1998) Field preservation of marine invertebrate tissue for DNA analyses. *Molecular Marine Biology and Biotechnology* 7: 145–152

81 Seutin G, White BN, Boag PT (1991) Preservation of avian blood and tissue samples for DNA analyses. *Canadian Journal of Zoology* 69: 82–90

82 Nickrent DL (1994) From field to film – Rapid sequencing methods for field-collected plant species. *BioTechniques* 16: 470

83 Shubin N, Tabin C, Carroll S (1997) Fossils, genes and the evolution of animal limbs. *Nature (London)* 388: 639–648

84 Kramer EM, Irish VF (1999) Molecular genetic studies in *Arabidopsis thaliana* and other higher-eudicot flowering plants have led to the development of the 'ABC' model of the determination of organ identity in flow. *Nature (London)* 399: 144–148

85 Eernisse DJ, Kluge AG (1993) Taxonomic congruence versus total evidence, and the phylogeny of amniotes inferred from fossils, molecules and morphology. *Molecular Biology and Evolution* 10: 1170–1195

86 Jacobs DK, Wray CG, Wedeen CJ, Kostriken R et al. (2000) Molluscan engrailed expression, serial organization and shell evolution. *Evolution and Development* 2: 340–347

87 Wasser SK, Houston CS, Koehler GM, Cadd GG et al. (1997) Techniques for application of fecal DNA methods to field studies of ursids. *Molecular Ecology* 6: 1091–1097

88 Frantzen MAJ, Silk JB, Ferguson JWH, Wayne RK et al. (1998) Empirical evaluation of preservation methods for fecal DNA. *Molecular Ecology* 7: 1423–1428

89 Shankaranarayanan P, Singh L (1998) A rapid and simplified procedure for isolating DNA from scat samples. *Current Science* 75: 883–884

90 Launhardt K, Epplen C, Epplen JT, Winkler P (1998) Amplification of microsatellites adapted from human systems in fecal DNA of wild Hanuman langurs (*Presbytis entellus*) *Electrophoresis* 19: 1356–1361

91 Dowd SE, Gerba CP, Enriquez FJ, Pepper IL (1998) PCR amplification and species determination of *Microsporidia* in formalin-fixed feces after immunomagnetic separation. *Applied and Environmental Microbiology* 64: 333–336

92 Jannett FJ Jr. (1989) Some tests of synthetic paper and polyethylene sacks for specimens preserved in fluids. *Curator* 32: 24–25

93 Blackwell M, Chapman R (1994) Collection and storage of fungal and algal samples. In: EA Zimmer, TJ White, RL Cann, AC Wilson (eds): *Methods in Enzymology Vol. 224, Molecular Evolution: Producing the Biochemical Data.* Academic Press, San Diego, CA, 65–77

94 Lee WL, Bell BM, Sutton JF (1982) *Guidelines for Acquisition and Management of Biological Specimens. A Report of the participants of a Conference on Voucher Specimen Management sponsored under the auspices of the Council of Curatorial Methods of the Association of Systematics Collections.* Association of Systematics Collections, University of Kansas Press, Lawrence, KS

95 Williams SL, Hawks CA (1986) Inks for documentation in vertebrate research collections. *Curator* 29: 93–108

96 Wood RM, Williams SL (1993) An evaluation of disposable pens for permanent museum records. *Curator* 36: 189–200

97 Baker RJ, Hafner MS (1984) Curatorial problems unique to frozen tissue collections. In: HC Dessauer, MS Hafner (eds): *Collections of Frozen Tissues: Value, Management, Field and Laboratory Procedures, and Directory of Existing Collections.* Association of Systematics Collections, University of Kansas Press, Lawrence, KS, 35–40

98 Jewett SL (1987) Byron Weston Resistall Ledger and other papers used for labels in aqueous preservatives. *Curation Newsletter* 9: 1–2

99 Van Guelpen L, McElman JF, Sulak KJ (1987) Additional label papers for specimens in aqueous preservatives. *Curation Newsletter* 9: 3–4

100 Walker S (1986) Investigation of the properties of Tyvek pertaining to its use as a storage material for artifacts. *The International Institute for Conservation, Canadian Group Newsletter* 12: 21–25

101 Gisbert J, Palacios F, Garcia-Perea R (1990) Labelling vertebrate collections with Tyvek® synthetic paper. *Collection Forum* 6: 35–37

102 Remsen JV Jr. (1977) On taking field notes. *American Birds* 31: 946–953

103 Herman SG (1980) *The Naturalist's Field Journal.* Buteo Books, Vermillion, SD

104 Friend M (1987) *Field Guide to Wildlife Diseases.* U.S. Department of the Interior, Fish and Wildlife Service Resource Publication No. 167, Washington, DC

105 Jong S-C, Atkins WB (1985) Conservation, collection, and distribution of cultures. In: S-C Jong, WB Atkins, DH Howard (eds) *Fungi Pathogenic for Humans and Animals.* Marcel Dekker, Inc., New York, 153–194

106 Cypess RH, Jong S-C (1997) Culture collections for equitable use of microbial germplasm. In: KE Hoagland, AY Rossman (eds): *Global Genetic Resources: Access, Ownership, and Intellectual Property Rights. Beltsville Symposia in Agricultural Research 21.* Association of Systematics Collections, Washington, DC, 235–244

107 Staines JE, McGowan VF, Skerman VAD (1986) *World Directory of Collections of Cultures of Microorganisms.* 3rd Edition. World Data Centre, University of Queensland, Brisbane

108 Miyachi S, Nakayama O, Yokohama Y, Hara Y et al. (1989) *World Catalogue of Algae.* 2nd Edition. Japan Scientific Societies Press, Tokyo

109 Edwards MJ (1988) *Microbes and Cells at Work: An Index to ATCC Strains with*

Special Applications. American Type Culture Collection, Rockville, MD

110 Maglott DR, Nierman WC (1991) *Catalogue of Recombinant DNA Materials*. 2nd Edition. American Type Culture Collection, Rockville, MD

111 Krickevsky MI, Fabricius BO, Sugawara H (1988) Information resources. In: DL Hawksworth, BE Kirsop (eds): *Living Resources for Biotechnology, Filamentous Fungi*. Cambridge University Press, New York, 31–53

112 Starr RC, Zeikus JA (1993) The Culture Collection of Algae at The University of Texas at Austin: 1993 list of cultures. *Journal of Phycology* 29 (suppl.): 1–106

113 Gardes M, White TJ, Fortin JA, Bruns TD et al. (1991) Indentification of 8 indigenous and introduced fungi in mycorrhizae by amplification of internal transcribed spacer. *Canadian Journal of Botany* 69: 180–190

114 Lee SB, Taylor JW (1990) Isolation of DNA from fungal mycelia and single spores. In: Innis MA, Gelfand DH, Sninsky JJ, White TJ (eds): *PCR Protocols: A Guide to Methods and Applications*. Academic Press, San Diego, CA, 282–287.

115 Bold HC, Wynne MJ (1985) *Introduction to the Algae*. Prentice-Hall, Englewood Cliffs, NJ

116 Shands HL, Stoner AK (1997) Agricultural germplasm and global contributions. In: KE Hoagland, AY Rossman (eds): *Global Genetic Resources: Access, Ownership, and Intellectual Property Rights. Beltsville Symposia in Agricultural Research 21*. Association of Systematics Collections, Washington, DC, 97–106

117 Hawtin G, Engels J, Siebeck W (1997) International germplasm collections under the Biodiversity Convention – Options for a continued multilateral exchange of genetic resources for food and agriculture. In: KE Hoagland, AY Rossman (eds): *Global Genetic Resources: Access, Ownership, and Intellectual Property Rights. Beltsville Symposia in Agricultural Research 21*. Association of Systematics Collections, Washington, DC, 247–262

118 Henderson DM , Prentice HT (1977) *International Directory of Botanical Gardens. Regnum Vegetabile. Vol. 95*. International Bureau for Plant Taxonomy and Nomenclature, Utrecht, The Netherlands

119 Raven PH, Johnson GB, Zenger VE (1986) *Instructor's Manual to Accompany Raven and Johnson Biology*. Times Mirror/Mosby College Publ., St. Louis, MO, 33

120 Connor KF, Towill LE (1993) Pollen-handling protocol and hydration-dehydration characteristics of pollen for application to long-term storage. *Euphytica* 68: 77–84

121 Dessauer HC, Hafner MS, Zink RM (1988) A national program to develop, maintain and utilize frozen tissue collections for scientific research. *Association of Systematics Collections Newsletter* 16: 3–10

122 Sheldon FM, Dittmann DL (1997) The value of vertebrate tissue collections in applied and basic science. In: KE Hoagland, AY Rossman (eds): *Global Genetic Resources: Access, Ownership, and Intellectual Property Rights. Beltsville Symposia in Agricultural Research 21*. Association of Systematics Collections, Washington, DC, 151–162

123 Baker R, Hager B, Monk R (1999) Issues concerning loan policies for destructive analysis of museum based frozen tissue and DNA samples. (Abstract). In: *Proceedings of Society for the Preservation of Natural History Collections, 14th Annual Meeting, June 27–2 July 1999*. National Museum of Natural History, Smithsonian Institution, Washington, DC, 20

124 Thomas WK, Pääbo S, Villablanca FX, Wilson AC (1990) Spatial and temporal continuity of kangaroo rat populations shown by sequencing mitochondrial DNA from museum specimens. *Journal of Molecular Evolution* 31: 101–112

125 Hagelberg E, Sykes B, Hedges R (1989) Ancient bone DNA amplified. *Nature (London)* 342: 485–486

126 Lee HC, Pagliaro EM, Berka KM, Folk NL et al. (1991) Genetic markers in human bones: DNA analysis. *Journal of Forensic Science* 36: 320–326.

127 Tuross N, Stathoplos L (1994) Ancient proteins in fossil bones. In: EA Zimmer, TJ White, RL Cann, AC Wilson (eds): *Methods in Enzymology Vol. 224, Molecular Evolution: Producing the Biochemical Data.* Academic Press, San Diego, CA, 121–129

128 Simpson TA, Smith RJH (1995) Amplification of mitochondrial DNA from archival temporal bone specimens. *Laryngoscope* 105: 28–34

129 Koshiba M, Ogawa K, Hamazaki S, Sugiyama T et al. (1993) The effect of formalin fixation on DNA and the extraction of high molecular weight DNA from fixed and embedded tissues. *Pathology Research and Practice* 189: 66–72

130 Savioz A, Blouin JL, Guidi S, Antonarakis SE et al. (1997) A method for the extraction of genomic DNA from human brain tissue fixed and stored in formalin for many years. *Acta Neuropathologica* 93: 408–413

131 Waller R, McAllister DE (1986) A spot test for distinguishing formalin from alcohol solutions. In: J Waddington, DM Rudkin (eds): *Proceedings of the 1985 Workshop on Care and Maintenance of Natural History Collections.* Life Sciences Miscellaneous Publications of the Royal Ontario Museum, Toronto, 93–99

132 Moore SJ (1994) What fluid is in this bottle? *Biology Curators' Group Newsletter* 6: 44–45

133 Criscuolo G (1992) Extraction and amplification of DNA from wet museum collections. *Ancient DNA Newsletter* 1: 12–13

134 Criscuolo G (1994) On the state of preservation of DNA from museum spirit collections. *Biology Curators' Group Newsletter* 6: 39–41

135 Bulat SA, Zakharov IA (1992) Identification of DNA in the material of entomological collections by means of polymerase chain reaction. *Zhurnal Obshchei Biologii* 53: 861–863

136 Rollo F, La Marca A, Amici A (1987) Nucleic acids in mummified plant seeds: screening of twelve specimens by gel-electrophoresis, molecular hybridization and DNA cloning. *Theoretical and Applied Genetics* 73: 501–505

137 Golenberg EM, Giannasi DE, Clegg MT, Smiley CJ et al. (1990) Chloroplast DNA sequence from a Miocene *Magnolia* species. *Nature (London)* 344: 656–658

138 Golenberg EM (1994) Antediluvian DNA research. *Nature (London)* 367: 692–692

139 Bruns TD, Fogel R, Taylor JW (1990) Amplification and sequencing from fungal herbarium specimens. *Mycologia* 82: 175–184

140 Swann EC, Saenz GS, Taylor JW (1991) Maximizing information content of morphological specimens: Herbaria as sources of DNA for molecular systematics. (Abstract). *Mycological Society of America Newsletter* 42: 36

141 Goff LJ, Moon DA (1993) PCR amplification of nuclear and plastid genes from algal herbarium specimens and algal spores. *Journal of Phycology* 29: 381–384

142 Douglas MP, Rogers SO (1998) DNA damage caused by common cytological fixatives. *Mutation Research – Fundamental and Molecular Mechanisms of Mutagenesis* 401: 77–88

143 Young DJ, Gilbertson RL, Alcorn SM (1982) A new record for longevity of *Macrophomina phaseolina sclerotia* soil-inhabiting fungi. *Mycologia* 74: 504–505

144 Shibata DK, Arnhem N, Martin WJ (1988) Detection of human papilloma virus in paraffin-embedded tissue using the polymerase chain reaction. *Journal of Experimental Medicine* 167: 225–230

145 Crisan D, Mattson JC (1993) Retrospective DNA analysis using fixed tissue specimens. *DNA and Cell Biology* 12: 455–464

146 Iwamoto KS, Mizuno T, Ito T, Akiyama M et al. (1996) Feasibility of using decades-old archival tissues in molecular oncology/epidemiology. *American Journal of Pathology* 149: 399–406

147 McEwen JE, Reilly PR (1994) Stored Guthrie cards as DNA banks. *American Journal of Human Genetics* 55: 196–200

148 Visvikis S, Schlenck A, Maurice M (1998) DNA extraction and stability for epidemiologic studies. *Clinical Chemistry and Laboratory Medicine* 36: 551–555

149 Frank TS, Svobodanewman SM, His ED (1996) Comparison of methods for extracting DNA from formalin-fixed paraffin sections for nonisotopic PCR. *Diagnostic Molecular Pathology* 5: 220–224

150 Morgan K, Lam L, Kalsheker N (1996) A rapid and efficient method for DNA extraction from paraffin wax embedded tissue for PCR amplification. *Journal of Clinical Pathology – Clinical Molecular Pathology Edition* 49: M179–M180

151 Pavelic J, Galltroselj K, Bosnar MH, Kardum MM et al. (1996) PCR amplification of DNA from archival specimens – A methodological approach. *Neoplasma* 43: 75–81

152 Sabrosky CW (1971) Packing and shipping pinned insects. *Bulletin of the Entomological Society of America* 17: 6–8

153 Piegler RS (1992) Shipping of pinned insects. *Collection Forum* 8: 73–77

154 McCoy CJ (1993) Packing fluid-preserved herpetological specimens for shipment. *Collection Forum* 9: 70–75

155 Mikoliczeak J, Stephens M, Simione FP Jr. (1987) Comparison of –130°C mechanical refrigeration with liquid nitrogen storage for biological specimens. *SIM News* 37: 74

156 Naber SP (1996) Continuing role of a frozen-tissue bank in molecular pathology. *Diagnostic Molecular Pathology* 5: 253–259

157 Jong S-C (1978) Conservation of the cultures. In: ST Chang, WA Hay (eds): *The Biology and Cultivation of Edible Mushrooms*. Academic Press, London, 119–135

158 Smith D (1988) Culture and preservation. In: DL Hawksworth, BE Kirsop (eds): *Living Resources for Biotechnology, Filamentous Fungi*. Cambridge University Press, New York, 75–99

159 Palmer WM (1974) Inexpensive jars for museum specimens. *Curator* 17: 321–324

160 Legler JM (1981) Inexpensive containers for large herpetological specimens. *Herpetological Review* 12: 9–10

161 De Moor FC (1990) Containers for wet collections – problems and solutions. In: EM Herholdt (ed): *Natural History Collections: Their Management and Value*. Transvaal Museum Special Publication No. 1, Transvaal Museum, Pretoria, 27–36

162 Hoebeke ER (1985) A unit storage system for fluid-preserved arthropods. *Curator* 28: 77–83

163 Holm-Hansen O (1973) Preservation by freezing and freeze-drying. In: R Stein (ed): *Handbook of Phycological Methods: Culture Methods and Growth Measurements*. Cambridge University Press, New York, 195–206

164 Jong S-C (1989) Microbial germplasm. In: L Knutson, AK Stoner (eds): *Biotic Diversity and Germplasm Preservation, Global Imperatives*. Kluwer Academic Publishers, Dordrecht, The Netherlands, 241–273

165 Stein JR (1973) *Handbook of Phycological Methods: Culture Methods and Growth Measurements*. Cambridge University Press, New York

166 Hall GS, Hawksworth DL (1990) *International Mycological Directory*. 2nd Edition. International Mycological Association, CAB International Mycological Institute, Oxon, UK

167 Arnett RH Jr., Samuelson GA (1986) *The Insect and Spider Collections of the World*. Fauna and Flora Publications, Gainesville, FL

168 Voss G. L (1976) *Seashore Life of Florida and the Caribbean*. E.A. Seamann Publishing, Miami, FL

169 Bland RG, Jaques HE (1978) *How to Know the Insects*. W. C. Brown, Dubuque, IO

170 Meinkoth NA (1981) *The Audubon Society Field Guide to North American Seashore Creatures*. Alfred A. Knopf, New York

171 Schultz GA (1969) *How to Know the Marine Isopod Crustaceans*. W.C. Brown, Dubuque, IO

172 Hayward PJ (1985) *Ctenostome Bryozoans*. Synopses of the British Fauna (New Series) No. 33. E.J. Brill, London

173 Sims RW (1980) *Animal Identification, A Reference Guide*; *Vol. 1, Marine and Brackish Water Animals*; *Vol. 2, Land and Freshwater Animals (not Insects)*; *Vol. 3, Insects*. British Museum (Natural History), London, and Wiley, Chichester

174 Huber JT (1998) The importance of voucher specimens, with practical guidelines for preserving specimens of the major invertebrate phyla for identification. *Journal of Natural History* 32: 367–385

175 Meester J (1990) The importance of retaining voucher specimens. In: EM Herholdt (ed): *Natural History Collections: Their Management and Value*. Transvaal Museum Special Publication No. 1, Transvaal Museum, Pretoria, 123–127

176 Robinson WH (1975) Taxonomic responsibilities of non-taxonomists. *Bulletin of the Entomological Society of America* 21: 157–159

177 Yates TL (1985) The role of voucher specimens in mammal collections: characterization and funding responsibilities. *Acta Zoologica Fennica* 170: 81–82

178 Dingerkus G (1982) Preliminary observations on acidification of alcohol in museum specimen jars. *Curation Newsletter* 5: 1–3

179 Tucker JW, Chester AJ (1984) Effects of salinity, formalin concentration, and buffer on quality and preservation of southern flounder (*Paralilchthys lethostigma*) larvae. *Copeia* 4: 981–988

180 Hughes GW, Cosgrove JA (1990) pH change in a formalin borax solution with inferences about uses of neutralized formalin in vertebrate collections. *Collection Forum* 6: 21–26

181 Owen G (1955) Use of propylene phenoxetol as a relaxing agent. *Nature (London)* 175: 434

182 Hardwick DF (1950) Preparation of slide mounts of lepidopterous genitalia. *Canadian Entomologist* 82: 231–235

183 Turner RD (1960) Mounting minute radulae. *The Nautilus* 73: 135–137

184 Gurney AB, Kramer JP, Steyskal GC (1964) Some techniques for the preparation, study, and storage in microvials of insect genitalia. *Annals of the Entomological Society of America* 57: 240–242

185 Smith EH (1979) Techniques for the dissection and mounting of the male (aedeagus) and female (spermatheca) genitalia of the Chrysomelidae (Coleoptera) *Coleopterists Bulletin* 33: 93–103

186 Mueller GJ (1972) *Field Preparation of Marine Specimens*. University of Alaska Museum, Fairbanks, AK

187 Lincoln RJ, Sheals JB (1979) *Invertebrate Animals: Collection and Preservation*. Cambridge University Press, Cambridge, UK

188 Hall ER (1962) Collecting and preparing study specimens of vertebrates. *University of Kansas Museum of Natural History Miscellaneous Publication* 30: 1–46

189 Wagstaffe R, Fidler JH (1968) *Preservation of Natural History Specimens, Vol. 2: Zoology/Vertebrates*. H.F. & G. Witherby Ltd., London

190 Savile DBO (1962) *Collection and Care of Botanical Specimens*. Canadian Department of Agriculture Publication 1113, Ottawa

191 Ketchledge EH (1970) *Plant Collecting: A Guide to the Preparation of a Plant Collection*. State University College of Forestry at Syracuse University, Syracuse, NY

192 Smith CE (1971) *Preparing Herbarium Specimens of Vascular Plants*. United States Department of Agriculture Information Bulletin 348, Washington, DC

193 Harris RH (1990) Zoological preservation and conservation techniques. *Journal of Biological Conservation* 1: 4–68

194 Knudsen JW (1972) *Collecting and Preserving Plants and Animals*. Harper and Row, New York

195 Herholdt EM (1990) *Natural History Collections: Their Management and Value*. Transvaal Museum Special Publication No. 1, Transvaal Museum, Pretoria

196 Stansfield G, Mathias J, Reid G (1994) *Manual of Natural History Curatorship.* Her Majesty's Stationery Office, London

197 Woodward SM, Hlywka WE (1993) A database for frozen tissues and karyotype slides. *Collection Forum* 9: 76–83

198 Zaldumbide K, Boulukos E, Pognonec P (2000) Virtual nitrogen tank to monitor frozen cell stocks. *BioTechniques* 29:122–126

199 McAllister DE, Murphy R, Morrison J (1978) The compleat minicomputer cataloging and research system for a museum. *Curator* 21: 63–91

200 Arnold TH (1990) Computerization of the curatorial and service functions of the National Herbarium, Pretoria. In: EM Herholdt (ed): *Natural History Collections: Their Management and Value.* Transvaal Museum Special Publication No. 1, Transvaal Museum, Pretoria, 81–87

201 Wingate LR (1990) Microcomputer use for mammal collection management at the Kaffrarian Museum. In: EM Herholdt (ed): *Natural History Collections: Their Management and Value.* Transvaal Museum Special Publication No. 1, Transvaal Museum, Pretoria, 89–104

202 Owen RD (1990) Database computerization and consortium development for vertebrate collections – a collection management perspective. In: EM Herholdt (ed): *Natural History Collections: Their Management and Value.* Transvaal Museum Special Publication No. 1, Transvaal Museum, Pretoria, 105–116

203 Lesser WH (1997) International treaties and other legal and economic issues relating to the ownership and use of genetic resources. In: KE Hoagland, AY Rossman (eds): *Global Genetic Resources: Access, Ownership, and Intellectual Property Rights. Beltsville Symposia in Agricultural Research 21.* Association of Systematics Collections, Washington, DC, 31–50

204 Stenson AJ, Gray TS (1999) *The Politics of Genetic Resource control.* St. Martins Press, London

205 Hoagland KE (1997) Access to specimens and genetic resources: an Associa-

tion of Systematics Collections position paper. In: KE Hoagland, AY Rossman (eds): *Global Genetic Resources: Access, Ownership, and Intellectual Property Rights. Beltsville Symposia in Agricultural Research 21.* Association of Systematics Collections, Washington, DC, 317–330

206 Holden C (1992) NSF bombs in ice capade. *Science* 255: 406.

207 Nicholson TD (1976) Systematics collections and the law. *Curator* 19: 21–28

208 Williams SL, Laubach R, Genoways HH (1977) A guide to the management of recent mammal collections. *Carnegie Museum of Natural History, Special Publication* 4: 1–105

209 Cato PS (1993) Institution-wide policy for sampling. *Collection Forum* 9: 27–39

210 Merritt E (1992) Conditions on outgoing research loans. *Collection Forum* 8: 78–82

211 Cato PS, Williams SL (1993) Guidelines for developing policies for the management and care of natural history collections. *Collection Forum* 9: 84–107

212 Hoagland KE (1994) *Guidelines for Institutional Policies and Planning in Natural History Collections.* Association of Systematics Collections, Washington, DC

213 Reid W (1997) Regulating the biotrade to promote the conservation and sustainable use of biodiversity. In: KE Hoagland, AY Rossman (eds): *Global Genetic Resources: Access, Ownership, and Intellectual Property Rights. Beltsville Symposia in Agricultural Research 21.* Association of Systematics Collections, Washington, DC, 303–315

214 Oring L, Able KP, Anderson DW, Baptista LF et al. (1988) *Ad Hoc* Committee on the Use of Wild Birds in Research, Report of the American Ornithologists' Union, Cooper Ornithological Society, Wilson Ornithological Society, Supplement to *Auk* 105

215 Baker S, Palumbi S (1994) Which whales are hunted? A molecular approach to monitoring whaling. *Science* 265: 1538–1539

216 Echandi CMR (1997) Biodiversity, research agreements, intellectual and

other property rights: the Costa Rican case. In: KE Hoagland, AY Rossman (eds): *Global Genetic Resources: Access, Ownership, and Intellectual Property Rights. Beltsville Symposia in Agricultural Research 21*. Association of Systematics Collections, Washington, DC, 213–218

217 Hoagland KE, Rossman AY (1997) *Global Genetic Resources: Access, Ownership, and Intellectual Property Rights. Beltsville Symposia in Agricultural Research 21*. Association of Systematics Collections, Washington, DC

Further reading

Dessauer, H. C. and Hafner, M. S. (1984). *Collections of Frozen Tissues: Value, Management, Field and Laboratory Procedures and Directory of Existing Collections*. Association of Systematics Collections, University of Kansas Press, Lawrence, KS.

This volume presents the proceedings of a Workshop on Frozen Tissue Collection Management, in which papers stressing the value of frozen tissue collections for studies in basic, applied and forensic sciences were presented, guidelines for collection management were discussed and a plan to promote and coordinate use, growth and funding of tissue collections was formulated. Contents include sections detailing the value of and need for, collections of frozen tissues, simple methods for collecting tissues under field conditions, knowledge on long-term stability of tissue components, U.S. federal regulations concerning tissue collection and transport and results of a worldwide survey of tissue collections holdings. This seminal volume is still of tremendous value to tissue collection managers, users and potential users and officials of all federal, state and private funding agencies. The directory of frozen tissue collections was updated in Dessauer et al. (1996), and recent websites are provided in Appendix 5 of this chapter.

Herholdt, E. M. (1990). *Natural History Collections: Their Management and Value*. Transvaal Museum Special Publication No. 1, Transvaal Museum, Pretoria.

This volume presents the proceedings of a symposium on the management and value of natural history collections held at the Transvaal Museum, Pretoria. Contributed papers deal with general aspects of the management of fluid collections, the conservation of natural history material, the computerization of collections, the importance of voucher specimens and the extensive uses of collections. Contents reflect the broad nature of collections management, the value of collections and the expertise involved in their curation. This volume serves both as a useful reference for museum curators and collections managers, by stimulating professionalism in collections management and as a primer on the value of natural history collections outside the scientific community.

Hoagland, K. E. and Rossman, A. Y. (1997). *Global Genetic Resources: Access, Ownership and Intellectual Property Rights. Beltsville Symposia in Agricultural Research 21*. Association of Systematics Collections, Washington, DC.

This volume presents the proceedings of the Beltsville Symposium on Global Genetic Resources, which explored issues relating to ownership of and access to genetic resources and biological specimens. Laws are being implemented in countries throughout the world that severely restrict access to these research resources, thereby affecting the ability of scientists to pursue their goal of providing knowledge to benefit the world's people. Contributed papers examine the status of the various treaties, national laws and agreements in effect around the world, present case studies demonstrating how research using international resources benefit the global community, explore models of the equitable use of genetic resources, and discuss potential solutions and mutually beneficial compromises. An Association of Systematics Collections position paper, based on the presentations, discussions, and an open session immediately following the symposium, is also presented.

Zimmer, E. A., White, T. J., Cann, R. L. and Wilson, A. C. (1994). *Methods in Enzymology Vol. 224, Molecular Evolution: Producing the Biochemical Data*. Academic Press, San Diego, CA.

This volume presents a compendium of excellent chapters, contributed by authorities, dealing with all aspects of data acquisition and analysis for studies in molecular evolution. Guidelines for the collection and storage of samples are arranged into chapters on land plants, vertebrates, invertebrates, fungi and algae. Each chapter presents methods for the collection and temporary storage of freshly collected tissues in the field, as well as suggestions for alternative sources of tissues. Procedures for the transportation, long-term storage and documentation of samples, including the collection and deposition of voucher specimens, are provided. Each chapter also offers general information on the regulations permitting collection and transportation of samples. Subsequent chapters present protocols for isolation of proteins and DNA from fresh and ancient samples, including advice on the selection of tissues.

12 DNA Isolation Procedures

Michele K. Nishiguchi, Phaedra Doukakis, Mary Egan, David Kizirian, Aloysius Phillips, Lorenzo Prendini, Howard C. Rosenbaum, Elizabeth Torres, Yael Wyner, Rob DeSalle and Gonzalo Giribet

Contents

Methods and Tools in Biosciences and Medicine
Techniques in molecular systematics and evolution, ed. by Rob DeSalle et al.
© 2002 Birkhäuser Verlag Basel/Switzerland

1   Introduction

Literally hundreds of protocols for DNA preparation from various sources of tissue
have been published over the last few decades. To display all of these preparations
would take volumes of manual space so instead we present in this chapter several
of the preparations that have been used successfully in our laboratories. We also
present a few "classical" procedures that are "tried and true" and nearly always
work. In addition the www is an excellent source for protocols. Some forums
exist for the dissemination of protocols for DNA and RNA isolation (DNA
isolation protocols forums: http://www.nwfsc.noaa.gov/protocols.html, http://
bric.postech.ac.kr/resources/rprotocol/; RNA isolation protocols forum: http://
www.nwfsc.noaa.gov/protocols/methods/RNAMethodsMenu.html).

Myriad permutations of the traditional phenol/chloroform extraction methods (e. g., [1–5]) are still in use because they reliably produce high-quality DNA.

For DNA fragment analysis [6, 7], we recommend using a cesium-chloride (CsCl)
gradient (which takes 3–4 days) to minimize the possibility of amplifying nuclear
mitochondrial sequences. Some investigators use salt-precipitation [8] before
phenol/chloroform extraction and others follow phenol/chloroform extraction
with further purification using a Centricon 30000 MW membrane (Amicon).

2   Materials

Organellar DNA
We include a CsCL preparation not merely out of tradition. Modern PCR
techniques have all but eliminated the need for CsCl gradient purification of
target DNA. The reason we include this procedure is that the CsCl gradient
method can also be used as a last resort when organellar DNA studies result in
the discovery of organellar DNA inserted into the nuclear genome. The CsCl
gradient can be used to purify the organellar DNA away from the inserted
organellar DNA that is contained within the nuclear genome, thus avoiding the
problem of spurious results from nu-mtDNA. These methods use more time, are
more susceptible to contamination because tubes are opened and closed more
frequently, and are unpleasant due to the exposure to toxic chemicals.

When separating organellar DNA from nuclear DNA, fluorescent dyes (either
Hoechst 33258 or Ethidium Bromide) are used to visualize the different types of
DNA using a CsCl gradient. Ethidium bromide (EtBr), similar to propidium

iodide, is an intercalating dye. Both dyes insert between the stacked purine and pyrimidine base pairs of double-stranded DNA. The intercalation of EtBr causes the DNA to become buoyant, resulting in DNA with a lower density (and therefore higher in the centrifuge tube) in a CsCl gradient. As supercoiled DNA binds with EtBr, it relaxes the supercoil, such that it rewinds in the opposite direction. During this rewinding, more EtBr is bound to the DNA, so that the strands cannot bind to each other, unless a nick is introduced into the strand to cause relaxation. Therefore, linear DNA can bind more EtBr than plasmid DNA and its buoyant density is less than plasmid DNA. This results in the characteristic plasmid gradient where the supercoiled plasmid DNA is below that of the linearized genomic or nicked DNA.

Contrasted to EtBr, Hoechst 33258 does not intercalate into the DNA. It interacts with the large groove of the DNA molecule by hydrogen bonding. This blue fluorescing dye interacts more with A and T nucleotides rather than G and C nucleotides. Hoechst dye also decreases the buoyant density of the DNA. Since many plastid and mitochondrial genomes have a much higher ratio of AT:GC residues, these genomes will bind more Hoechst dye than the nuclear DNA which characteristically has lower AT:GC ratios (although this will vary depending on what organism is being sampled). The lower density mitochondrial DNA (mtDNA) and plastid DNA (pDNA) will migrate higher during centrifugation and can be separated from the nuclear DNA in this manner.

Plants and Algae

Plant and algal DNA isolation also present particular problems that oftentimes require the use of the traditional methods. The isolation of nucleic acids from plants and algae differs from most modern and generic techniques used for animal tissues due to the cellular structure of plant material *versus* animal tissues. Plants and macroalgae have cell walls mostly comprised of cellulose or some other complex polysaccharide, and the degree to which they must be separated from the nucleic acid material (particularly RNA) depends on the intended use of the nucleic acids. For example, RNA that will eventually be used to make cDNA library material must be completely free of any complex polysaccharides which decrease the amount of mRNA yields following the initial separation. The use of herbaria-preserved material has also proven to be valuable for obtaining DNA from rare or unique specimens. We include a protocol in this chapter that has been quite successful at extracting DNA from preserved plant specimens.

Interestingly enough, plant and macroalgal material can yield large quantities of nucleic acids, primarily due to their large genome sizes. In macroalgae, nuclear DNA per cell varies over four orders of magnitude (200–0.2 pg) and in algal species with smaller genomes, there can be a 1000-fold difference between the size of the nuclear genome and that of the plastid genome [9]. Nuclear DNA can exist in one to four copies per cell, depending on what stage it is in its life history. However, specific sub-regions within the nuclear genome are comprised of identical tandem repeats of the same sequence. One example of this is the region of the nuclear rDNA that contains the genes for the large- and the small- subunit ribosomal RNAs plus spacer regions, both transcribed and non-transcribed regions. Certain angiosperm nuclear rDNA repeats can range from 9–12 kb in length and include large, non-transcribed regions that can vary even within a species (they are not conserved and they can vary in the number of tandem repeats [10].

Plants and macroalgae also contain two other genomes: mitochondrial and plastid. In both terrestrial plants and macroalgae, the plastid DNA genome is a double-stranded DNA circle, which contains the genes for plastid rRNA, tRNA and some other proteins. Plant genomes can range from 120–217 kilobases [11], in contrast to macroalgae, which range from 73 to over 400 kb. Plastid genomes of angiosperms have two copies of the region containing the ribosomal DNA (rDNA) genes, and are on opposite sides of the circular DNA molecule, but in reverse orientation (called an inverted repeat). Macroalgae may contain variations of this organization, with strings of tandem repeats of the rDNA, or have only a single copy of the ribosomal gene [10].

Eukaryotic mitochondrial genomes have a much greater variety than plastid genomes in size. While metazoan mitochondrial genomes are small circles (16–40 kb) of conserved gene content [12, 13], angiosperm mitochondrial genomes are some of the largest known (200–>2000 kb), with the greatest variation even within a genus [14]. Even certain species have several sizes of mitochondrial DNA molecules within an individual. Macroalgae, in contrast, have small mitochondrial genomes ranging from 100–500 kb pairs. They are found in both circular and linear forms, similar to fungi and protozoa. As with plastid DNA, there are identical copies of the mitochondrial DNA per cell, since each mitochondrion has one or more copies. Similar to metazoans, uni-parental inheritance is observed [15].

Microscopic Organisms
Because many of the traditional DNA isolation preparations for animals were originally developed for vertebrates and insects, microscopic organisms such as protozoa and extremely small animals pose difficult problems for DNA isolation. Because of the wide range of animals and microscopic organisms, we will focus on several protocols that have been developed for rapid and efficient isolation of DNA.

Isolation of DNA from museum-preserved specimens has always been diffi-cult due to the nature of the liquids in which specimens are preserved. Previous fixatives like formaldehyde and other aldehyde mixtures work well in preser-ving macro- and ultra- structural components of the specimens. Unfortunately, because these types of chemicals bind tightly to the tissue matrix as well as to the nucleic acids retained inside them, processing the material for DNA can be difficult. Also, specimens may have initially been preserved in formalin or some other strong fixative, and then eventually placed in ethanol, which produces a sample from which it is difficult to obtain and therefore isolate DNA. The method described here for these types of "fixed" specimens may not always be fruitful, but in many instances has proved to be successful for obtaining PCR templates of various loci used in phylogenetic analysis [16]. Due to the special nature of ancient specimens we dedicate an entire section to that subject.

Solutions
In this section we explain how to make the common stock solutions for Molecular Biology laboratories in the protocol boxes in this chapter. Other specific solutions are described in the relevant sections of this manual.

- 5 M ammonium acetate (NH_4Ac)
Dissolve 335 g of ammonium acetate (CH_3COONH_4; M.W. 77.08) in 800 mL of H_2O. Adjust the volume to 1 liter with H_2O.
Sterilize by filtration.

- 1 M ammonium sulfate
Dissolve 132.14 g of ammonium sulfate [$(NH_4)_2 SO_4$; M.W. 132.14) in 800 mL of H_2O. Adjust the volume to 1 liter with H_2O.

- BCIP (5-bromo-4-chloro-3indolyl-phosphate, Boehringer-Mannheim) solution
Dissolve BCIP to a concentration of 50 mg/ml in dimethyl formamide.
Store aliquots at −20°C.

- Carlson lysis buffer
For 100 mls:
100 mM Tris, pH 9.5 (1.21g Tris base in 70 ml ddH2O, pH to 9.5 with HCl)
20 mM EDTA (0.76g)
1.4 M NaCl (8.18g)
2% CTAB (2 g)
1% PEG 8000 or 6000 (1 g)
Stir overnight. Bring up to 100 ml volume. Add 2μl β-mercaptoethanol per 1 ml of buffer just before use.

- Chelex® 100 buffer
Chelex 100 buffer-0.001 M Tris, pH 8.0
0.05 mM EDTA
5% (w/v) Chelex® 100 resin

- 2x CTAB (cetyltrimethylammonium bromide) extraction buffer
100 mM Tris-HCl (pH 8.0)
1.4 M NaCl
20 mM EDTA
2% (w/v) CTAB
0.1% (w/v) PVPP (polyvinyl polypyrrolidine)
0.2% (v/v) B-mercaptoethanol (add directly before use, but do not store above with this).

- CTAB/NaCl solution (10% CTAB, 0.7 M NaCl)

Dissolve 10 g of CTAB in 80 mL of 0.7 M NaCl solution. Stir on a hot magnetic stirrer until the CTAB has dissolved. Adjust volume to 100 mL with 0.7 M NaCl solution.

- DEPC (diethyl pyrocarbonate) – treated H_2O

0.5% DEPC in H_2O, stir vigorously in a fume hood for an hour and let sit for 2 hours prior to autoclaving.

- Digestion buffer

100 mM NaCl
10 mM Tris pH 8.0
25 mM EDTA
Prepare aliquots of 100 mL

- DMSO (dimethylsulfoxide) buffer

From the new Sambrook & Russell [30] Purchase a high grade DMSO (HPLC grade or better). Divide the contents of a fresh bottle into 1 mL aliquots in sterile tubes. Close the tubes tightly and store at –20°C. Use each aliquot only once and then discard.

- DTAB (dodecyltrimethylammonium bromide) solution

8% DTAB
1.5 M NaCl
100 mM Tris (pH 8.8)
50 mM EDTA

- 0.5 M EDTA pH 8.0 (ethylenediaminetetraacetic acid)

Add 168.1 g of disodium ethylenediaminetetraacetate·$2H_2O$ to 800 mL of H_2O. Stir vigorously on a magnetic stirrer. Adjust the pH to 8.0 with NaOH (approximately 20 g of NaOH pellets). Adjust the volume to 1 liter with H_2O.
NOTE: The disodium salt of EDTA will not go into solution until the pH is adjusted to approximately 8.0.
Sterilize by autoclaving.

- Ethidium bromide solution (10 mg/mL)

Add 200 mg of ethidium bromide to 20 mL of H_2O.
Stir on a magnetic stirrer for several hours to ensure that the dye has dissolved. Store in a light-proof container (e. g., in a falcon tube wrapped in aluminum foil) at room-temperature.

- Glycine

2mg/ml of glycine in PBT.
Store at –20°C.

- 5 M guanidinium thiocyanate

Dissolve 59 g of guanidinium thiocyanate to 100 mL of H_2O. Heat at 65°C until dissolved in a water bath. Filter the solution through a Whatman No. 1 filter or equivalent (i. e., a Nalgene Filtration Unit).

- Guanidinium iso thiocyanate (GITC) homogenization buffer

4 M guanidinium iso thiocyanate ($CH_5N_3 \cdot CHNS$; M.W. 118.16)

0.1 M Tris-HCl (pH 7.5)

1% β-mercaptoethanol

Dissolve 50 g of guanidinium iso thiocyanate in 10 mL of 0.1 M Tris-HCl (pH 7.5) and RNAse-free, DNAse-free H_2O to 100 mL. Heat at 65°C until dissolved in a water bath. Filter the solution through a Whatman No. 1 filter or equivalent (i. e., a Nalgene Filtration Unit).

To avoid filtration, the solution may be prepared for half volume (25 g of guanidinium iso thiocyanate in 5 mL of Tris-HCl, and H_2O to 50 mL) in a sterile 50 mL Falcon tube using sterile spatulas for pouring the guanidinium iso thiocyanate, and sterile reagents.

This solution is stable and can be stored indefinitely at room temperature. Just before use, add β-mercaptoethanol to a final concentration of 1% (0.14 M).

- HM

1 mM $CaCl_2$

1.5 mM $NaHCO_3$

0.1 mM $MgCl_2$

0.08 mM $MgSO_4$

0.03 mM KNO_3

in Arrowhead spring H_2O

- Liftons buffer

100 mM EDTA

25 mM Tris-HCl pH 7.5

1% SDS

- 4M lithium chloride (LiCl)

Dissolve 169.56 g of lithium chloride (LiCl; M.W. 42.39) in 800 mL of H_2O, and adjust the volume to 1 liter with H_2O.

- 1 M Levamisole

Dissolve 60 mg of levamisole in 250 µl of RNase-free H_2O.

Make fresh stock for every use.

- Lysis buffer

100 mM EDTA

10 mM Tris-HCl pH 7.5

- MAB

100 mM maleic acid

150 mM NaCl, pH 7.5

- 1 M magnesium chloride (MgCl$_2$)

Dissolve 203.31 g of magnesium chloride hexahydrate (MgCl$_3$·6H$_2$O; M.W. 203.31) in 800 mL of H$_2$O, and adjust the volume to 1 liter with H$_2$O.

Dispense in aliquots and sterilize by autoclaving.

NOTE: MgCl$_2$ is extremely hygroscopic. Buy small bottles and do not store opened bottles for long periods of time.

- MOPS [3-(N-morpholino)propanesulfonic acid] buffer

0.1 M MOPS (pH 7)

0.5 M NaCl

0.1% Tween-20.

- NBT (4-nitroblue tetrazolium chloride, Boehringer Mannheim)

Dissolve NBT at 75mg/ml in 70% dimethyl formamide.

Store aliquots at −20°C.

- NTMT

100 mM NaCl

100 mM Tris HCl (pH 9.5)

50 mM MgCl$_2$

0.1% Tween-20

2 mM levamisole (add on day of use).

- 4% Paraformaldehyde

Dissolve 10 g paraformaldehyde in 200 ml of DEPC-treated H$_2$O at 65°C in a fume hood and cool on ice. Adjust pH to 7.5 with 5–10 μl NaOH. Add 25 ml 10× PBS and make volume up to 250 ml with DEPC treated H$_2$O.

Aliquots can be stored for several months at −20°C.

- PCI (phenol:chlorophora:isoamyl alcohol)

This is a solution of phenol, chloroform, and isoamyl alcohol, in a ratio of 25:24:1.

We recommend the use of commercially mixed PCI at a pH of 7.5–8.0, which avoids the hassle of handling phenol solutions.

- 5 M potassium acetate (KOAc) pH 7.5

Dissolve 49.1 g of potassium acetate (CH$_3$COOK; M.W. 98.15) in 90 mL of RNAse-free, DNAse-free H$_2$O. Adjust the pH to 7.5 with 2 M acetic acid. Adjust the volume to 100 mL with H$_2$O.

Aliquot and store at −20°C.

- 10× PBS (phosphate-buffered saline solution)

Dissolve 80 g of sodium chloride, 2 g of potassium chloride, 14.4 g of sodium phosphate, and 2.4 g of potassium phosphate in 800 mL of H_2O. Adjust the pH to 7.4 with HCl. Adjust the volume to 1 L with H_2O.
Aliquot and sterilize by autoclaving on liquid cycle.
Store at room temperature.

- PBT (phosphate buffered saline, Triton X-100)

0.1% of Tween-20 in 1×PBS buffer.

- 10× PCR buffer

Dissolve 8.116 g of Tris Base (Molecular biology grade), 0.610 g of $MgCl_2$, and 2.227 g of ammonium sulfate in 90 mL of HCl. Stir until dissolved. Adjust the volume to 100 mL with H_2O.
Sterilize by autoclaving and aliquot in 1 mL eppendorf tubes.

- PCR buffer w/non-ionic detergents

50 mM KCl
10 mM Tris-HCl, pH 8.3
2.5 mM $MgCl_2$
0.1 mg/mL gelatin
0.45% NP-40
0.45% Tween-20
Prepare aliquots of 10 mL in sterile 10 mL falcon tubes using sterile reagents since this buffer cannot be filtered or autoclaved. Use 0.5 mL of 1 M potassium chloride, 0.1 mL of 1 M Tris-HCl (pH 8.3), 25 μL of 1 M magnesium chloride, 1 mg of gelatin, 450 μL of NP-40, and 450 μL of Tween-20 and bring the volume up to 10 mL in RNAse-free, DNAse-free d H_2O.

- PK buffer

Dilute 10 ml 1M Tris-HCl and 2 ml 0.5 M EDTA in RNase-free H_2O, and adjust volume to 200 ml with RNase free H_2O.

- 1 M potassium chloride (KCl)

Dissolve 7.45 g of potassium chloride (KCl; M.W. 74.55) in 80 mL of H_2O, and adjust volume to 100 ml with H_2O.

- Proteinase K (20mg/mL)

Use at a concentration of 50–60 μg/mL with a reaction buffer containing 0.01 M Tris (pH 7.8), 0.005 M EDTA, and 0.5% SDS.
Incubate at 37–56°C
Store Proteinase K at –20°C.

- RNA preparation binding buffer

0.5M NaCl

10mM TrisCl pH 7.4
1mM EDTA pH 8.0,
0.1% SDS
Mix appropriate amounts of RNAse-free stock solutions of Tris-HCl, NaCl and EDTA. Autoclave the mixture and allow to cool to 65°C. Then add appropriate amount of SDS from a 10% stock solution.

• RNA preparation elution buffer
10 mM TrisCl pH 7.4
0.1 mM EDTA pH 8.0
Mix appropriate amounts of RNAse free stock solutions of Tris-HCl and EDTA. Autoclave the mixture and allow to cool to 65°C.

• RNA preparation washing buffer
0.1M NaCl
10mM TrisCl pH 7.4
1mM EDTA pH 8.0
0.1% SDS
Mix apprpriate amounts of RNAse free stock solutions of Tris-HCl, NaCl and EDTA. Autoclave the mixture and allow to cool to 65°C. Then add appropriate amount of sodium lauryl sulfate sarcosinate from a 10% stock solution.

• 5% sarkosyl buffer
Dissolve 25 g of sarkosyl in 15 mL of 5 M sodium chloride, 25 mL 1 M Tris-HCl (pH 8.0), 15 mL 0.5 M EDTA, and 400 mL of H_2O. Adjust the volume to 500 mL with H_2O. Do not refrigerate or autoclave.

• 4% SDS buffer
Dissolve 20 g of SDS (Sodium Dodecyl Sulfate; also called Sodium Lauryl Sulfate) in 30 mL of 5 M sodium chloride, 25 mL 1 M Tris-HCl (pH 8.0), 100 mL 0.5 M EDTA, and 300 mL of H_2O. Adjust the volume to 500 mL with H_2O. Do not refrigerate or autoclave.

• 10% SDS
Dissolve 10 g of SDS (Sodium Dodecyl Sulfate) in 80 mL of H_2O, and adjust the volume to 100 mL with H_2O.

• SDS/Sarkosyl lysis buffer
3 volumes of 4% SDS buffer
1 volume of 5% Sarkosyl Buffer
5 µg/mL proteinase K
0.12% of β-mercaptoethanol
Add 200 µL of Proteinase K (20mg/mL) and 100 µL of β-mercaptoethanol to 60 mL of 4% SDS buffer, and 20 mL of 5% sarkosyl buffer.
Do not refrigerate or autoclave.

- 20× SSC (pH 4.5)
3 M NaCl
0.3 M sodium citrate

- Silica solution
Weigh out 4.8 grams silica dioxide into a 50 ml polypropylene tube
Add 40 ml distilled water and agitate
Let stand 24 hrs
Pipette 35 ml off
Bring up to 40 ml with distilled water and agitate
Let stand 5 hrs
Pipette 36 ml off
Add 48 ml HCl
Aliquot into 1.5 ml tubes and store in dark

- 3 M sodium acetate (NaOAc) (pH 5.2 and 7.0)
Dissolve 40.82 g of sodium acetate trihydrate ($CH_3COONa \cdot 3H_2O$; M.W. 136.08) in 80 mL of H_2O. Adjust the pH to 5.2 with glacial acetic acid or adjust the pH to 7.0 with dilute acetic acid. Adjust the volume to 100 mL with H_2O.
Sterilize by autoclaving.

- 2 M sodium acetate (pH 4.0)
Dissolve 27.22 g of sodium acetate trihydrate ($CH_3COONa \cdot 3H_2O$; M.W. 136.08) in 80 mL of H_2O. Adjust the pH to 4.0 with glacial acetic acid. Adjust the volume to 100 mL with H_2O.
Sterilize by autoclaving.

- 5 M sodium chloride
Dissolve 292.2 g of sodium chloride (NaCl; M.W. 58.44) in 800 ml of H_2O. Adjust the volume to 1 liter with H_2O.
Sterilize by autoclaving.

- Solution I
50 mM glucose
25 mM Tris-HCl (pH 8.0)
10 mM EDTA (pH 8.0)
Solution I can be prepared in batches of 100 mL. Autoclave for 15 minutes on liquid cycle. Store at 4°C.

- STE (Sodium Chloride-Tris-EDTA) buffer
0.1 M NaCl
10 mM Tris-HCl (pH 8.0)
1 mM EDTA (pH 8.0)

- 50× TAE (Tris-Acetate Buffer)

Dissolve 242 g of Tris base, 57.1 mL of glacial acetic acid, and 100 mL of 0.5 M EDTA (pH 8.0) in H_2O up to 1 liter.

The 50× TAE is the concentrated stock solution. Use 1× TAE as working solution (0.04 M Tris-acetate, 0.001 M EDTA).

- 5× TBE (Tris-Borate/EDTA buffer)

Dissolve 54 g of Tris base, 27.5 g of boric acid, and 20 mL of 0.5 M EDTA (pH 8.0) in H_2O up to 1 liter. Stir until dissolved.

The 5× TBE is the concentrated stock solution. Use 0.5× TBE (0.045 M Tris-borate, 0.001 M EDTA) as electrophoresis buffer.

NOTE: A precipitation forms when concentrated solutions of TBE are stored for long periods of time. Discard any batches that develop a precipitate.

NOTE: 10× TBE buffer is commercially available, and it constitutes a good solution for laboratories not using much TBE buffer, or for laboratories with high budgets.

- TBST solution

135 mM NaCl

2.7 mM KCl

25 mM Tris HCl (pH 7.5)

0.1% Tween-20

2 mM levamisole (add on day of use).

- TE Buffer Solution (pH 7.4, 7.6, 8.0)

10 mM Tris-HCl, pH 7.4

1 mM EDTA, pH 8.0

10 mM Tris-HCl, pH 7.6

1 mM EDTA, pH 8.0

10 mM Tris-HCl, pH 8.0

1 mM EDTA, pH 8.0

- 10 mM Tris-HCl (pH 7.4, 7.6 and 8.0)

Dissolve 121.1 g of Tris base in 800 ml of H_2O. Adjust pH to the desired value by adding concentrated HCl.

pH 7.4 add 70 ml

pH 7.6 add 60 ml

pH 8.0 add 42 ml.

Other pHs desired can be obtained by titrating the HCl.

Adjust the final volume of the solution to 1000 ml with H_2O.

3   Methods

3.1   Kits

Various commercially available DNA extraction kits and systems are becoming increasingly popular because of their ease of use, limited labor, and ability to consistently produce high-quality DNA. Because of proprietary considerations of the manufacturer, the composition of some components in these kits is not revealed to the user. We describe a few below that we have used with success in our laboratories. Not all kits are economically favorable to use. Qiagen (www.qiagen.com) produces several kits (e. g., QIAamp DNA Mini Kit #51304, #51306; DNeasy Tissue Kit #69504, #69506) which seem to be economically feasible for extracting large numbers of samples. These kits contain a stable proteinase K solution, unidentified non-phenol/chloroform buffers, a silica-gel-membrane spin-column to isolate and purify high-quality DNA for PCR, and easy-to-follow instructions. These kits cost approximately $1.50 to $1.80 per sample. Other kits, which are expensive but adequate for large numbers of samples include: Nucleospin (www.clonetech.com) and Isoquick (Microprobe Corp.). Bio 101 (www.bio101.com) sells a Kit which employs special tubes containing a ceramic lysing matrix, a machine that violently shakes samples (twelve at a time), and a subsequent purification process. This system is ideal for small samples or organisms with sturdy cell walls like plants.

3.2   The Generic DNA Preparation

Most DNA preparation protocols outline the steps involved in three major phases of DNA isolation. The first phase (in *italic*) contains preparatory procedures and the solutions used during these steps. We also include in this phase of the preparation any steps that involve maceration or dispersion of tissue. The second phase of the protocols (in **bold**) is the actual isolation steps required to separate the nucleic acids from the rest of the cellular proteins and debris. The third phase (in normal font) in these preparations includes steps required to purify nucleic acids from impurities that will interfere with subsequent enzymatic manipulations. It should be noted that when protocols fail, the phenol/chloroform prep will usually work on any remainder of the sample. Resuspension of the isolated DNA should be accomplished so that a DNA collection is standardized for stock concentration. We routinely attempt to resuspend our samples to a concentration of 1μg DNA/μl of solution. The DNA is resuspended in dH$_2$O or in TE Buffer Solution. Dilutions of the stock solutions can then be made to accommodate the concentrations used for PCR or cloning purposes.

Since some of the preparations we present in this chapter are organism-dependent, we list them in a phylogenetic context. Next to the major groups of organisms in Figure 1 we list the protocols that work best for that group. This does not mean that a preparation listed for bacteria will not work on other organisms; some mixing and matching of protocols can be accomplished to maximize efficiency and yield of DNA preparation.

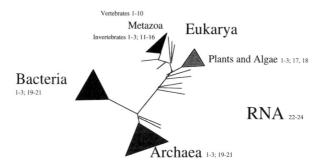

Figure 1 Highly condensed "tree of life" showing the various DNA isolation techniques listed in this chapter next to the group of organisms most appropriate for the protocol.

4   Protocols

Protocol 1   Traditional phenol/chloroform extraction for vertebrates or invertebrates.

(Time: 2 days)

1. *Different starting materials will require different maceration and dispersion methods.*
2. *Liquid nitrogen powdering is a common method for dispersing tissues where large starting amounts are available.*
3. *Smaller amounts of starting material can be dounced with a commercial homogenizer or with a homemade douncer.*
4. *The macerated dispersed material is placed in an extraction buffer in a ratio of 3 extraction buffer volumes to 1 sample volume. A common extraction buffer is the Lifton buffer, among many others.*
5. *Add 1/50 volume of Proteinase K (10 mg / mL) and incubate the sample at 65°C for 2 hrs to overnight.*
6. *Add 1/10 volume of 5 M potassium acetate and mix well by inverting. Incubate on ice for 30 min.*
7. *Spin tubes at 10,000 rpm for 10 min and transfer supernatant to new tube.*

8. **Add an equal volume of PCI.**

9. **Mix the solution by inverting the tube several times.**

10 **Spin tubes at 3000 to 5000 rpm for 5 min. The upper aqueous phase containing nucleic acids and the lower organic phases should separate. Transfer upper phase to a new tube.**

11. **If a sample is especially rare, a back extraction can be performed on the phenol phase by adding an equal volume of TE buffer solution and mixing, followed by spinning tubes at 3000 to 5000 rpm. The aqueous phase from this separation and the aqueous phase from the previous spin can then be combined.**

12. **Add an equal volume of chloroform to the tube and mix by inverting the tube several times.**

13. **Spin the tubes again at 3000 to 5000 rpm for 5 min.**

14. **Remove upper aqueous layer to transfer a new tube.**

15. Add two volumes of ice-cold 95–100% ethanol, invert the tubes several times and allow the DNA to precipitate at −20°C for 10 min.

16. The DNA can be collected by spinning the solution at top speed in a microcentrifuge.

17. The ethanol is gently poured off and the pellet is resuspended in 0.5 mL of TE buffer solution.

18. 0.3 ml of 5M ammonium acetate and 1 ml of 100% ethanol are added to the resuspended DNA for a second precipitation. The solution can be placed at −20°C for 10 min.

19. The tubes are spun at top speed in a microcentrifuge and the ethanol supernatant is poured off gently.

20. The resulting pellet is resuspended in TE buffer solution or dH_2O.

Protocol 2   Crude total cellular miniprep

(Time: 3 hours)

This procedure is highly recommended when large numbers of individuals are to be processed for systematic or population genetic studies.

1. *The equivalent of no more than 100 µL of tissue is placed in a microcentrifuge tube.*

2. *Add 200 µL of grinding buffer and dounce the tissue using a commercial or homemade dounce.*

3. *Add 200 µL of mini lysis solution.*

4. *Place tubes at 65°C for 30 min.*

5. **Add 60 µL of 5M potassium acetate and mix the solution by inverting.**

6. **Place tubes on ice for 30 min.**

7. **Spin tubes at max speed in a microcentrifuge and transfer supernatant to a new tube.**

8. Add 1 mL of ice-cold 95–100% ethanol and mix solution by inverting. Let tubes sit for 30 min at room temperature

9. Pellet the DNA by spinning at maximum speed in a microcentrifuge and remove ethanol supernatant

10. Resuspend pellet in 100 µL of 0.5M ammonium acetate (NH_4OAc).

11. Add 400 µL of 70% ethanol and mix the tubes by inverting. Let tubes sit at room temperature for 10 min.

12. Pellet the DNA by spinning at maximum speed in a microcentrifuge for 10 min.

13. Pour off ethanol and allow pellet at bottom of tube to air dry for 30 min to one hour.

14. Resuspend the pellet in 50 to 100 µL of dH_2O or TE buffer solution.

Protocol 3   Separation of nuclear and organellar DNAs using cesium chloride gradients

(Time: 2 to 3 days)

This protocol describes steps after obtaining high molecular weight nucleic acids most typically through phenol chloroform preps.

1. **Measure the volume of solution of total DNA that will be used for the cesium chloride gradients. For every mL of solution, add 1 g of finely ground cesium chloride (using a mortar and pestle) to the solution slowly, keeping it warm by placing the tube in one's hands. Agitate the tube gently.**

2. **While the cesium chloride is dissolving, add 15 mL of ethidium bromide (10 mg/mL stock) to a Beckman 13 mL quickseal tube.**

3. **Add the DNA/cesium chloride solution to the quickseal tube till the volume reaches the lower part of the neck of the seal.**

4. **Balance tubes, and seal.**

5. **Centrifuge at 40,000 rpm for 40–48 hrs. Once centrifugation is complete, carefully remove tubes (do not disturb the gradient!) and place in a rack.**

6. **Use a small, hand-held UV lamp to visualize the bands. There should be at least 4 bands in the tubes. The lowest density band (and therefore the highest one) is the mitochondrial DNA, the next one is the plastid DNA. The third lower band is sheared rDNA and the bottom band (and this should contain the most DNA) is the nuclear DNA.**

7. **Using a large gauge needle (18 is okay), puncture a hole next to the fraction that is to be isolated. If all 3 fractions are to be isolated, then start with the topmost band (mitochondrial) and work down the layers.**

8. **Slowly draw up the band until everything is collected, and place in separate collection tubes.**

9. Extract DNA with 100% iso-butanol two to three times to remove the ethidium bromide by mixing an equal volume of iso-butanol and inverting the tubes several times. The aqueous phase (top) containing the DNA should lose the pink color of ethidium bromide with each wash. Extract one extra time after all ethidium bromide appears to be removed from aqueous phase.

10. After collecting all 3 types of DNAs, the solutions must be dialyzed in TE buffer solution (usually several changes of 1–2 L overnight) at 4°C. This will remove all the cesium chloride.

11. After dialysis, the DNA can be used for several techniques (PCR, restriction digest, Southern and Northern blot analysis, genomic libraries). This solution must be kept at 4°C (freezing the DNA will shear the longer strands).

Protocol 4   Standard vertebrate isolation protocol using CHELEX®

(Time: approximately 15 min)

Chelating resin (available from Sigma as Chelex, [17]) works by denaturing proteins and removing inhibitory ions [18]. There are several reasons to prefer this method over others. In as little as 12 min, one can have DNA suitable for PCR. The risk of contamination resulting from opening and closing of tubes is greatly reduced over other methods, because the tube is opened just one time to add the tissue sample. It is inexpensive ($1/gm), which is important when processing large numbers of samples. It requires little tissue; all that is needed for a successful reaction (300 µL), is a piece of tissue approximately the size of the period at the end of this sentence. Chelex samples work with a variety of tissue preparations including blood, muscle and liver. Finally, Chelex is a very successful method; only a small percentage of reactions do not work with this preparation. Following is a minimum-step method:

1. *Add a small piece of tissue (< 1 mg) to 300 µL of 5% Chelex® (5% weight to volume chelating resin in ultrapure water).*
2. *Incubate at 100°C for 12 min.*
3. *Vortex once or twice during incubation.*

4. **Use 0.8–1.0 µL in a 10–20 µL PCR reaction.**

5. Store at –20 or –80°C.
6. Some of the modifications of the Chelex® protocol include: a longer or shorter incubation time (e.g., 10–15 min); more complicated incubation profiles: a) 100°C for 10 min, then 70°C for 20 min; b) 5 hr at 56°C; vortex; then 15 min at 95°C; 20% Chelex; boil for 8 min; vortex; spin.
7. Brief homogenization of tissue in DMSO buffer before adding it to the Chelex improves extraction and PCR amplification.
8. Spin the samples briefly (e.g., 5 min at 13,000 rpm) before drawing off supernatant for PCR.

9. In each case, reactions can be left at room temperature overnight after incubation. Vortexing one or more times after the sample has had time to extract seems to be essential for better PCR.

10. A common problem with reactions that do not produce a PCR product on the first attempt is too much DNA. This is often remedied by using some portion of the supernatant in a subsequent "re-chelexing" reaction. A disadvantage of this method is that the DNA product is not readily quantifiable using ethidium-bromide-infused agarose gels because of the small amount of DNA recovered.

Protocol 5   Isolation of DNA from museum-preserved specimens-formalin

(Time: 3 hours to 1 day)

1. *Initially, the specimen or tissue sample is placed in a TE buffer to rid the tissue of any remaining fixative.*

2. *This pre-incubation can be from one hour to overnight, depending on the specimen size and the type of fixation that it has undergone (if it comes directly from formalin, or if it has been preserved in ethanol following formalin fixation).*

3. *Place each sample in an eppendorf tube, add 500 µL of STE buffer containing 0.2% SDS and 250 µL of 10 M ammonium acetate.*

4. *Place tube in a heating block at 55°C to preheat.*

5. **Place a small amount of tissue (the size of a match head) into each tube and grind the tissue with a sterile teflon eppendorf grinder (Kontes).**

6. **After tissue has been ground, return to the heating block and incubate for 1 hour.**

7. **Centrifuge tubes in a microcentrifuge at 14,000 rpm for 5 min to pellet the cell debris and precipitated proteins.**

8. Transfer supernatant to a new tube and add 2 volumes of ice-cold 100% ethanol.

9. Mix gently by inverting tubes.

10. Place tubes at –20°C overnight, or at –80°C for 1 hour until DNA precipitates.

11. Centrifuge tubes at 4°C at 14,000 rpm for 15 min.

12. Remove supernatant and add the same volume of cold 70% ethanol.

13. Spin tubes at 4°C at 14,000 rpm for 10 min.

14. Pour off ethanol and dry tubes completely.

15. Resuspend the pellet in 50 µL of TE buffer overnight at 4°C or for 30 min at 40°C.

16. Use between 1–5 µL of solution for PCR reactions (depending on PCR reaction volume).

Protocol 6   Enriched cytoplasmic nucleic acid preparation from animals

Minced, ground, powdered or dounced tissue is brought up to 5 mL using any homogenization buffer.

1. *Add 100 μL of 10% stock of Nonidet P-40 or Triton X-100 and incubate on ice for 5 min.*
2. *Spin solution at 2,500 rpm in SS34 rotor. Recover supernatant, as nuclei will pellet.*
3. *Several low speed (2,500 rpm) spins may be necessary to remove majority of nuclei.*

4. **Add an equal volume of PCI.**
5. **Mix the solution by inverting the tube several times.**
6. **Spin tubes at 3,000 to 5,000 rpm for 5 min. The upper aqueous phase containing nucleic acids and the lower organic phases should separate.**
7. **If a sample is especially rare, a back extraction can be performed on the phenol phase by adding an equal volume of TE buffer and mixing, followed by spinning tubes at 3,000 to 5,000 rpm. The aqueous phase from this separation and the aqueous phase from the previous spin can then be combined.**
8. **Add an equal volume of chloroform to the tube and mix by inverting the tube.**
9. **Spin the tubes again at 3,000 to 5,000 rpm for 5 min.**
10. **Remove upper aqueous layer.**

11. Add two volumes of ice-cold 95–100% ethanol and allow the DNA to precipitate at –20°C for 10 min.
12. The DNA can be collected by spinning the solution at top speed in a microcentrifuge.
13. The ethanol is gently poured off and the pellet is resuspended in 0.5 mL of TE buffer.
14. 0.3 mL of 5 M ammonium acetate and 1 mL of 95–100% ethanol are added to the resuspended DNA for a second precipitation. The solution can be placed at –20°C for 10 min again.
15. The tubes are spun at top speed in a microcentrifuge and the ethanol supernatant is poured off gently.
16. The resulting pellet is resuspended in TE buffer or dH$_2$0.

Protocol 7   Plucked feathers using CHELEX®

1. *Using a sterile razor blade, cut off approximately 5 mm from the base of a plucked feather (calamus)*

2. **Place in a 0.8 mL tube containing 250 μL sterile 5% Chelex® (Bio-Rad).**

3. Incubate at 100°C for 15 min; vortex 2x's for 15 sec during incubation period .

4. Allow solution to cool to room temperature and spin for 30 sec at maximum speed.

5. Transfer supernatant to fresh, sterile 0.8 mL tube.

6. Store at 4°C, or at −20°C for long-term storage.

Protocol 8   Preparation for caviar and other fish tissues.

This method works well for fish tissue samples, fin snips, and eggs, even when the latter are processed (salted and pasteurized).

1. *Place a single egg or small piece of tissue in about 200 μL of Lifton buffer and add 20 μL of a 20 mg/ml proteinase K solution. If an egg sample is being processed, use the pipette tip to crush the egg. Incubate at 55–65°C for a minimum of one hour with gentle rocking. The yield of the extraction increases with the amount of incubation time. For caviar that is degraded or spoiled, or suspicious tissue samples, incubate overnight. A general rule is that once the tissue is completely dissolved, you can proceed to the next step.*

2. **Add 300 μL of phenol and 25 μL 5M potassium acetate, pH 4.8 to the above solution (after one hour to overnight incubation) and shake vigorously. Allow about 5–10 min incubation time with gentle rocking. Spin 14,000 rpm for 5 min. At this point, the aqueous phase is on top. The whitish material at the interface contains denatured proteins and carbohydrates and should be avoided. Remove top layer (aqueous phase) and transfer to new tube. Save the remaining solution.**

3. **Add 300 μL of chloroform and gently mix, incubate, and spin as above in step 2. If separate layers are not evident after spinning, the wrong phase was removed during step 2 (occasionally the phases flip because the DNA is heavier than the phenol). If so, use the saved solution from step 2 and repeat step 3 (i.e., add the chloroform to this layer, mix, incubate, and spin). Remove top layer (aqueous phase) and transfer to a new tube. Again, the interface should be avoided.**

4. **Add 1 mL of 100% ethanol to the aqueous phase removed in step 3 and mix. Place at −20°C for a minimum of 30 min. The extraction can also be left at −20°C overnight at this point if necessary.**

5. Spin at 14,000 rpm for 5 min. Pour off supernatant, being careful not to lose the pellet (sometimes the pellet will not be visible). Resuspend the pellet in 200 μL of RNAse free, sterile water and then add 500 μL 100% EtOH and 25 μL 7.5M NH$_4$OAc. Place at −20°C for 20 min.

6. Spin at 14,000 rpm for 5 min. Pour off supernatant and allow inverted tube to dry completely or centrifuge under speed vacuum without heat for 1–2 min. Resuspend pellet in 50 μL sterile H₂O.

7. If a high-quality extraction is obtained, use approximately 1μL of a 1:100 dilution of the extraction for a 25 μL or 50 μL total PCR reaction. A dilution series for the appropriate amount for PCR can be performed.

8. If PCR reactions do not work using several dilutions, they may contain impurities or inhibitors in the extraction. This often happens with caviar egg extractions. In this case, try using the GeneClean (BIO 101) kit following manufacturer's specifications (extractions should be resuspended in 30 μL RNAse-free sterile H₂O). Be sure to keep solutions sterile at this step (by UV-treating everything) as this can introduce contaminants.

Protocol 9 Avian tissue and feathers from museum skins using the QIAgen DNAeasy® Tissue Kit

(Time: 3 hours; also used for molted feathers for maximum yield)

1. *Working in a UV hood or clean area restricted to ancient DNA extractions and using a sterile razor blade, shave off a thin tissue fragment (~1.5 × 1.5 × 3 mm) from ventral side of the appropriate digit of the foot (varies among avian groups) or approximately 5 mm from the base of a plucked feather (calamus) from each specimen [19].*

2. *Place approximately one-half of tissue sample or entire feather sample in a sterile 1.5 mL tube. Negative extraction controls should be carried out throughout the entire procedure.*

3. *Add 180 μL of Buffer ATL and 20 μL Proteinase K (20 mg/mL) to 1.5 mL tubes containing tissue.*

4. *Incubate at 55°C rotating until tissue is completely dissolved (may take up to 48 hours or more).*

5. *In order to guarantee full digestion, it may be necessary to add additional 20 μL aliquots of Proteinase K to the sample.*

6. *If additional Proteinase K is added, the volumes of subsequent solutions of Buffer AL and ethanol must be scaled up accordingly.*

7. After tissue is completely digested, add 200 μL Buffer AL (or scaled up volume).

8. Incubate at 65°C for 15 min.

9. Add 200 μL of 95–100% ethanol (or scaled-up volume) and mix thoroughly by vortexing.

10. Incubate at 4°C for 1 hour.

11. Place the provided spin column inside 2 mL collection tube and add entire volume of sample to spin column and centrifuge for 1 min at 8,000 rpm; discard filtrate and collection tube.

12. Place spin column in fresh 2 mL collection tube and add 500 μL of Buffer AW1 and centrifuge for 1 min at 8,000 rpm; discard filtrate and collection tube.

13. Place spin column in fresh 2 mL collection tube and add 500 μL of Buffer AW2 and centrifuge for 3 min at maximum speed to dry DNeasy membrane; discard filtrate and tube.

14. Place spin column in fresh 1.5 mL tube and add 50 μL Buffer AE preheated to 70°C [19].

15. Incubate at room temperature for 45 min.

16. Centrifuge for 3 min at 8,000 rpm.

17. A second elution may be performed by transferring spin column to a fresh 1.5 mL tube and repeating step 12; incubate sample for 10 min at room temperature and repeat step 14.

18. Store at 4°C, or at −20°C for long-term storage.

Buffers ATL, AL, AW1, AW2, and AE are supplied with the DNAeasy® Tissue Kit.

Protocol 10   Fecal samples using the QIAamp® DNA Stool Mini Kit

(Time: 4 to 6 hours)

1. *Using a sterile razor blade, cut 180–220 mg from surface of stool (thin slices) and place in a sterile 2 ml tube.*

2. *Add 1 mL of Digestion buffer (100 mM NaCl, 10 mM Tris-HCl pH 8.0, 25 mM EDTA), 2% SDS, and 20 μL Proteinase K (20 mg/mL) to each 2 mL tube containing stool; above Digestion buffer not provided by QIAgen; substituted for Buffer ASL.*

3. *Incubate at 65°C overnight or until well digested (solids may still be present); mix by vortexing.*

4. *Centrifuge sample for 1 min at full speed to pellet stool particles.*

5. *Transfer supernatant to a fresh 2 mL tube and discard pellet.*

6. *Add 1 InhibitEX tablet to each sample and vortex immediately and continuously for 1 min; incubate suspension at room temperature for 5 min.*

7. *Centrifuge sample for 3 min at full speed to pellet stool particles and InhibitEX.*

8. *Immediately pipet supernatant into a fresh 1.5 mL tube (this step may be repeated as needed in order to ensure maximum recovery of sample; discard pellet when all spins completed).*

9. *Pipet 600 μl supernatant from step 7 into a fresh 2 mL tube containing 25 μL Proteinase K (20 mg/mL); (if greater than 600 μL recovered from step 7, carry the remaining solution through the following procedure in separate tubes using scaled volumes of all solutions; samples will be consolidated at step 15).*

10. Add 600 μL Buffer AL and vortex for 15 sec (QIAgen notes that it is essential that the sample and Buffer AL are thoroughly mixed to form a homogenous solution).

11. Incubate at 70°C for 10 min.

12. Add 600 μL of ethanol (95–100%) to the lysate and vortex.

13. Place the provided spin column inside 2 mL collection tube and add 600 μL lysate from step 11 directly on to the QIAamp spin column.

14. Centrifuge for 1 min at full speed; place the spin column in a fresh 2 mL collection tube and discard filtrate.

15. Add a second aliquot of 600 μL lysate and centrifuge for 1 min at full speed; place the spin column in a fresh 2 mL collection tube and discard filtrate.

16. Repeat step 14 with a third aliquot of 600 μL lysate; if greater than 600 μL recovered from step 7, then add parallel sample to the spin column and repeat step 14.

17. Place spin column in fresh 2 mL collection tube and add 500 μL Buffer AW1 and centrifuge for 1 min at full speed; discard filtrate and collection tube.

18. Place spin column in fresh 2 mL collection tube and add 500 μL Buffer AW2 and centrifuge for 2 min at full speed; discard filtrate and collection tube.

19. Place spin column in fresh 2 mL collection tube and centrifuge for 2 min at full speed to dry column.

20. Place spin column in fresh 1.5 mL tube and add 50 μL Buffer AE preheated to 70°C.

21. Incubate at room temperature for 45 min.

22. Centrifuge for 3 min at 8,000 rpm; a second elution may be performed by transferring spin column to a fresh 1.5 mL tube and repeating step 19; incubate sample for 10 min at room temperature and repeat step 21.

23. Store at 4°C or –20°C for long-term storage.

InhibitEX, and buffers AL, AW1, AW2 and AE are supplied with the QIAamp® DNA Stool Mini Kit.

Protocol 11   Quick DNA extraction for invertebrates and arthropods

(Time: 2–3 hours)

This method has been used with fresh, frozen or ethanol-preserved tissues of many different metazoan phyla (Porifera, Cnidaria, Ctenophora, Echinodermata, Hemichordata, Priapula, Arthropoda, Platyhelminthes, Phoronida, Brachiopoda, Bryozoa, Nemertea, Mollusca, Annelida, Sipuncula, Echiura and Pogonophora). It is especially recommended for organisms that have mucous-type secretions such as molluscs and platyhelminths. This method does not yield good PCR products for specific organisms that present certain types of body pigmentation, especially Onychophora and some Diplopoda (Arthropoda, Myriapoda). Tissues are homogenized in a solution of guanidinium thiocyanate following a modified protocol for RNA extraction from Chirgwin et al. [20].

1. *Homogenize the tissue sample in 1 volume (e. g., 400 µL) of 4M guanidinium thiocyanate homogenization buffer and 0.1M beta-mercaptoethanol.*
2. *If the tissue sample cannot be disrupted easily, grind the sample to a fine powder in liquid nitrogen with a mortar and pestle and pour the powder into the guanidinium thiocyanate buffer.*

3. **Mix well, shaking for 10 min to 1 hour at room temperature. Add 1 volume of PCI and vortex.**
4. **Centrifuge for 4 min at 12,000 rpm in a microcentrifuge.**
5. **Carefully remove the upper (aqueous layer) with a micropipette and transfer to a clean tube. Be careful not to disturb the debris on the interface.**
6. **Add 1 volume of chloroform:isoamyl-alcohol (24:1) and vortex. Centrifuge for 4 min at 12,000 rpm in a microcentrifuge.**
7. **Carefully remove the upper layer with a micropipette and transfer to a clean tube. Repeat steps 6–8 if the sample is dirty.**
8. **Add 0.1 volumes (ca. 40 µL) of 3 M sodium acetate pH 5.2 and 2 volumes (ca. 800 µL) of ice-cold 95% ethanol (or 100% ethanol).**
9. **Precipitate the DNA at –80°C for 20 min (or from 2 hours to overnight at –20°C).**
10. **Centrifuge the precipitate for 20 min at 12,000 rpm in a microcentrifuge.**
11. **Discard the ethanol solution by decantation and wash the pellet with 1 mL of 70% ethanol.**
12. **Centrifuge the precipitate for 5 min at 12,000 rpm in a microcentrifuge.**
13. **Discard the ethanol solution and dry in a vacuum centrifuge (or at 55°C).**
14. **Resuspend the pellet in 50–100 µL of TE buffer (pH 7.6). Incubating the sample at 45–60°C can facilitate dissolution of the pellet.**
15. **Use 1–2 µL for PCR.**

It is recommended to use small amounts of tissue (i. e., a leg of a small spider is enough). The dirtier the sample, the worse the PCR works. But do not expect to see a band in an agarose gel after the extraction if small amounts of tissue are used. For long-term storage of the samples, proteinase-K and RNAse treatments are necessary.

Protocol 12   DNA isolation from small insects and crustaceans

(Time: 4–24 hours)

This protocol has been used to obtain DNA samples from frozen, live, or ethanol-preserved crustaceans and insects (1–3 mm in length). It can be used to extract DNA from small portions of larger crustaceans or insects. All steps are performed at room temperature unless specified.

1. For grinding specimens, use a razor blade to cut off the cap and rim of a 0.67 mL microcentrifuge tube. Make a pestle by inserting a 1.5 mL blue pipette tip into the smaller 0.67 mL microcentrifuge tube (this is a cheaper and time-saving alternative to using and resterilizing ground glass homogenizers).

2. *Rinse alcohol-preserved specimens in sterile distilled water briefly to remove excess alcohol. Frozen or live animals can be ground directly.*

3. *Grind animal on inside wall of larger tube (with pestle) in 20–30 µL of Lysis buffer. Once sufficiently ground, add Lysis buffer to a total volume of 1 mL. Add 100 µL of 10% SDS and 4 µL of Proteinase K (20 mg/mL) [keep enzyme mixture on ice] and mix.*

4. *Incubate at 55°C with rotation/shaking for a minimum of 4 hours to over-night.*

5. **Add 500 µL of equilibrated phenol (pH 8.0) and shake tube well for 5–10 min.**

6. **Centrifuge for 10 min at 12,000 rpm in a microcentrifuge.**

7. **Remove aqueous phase (usually the top layer) and place into fresh tube. Try not to collect any portion of the bottom layer when sample is transferred. Repeat steps 3–5 on the aqueous sample once (or until interface between lower organic phase and upper aqueous phase is clean).**

8. **Add 500 µL of chloroform, place on rotator for 5–10 min, and centrifuge for 10 min. Remove aqueous phase (top layer) and place in a new tube.**

9. **At this point you can stop and continue with the DNA precipitation steps at a later time, or amplify directly from the sample (if concerned about losing DNA at precipitation steps, storage is not recommended for long term).**

10. Add cold 100% ethanol to sample tube from step 6 above (fill to top).

11. Place at –80°C for 15 min.

12. Centrifuge for 25 min at 12,000 rpm in a microcentrifuge.

13. Discard alcohol carefully by pipetting, without disturbing pellet (if pellet is visible). Check pipette tip before ejecting to ensure that pellet was not drawn into pipette tip during alcohol removal. If pellet is inside pipette tip, return pellet to tube and spin again.

14. Add 1000 µL of 70% ethanol and vortex for a few sec.

15. Centrifuge at 12,000 rpm in a microcentrifuge for 10 min. Check for pellet as above.

16. Air dry at room temperature or place in a vacuum centrifuge without heat.

17. Resuspend DNA pellet in 100 µL sterile water.

18. Place DNA sample in a 50–55°C water bath for 5–10 min to dissolve DNA pellet before amplification. Thoroughly mix DNA sample before any further procedures. Use 1–2 µL of sample for PCR.

19. Keep a working DNA sample in the refrigerator (i. e., 30 µL aliquot) to avoid freeze/thaw cycles and place remainder of DNA sample at –20°C for future use.

Protocol 13   DTAB – CTAB preparation

(Time: 1 hour)

This preparation works well on a wide variety of tissues. Phillips and Simon [21] have used this preparation to isolate DNA from pinned insects. The preliminary step in this approach is to poke holes in the pinned insect exoskeleton prior to the first step below. After the DTAB soak, the pinned insect can be washed in chloroform and returned to its place in the collection. We have used the preparation on pinned insect body parts such as leg and wing tissues. We find that the insect parts should be broken into small fragments prior to placing them in the DTAB solution. The full preparation is given below:

1. *The tissue is soaked in 600 µL of DTAB solution (8% DTAB, 1.5 M NaCl, 100 mM Tris-HCl (pH 8.8), and 50 mM EDTA) at 68°C for at least one hour or preferably overnight.*

2. **An equal volume of chloroform:isoamyl alcohol (24:1) is added to the specimen in DTAB and mixed by inverting.**
3. **This mixed solution is spun at 10,000 g for 2 min.**
4. **The aqueous layer is removed and placed into another tube. An equal volume of chloroform:isoamyl alcohol (24:1) is added to the aqueous phase for a second extraction.**
5. **The mixed solution is centrifuged at 10,000 g for 2 min and the aqueous layer is removed and placed into a new tube.**
6. **Add 900 µL of water and 100 µL of CTAB (5% CTAB, 0.4 M NaCl) to the aqueous layer from above.**
7. **Gently mix by inverting tubes and leave at room temperature for 2 min.**
8. **Spin tubes at 10,000 g for 10 min.**
9. **Pour off supernatant and resuspend pellet in 300 µL of 1.2 M NaCl. This step exchanges CTAB.**

10. Precipitate the DNA in 750 µL of 100% ethanol. Mix and leave at room temperature for 10 min.
11. Spin at 10,000 g for 10 min. Pour off ethanol supernatant.
12. Wash pellet in 70% ethanol.
13. Centrifuge at 10,000 g for 10 min and air dry.
14. Resuspend pellet in 20–40 µL of dH₂O or TE buffer.

Protocol 14   DNA isolation from microscopic animals

(Time: 1 hour)

This protocol has been used to obtain DNA samples of organisms smaller than 1 mm in length, such as *Macrobiotus* (Tardigrada), and Pauropoda (Arthropoda, Hexapoda). The method is based on a direct lysis of the animal tissues, a modification of the protocol by Higuchi [22]. Method described in Giribet et al. [23].

1. *Pour 400 μL of PCR buffer with non-ionic detergents (50 mM KCl, 10 mM Tris-HCl pH 8.3, 2.5 mM MgCl$_2$, 0.1 mg/mL gelatin, 0.45% NP-40, 0.45% Tween-20) in an eppendorf tube.*
2. *Add 0.6 μL of Proteinase K (10 mg/mL) per 100 μL buffer (a total of 2.4 μL). Add one to three live individuals.*
3. *Incubate at 60°C for 1 hour with occasional vortexing.*

Use 10 to 25 μL for PCR. This method is not suitable for long-term DNA storage, and will yield DNA for just a few PCR reactions.

Protocol 15   Insect preparation – quick and dirty

This preparation can be used on small insects. Storage of DNA made from this preparation is not recommended.

1. *Whole small insect (or part of a larger insect, such as a leg) is ground with a pipette tip in 100 μL of "quick and dirty" extraction buffer (1X PCR Buffer) in a 1.5 mL tube.*

2. **The ground insect in solution is placed in a boiling water bath for 15 min.**
3. **The tube is centrifuged at 14,000 rpm to get rid of debris.**
4. **The supernatant containing crude DNA preparation is ready for PCR and can be transferred to a new tube.**

5. The crude DNA solution is stored at –20°C.

Different concentrations of the crude preparation need to be tested for PCR amplification.

Protocol 16   Centricon 30 concentration and purification (Millipore)

(Time: 2 hours)

Useful for tiny or very rare specimens, but is more expensive than ethanol precipitation. Follow general directions as provided with concentrators. In general, the Centricon step is used as a substitute for ethanol precipitation.

1. **The samples that are purified using this approach can be prepped to this stage by most of the protocols mentioned in this chapter.**

2. Add samples to filter tubes and cap.
3. Spin for 10 min at 5,000 rpm in standard fixed-angle centrifuge.
4. Add sterile water to the fill line on the Centricon tube.
5. Spin for 30 min (Repeat steps 3 & 4, twice).

6. Invert the tubes and spin again for 3 min. The sample will be collected at the bottom of the lid.
7. Transfer the samples into labeled tubes and add 100 μL of water.
8. Use 1–2 μL of your sample for PCR. Keep a working DNA sample in the refrigerator (i. e., 30 μL aliquot) to avoid freeze/thaw cycles and place the remainder of DNA sample at –20°C for future use.
9. For both precipitation protocols, an assay gel for DNA extractions is not performed since the amount of DNA is small and usually not detectable (however, the amount isolated is usually sufficient for PCR).

Protocol 17   DNA isolation from plants and algae

(Time: 6–8 hours)

Obtaining chemically pure DNA from plant and macroalgal material is usually a difficult and tedious process. If pure cultures can be obtained, they are the most ideal material used for nucleic acid purification. If field-collected specimens are the only sources available, then they must be cleaned and all epiphytic material (animal, plant or fungi) must be removed. Optimal materials are the tissues of actively growing parts, or those producing gametes or spores. A good example is the reproductive structure in red algae, the cystocarp, which produces large amounts of nuclei and organelles relative to other parts of the organism, and is usually devoid of epiphytic material. Most material obtained in this manner yields small amounts of DNA, but the nucleic acid material is pure.

1. *After cleaning, the plant/macroalgae is cut into smaller pieces, so that the tissue can be easily frozen with liquid N_2, and then ground to a fine powder with a mortar and pestle.*
2. *The material must be kept cold at all times during this process, particularly if the separated nucleic acids will be used for genomic analysis (library construction, restriction mapping, etc.).*
3. *Combine the ground tissue in an insulated container and keep covered with liquid nitrogen until all the material has been pulverized. This can be done by placing the powder in a plastic beaker, covering to prevent contamination, and placing it in a –80°C freezer until all material is crushed.*
4. *After all the material has been ground to a fine powder, it can be added to a number of various standard lysis solutions.*

5. **Add the powder slowly to a prewarmed (50°C) SDS/Sarkosyl lysis buffer, so that it does not freeze the buffer, keeping it in an incubated water bath. It is best to start with approximately 50–100 mL of buffer, and gradually add tissue and more buffer when necessary. If the solution is still thick once it reaches 50°C, add more buffer, but try to keep the total volume at a minimum to reduce the amount of reagents (and therefore cost).**

6. Continue to incubate this solution at 50°C for 1.5–2 hours. If the extraction solution is thick, or there are large amounts of tissue in the slurry, you can filter the solution through cheesecloth. It is necessary to remove all the liquid from the cheesecloth (which contains DNA) so that the yield is high. Alternatively, the solution can be centrifuged (5,000–10,000 g) for 10 min at 4°C. Both of these steps can be skipped if there is not much undigested material in the solution.

7. Once the digestion is complete, the filtered (or unfiltered) solution can be put into 35 mL Oakridge or Sepcor centrifuge tubes (these must be phenol- and chloroform-resistant, and are tolerant of high-speed centrifugation).

8. Fill the tubes half full, and to this add an equal amount of room-temperature PCI.

9. Mix the solution well by inverting the tubes several times (no vortexing) and centrifuge at 10,000 g for 20 min at room-temperature.

10. Transfer the aqueous phase with a wide-bore pipet (a 10 mL plastic pipet with the end cut off is perfect), making sure that the interface is not transferred to the new tube.

11. Fill the new tube only half full, and add an equal amount of chloroform:isoamyl alcohol (24:1) to the sample.

12. Repeat mixing and centrifugation. Transfer the aqueous layer to a large glass beaker.

13. Slowly add 0.2–0.3 volumes of 95–100% ice-cold ethanol while constantly swirling. A large quantity of polysaccharide may precipitate. The level of precipitate may vary among species, but generally marine species produce large amounts of polysaccharides. After this step, the solution should be slightly more viscous.

14. Transfer the supernatant to new Oakridge/Sepcor tubes and add an equal volume of PCI at room temperature.

15. Invert the mixture several times for 10–15 min.

16. Centrifuge for 15 min at 10,000 g.

17. Remove the aqueous layer with a wide-bore pipet (reduces shearing of the DNA). Pool the fractions from all the tubes and place in a beaker on ice. Keep the solution cold from this step till the end. Add this solution to pre-chilled Oakridge/Sepcor tubes, only filling half full.

18. Add an equal volume of chloroform:isoamyl alcohol (24:1).

19. Invert tubes to mix for 10 min.

20. Centrifuge at 10,000 g for 10 min.

21. Remove the aqueous layer, place in a chilled beaker on ice, and measure the total volume.

22. Add exactly 1/10th the volume of 5 M NaCl.

23. Mix gently and add exactly two volumes of ice-cold 95% ethanol and mix by pouring between two beakers. The DNA will precipitate out into long, stringy threads.

24. Spool the DNA using a bent glass rod and transfer the DNA to a small beaker containing ice-cold 70% ethanol. The DNA will still contain a considerable amount of carbohydrates which have co-precipitated with the DNA.
25. Cover and place at –20°C overnight.
26. Following the 70% ethanol wash, take the DNA out of the alcohol and place in a 50 mL falcon tube.
27. Drain off most of the ethanol completely. Add TE buffer (pH 8.0), until the DNA is completely dissolved.
28. Depending on how large the pellet is, this may take 1–2 days. Make sure that the solution is kept cold (in the refrigerator or on ice).

Protocol 18   Plant DNA isolation from herbarium and fresh collected specimens.

Modified from Struwe et al. [24].

1. *Place 0.5 to 1.0 cm square piece of plant tissue in a tube and rapidly macerate tissue. The most efficient way to do this is to use a FastPrep machine.*
2. *Mix 500 µL of Carlson lysis buffer and 75 µL of beta-mercaptoethanol into each tube. Samples can be left at 4°C for extended periods of time (1 week) in this solution.*
3. *Incubate extraction at 74°C for 20 min to 1 hour.*

4. **Remove samples to room temperature and let cool.**
5. **Add 575 µL of PCI to each tube and mix gently.**
6. **Spin tubes at 14,000 for 2 min to separate organic and aqueous phases.**

7. While tubes are spinning prepare GENECLEAN® (Bio 101) glass milk by vortexing to get glass beads into solution.
8. Remove 300 µL of upper aqueous layer from step 6 to a new eppendorf tube.
9. Add 900 µL of sodium iodide (NaI; concentration not specified in the Kit) from the GENECLEAN® kit and 20 µL of glass milk per tube.
10. Mix gently for 10 min at room temperature.
11. Spin tubes at 14,000 rpm for 30 sec to pellet glass milk.
12. Pour off supernatant and add 900 µL of New™ Wash concentrate.
13. Resuspend glass milk pellet by gently breaking the pellet with a pipette tip and shaking the tube.
14. Spin tubes at 14,000 rpm for 30 sec to pellet glass milk.
15. Repeat steps 12–14 twice more.
16. Use a pipette to remove the last 10 or 20 µL of New™ Wash concentrate without disrupting glass milk pellet.
17. Resuspend glass milk pellet in water or TE buffer and place at 50°C for 10 min to elute DNA away from glass milk beads.
18. Spin tubes at 14,000 rpm for 2 min and transfer supernatant with DNA to a new tube.
19. Store DNA in freezer.

Protocol 19   Bacterial genomic DNA preparation

(Time: 2 hours)

This protocol is probably generally applicable to a wide variety of bacterial species.

1. *Spin down 5 ml of saturated cell culture in eppendorf tube.*
2. *Resuspend pellet in 500 µL of ice-cold Solution I ([5]; used for small-scale preparations of plasmid DNA by lysis by alkali) containing lysozyme (final concentration 2mg/mL). Incubate on ice for 10 min.*

3. **Add 50 µL of 10% SDS buffer. Incubate at 37°C for 5–10 min until clear and viscous.**
4. **Put into a new tube (or a phase lock gel; phase lock gel is not necessary but makes life a lot easier).**
5. **Add 550 µL of phenol (freshly equilibrated with an equal volume of 0.3 M NaOAc).**
6. **Mix gently by inversion and centrifuge at 4°C for 15 min.**
7. **Transfer top layer to a new tube (or phase lock gel) and repeat step 5.**
8. **Transfer top layer to eppendorf and add one-tenth volume of 3 M NaOAc.**
9. **Spin for 3 min and transfer supernatant.**

10. Add 2 volumes of 100% ethanol and mix by inverting.
11. Cool sample at –80°C for 5 min.
12. Centrifuge for 15 min at 4°C.
13. Remove and discard the supernatant.
14. Vacuum dry the pellet and resuspend in 50–100 µL of water.

This is a modified version of a protocol from Current Protocols in Molecular Biology [25].

Protocol 20   Isolation of DNA from prokaryotes: CTAB

(Time: 2–3 hours)

The isolation of prokaryotic nucleic acid is much less work-intensive than those described for plants and macroalgae. Most Bacteria and Archaea have cell walls that can be easily broken through and lysed for the isolation of DNA and RNA. Ideally, the material used should be grown from pure culture. Depending on the type of organism used for genetic material, either agar or liquid cultures can yield similar results, as long as there is no contamination in the process. Extra care must be taken when using nutrient rich media, since most airborne bacteria can grow in this as well.

1. *For liquid cultures, spin down 1–3 mL of culture and remove the media.*
2. *Add 567 μL of TE buffer to the pelleted cells. If agar media are used, choose one colony from the petri dish (using a sterile toothpick) and place the toothpick with the colony into a sterile eppendorf tube containing the 567 μL of TE buffer.*
3. *Resuspend the pellet by repeated pipetting, or by gently vortexing the toothpick with the solution so that the cells become resuspended.*
4. *Add 30 μL of 10% SDS and 3 μL of a 20 mg/mL solution of proteinase K. Mix and incubate for 1 hour at 37°C.*

5. **After incubation, add 100 μL of 5 M NaCl and mix.**
6. **Afterwards add 80 μL of a CTAB/NaCl solution (0.7 M NaCl, 10% CTAB).**
7. **Incubate this solution at 65°C for 10 min.**
8. **After incubation, add an equal volume of chloroform: isoamyl alcohol (24:1) and mix.**
9. **Centrifuge 5 min, and transfer the aqueous solution to a new tube. Be careful not to transfer the interface. Add another equal volume of PCI and mix well.**
10. **Centrifuge at 14,000 rpm for 5 min and transfer supernatant to a new tube.**
11. **Repeat first extraction again (chloroform: isoamyl alcohol alone).**

12. Add 0.6 volumes of isopropanol and mix gently until the DNA precipitates.
13. Centrifuge and remove isopropanol. Add 1 mL of 70% ethanol to wash the salt away from the DNA.
14. Centrifuge, and discard the ethanol, drying on the benchtop at room temperature.
15. Resuspend the pellet in 50–100 μL of TE buffer and keep at 4°C.

This DNA can be used for restriction digests, Southern and Northern blot analysis, genomic library construction, and PCR.

Protocol 21   Isolation of DNA from prokaryotes: Chelex®

Isolation Time: 1 hour

Another quick method to isolate DNA from prokaryotes for PCR amplification via the Chelex® method.

1. *Bacterial cultures are pelleted and placed in 200 μL of Chelex® 100 buffer (Chelex 100 buffer-0.001 M Tris-HCl, pH 8.0; 0.05 mM EDTA; 5% (w/v) Chelex® 100 resin).*
2. *Grind the cells with the Chelex(r) 100 buffer (the resin helps this process) with a sterile teflon eppendorf grinder (Kontes). Resuspending the cells helps break open the cells during this process.*

3. **Heat extract the DNA by incubating the tubes at 80°C for 25 min or longer, and then boiling for 10 min.**
4. **Another heat extraction method is to autoclave the cells for 10 min on slow exhaust, making sure a small hole is punctured through the top of the eppendorf tube to allow it to equilibrate in the autoclave.**

5. Once the heat extraction is complete, centrifuge the tubes and avoid cell debris when pipetting.
6. Use from 1–5 µL of extract for PCR. Extracts can be stored at –20°C for several months.

This method can also be used for isolating template DNA from dried botanical museum specimens, particularly the spores from any reproductive organ that is still intact [26].

Protocol 22   Isolation of total RNA from tissues and cultured cells

The following protocol was modified from the guanidinium thiocyanate/phenol DNA purification method described by Chomzynski & Sacchi [27] and Chomzynski [28]. Guanidinium iso thiocyanate is a very powerful protein denaturant, which inactivates RNases during the extraction procedure. This protocol allows the isolation of total RNA within 4 hours and provides both high yield and purity of undegraded RNA preparations. It is especially recommended for full-length cDNA synthesis and RT-PCR reactions. This method also permits recovery of total RNA from small quantities of tissues or cultured cells.

From Tissues

1. *Freeze the tissue in liquid nitrogen or on dry ice immediately after dissection. It is possible to store tissue at –80°C for several months.*
2. *Cut the frozen tissue into small pieces (less than 0.5 cm cubes) and transfer in an appropriate tube according to the amount of tissue (do not use polycarbonate tubes with guanidine iso thiocyanate). Add 1 mL of GITC solution (4 M guanidinium iso thiocyanate, 25 mM sodium citrate, pH 7.0, 0.5% sarkosyl, 0.1 M 2-mercaptoethanol, stored in light-proof vessel) per 100 mg tissue.*
3. *Homogenize small amounts of tissue by passing pieces through a syringe fitted with a 18–21 gauge needle. Break up the tissue by grinding it against the side of the tube with the needle. If the tissue is difficult to homogenize in this manner, or if larger amounts will be processed, a Dounce or power homogenizer may be used. It is also possible to grind up the frozen tissue with a mortar and pestle with liquid nitrogen.*

From Cultured Cells

1. *Wash the cells in ice-cold phosphate buffered saline (PBS) solution and transfer to a sterile microcentrifuge tube.*

2. *Pellet the cells by centrifugation (10 min at 1,000 rpm) and add the appropriate amount of GITC solution (1 mL per 10⁷ cells). Pass the suspension through a plastic syringe fitted with an 18–21 gauge needle at least 3 times.*

3. *Proceed with the RNA isolation.*

4. **Add the following solutions to the homogenate and mix thoroughly by inverting the tube after the addition of each reagent (all quantities and volumes in this protocol are calculated for 100 mg tissue or 10⁷ cells): 0.1 mL of 2 M sodium acetate (pH 4.0), 1 mL of acid phenol (pH 5.0) and 0.2 mL of chloroform.**

5. **Shake the suspension vigorously for 30 sec.**

6. **Incubate on ice for 15 min.**

7. **Centrifuge the samples at 10,000 rpm for 20 min at 4°C.**

8. **Transfer the upper aqueous phase (containing the RNA) in a fresh tube. Be careful not to disturb the interphase containing proteins and DNA.**

9. Add 1 mL 100% isopropanol to the aqueous phase, mix and precipitate the RNA by incubating the solution at –20°C for 1 hour.

10. Centrifuge at 14,000 rpm for 20 min at 4°C.

11. Resuspend the RNA pellet in 0.5 mL GITC solution and transfer to a fresh 1.5 mL microcentrifuge tube.

12. Precipitate RNA by adding 0.5 mL 100% isopropanol and incubating at –20°C for 1 hour.

13. Centrifuge for 10 min at 4°C (14,000 rpm), wash the RNA pellet twice with 70% ethanol and air dry the precipitant.

14. Resuspend RNA in 50 μL RNase-free water or TE buffer. If mRNA will be isolated, resuspend in 5 mL STE/0.5 M NaCl/Proteinase K (200 μg/mL) and proceed with the poly(A)+ selection procedure (see below). The RNA can be stored at –80°C for several months. For long-term storage, RNA is best kept in formamide [29].

Note: Since guanidinium iso thiocyanate does not irreversibly inactivate RNases, it is important to carefully separate the RNA-containing aqueous phase from the interphase and lower organic phase. If residual RNase contaminates the sample after deproteinization, these enzymes can become active again following the removal of the denaturant.

Protocol 23   Isolation of poly(A)+ RNA: selection of polyadenylated RNA from total RNA by Oligo-deoxythymidine (DT) Cellulose Chromatography

Poly(A)+ RNA can be isolated from total RNA preparations using oligo(dT) cellulose selection or directly from tissue and cell culture samples. The purification of mRNA from total RNA is recommended for tissue sources and cell lines rich in RNases in order to minimize possible RNA degradation during the extraction process. Total RNA is first isolated from tissue or cultured cells and mRNA is isolated by poly(A)+ selection using oligo(dT) cellulose (Boehringer or equivalent).

1. Prepare oligo(dT) cellulose (Boehringer or equivalent) in the following way:
2. Wash 1 g of oligo(dT) cellulose in 50 mL of 0.2 M sodium hydroxide in sterile water for 30 min at room temperature on a rotating wheel. Neutralize by washing twice in 50 mL of 500 mM Tris-HCl (pH 7.4), followed by six changes of sterile water. Wash in 50 mL of RNA preparation binding buffer to equilibrate the salt concentration, recover by centrifugation and then resuspend in 50 mL of RNA preparation binding buffer.
3. Resuspend total RNA from Protocol 22 in 5 mL STE/0.5 M NaCl/Proteinase K 200 µg/mL and add 0.25 mL of oligo(dT) cellulose. Allow the RNA to bind, incubating for at least 1 hour on a rotating wheel at room temperature.
4. Spin down oligo(dT) cellulose in lysate at 2,000 rpm for 1 min. Remove the supernatant and resuspend in 15 mL binding buffer. Repeat this step twice in order to remove unbound (mainly ribosomal) RNA.
5. Transfer oligo(dT) cellulose to alkali-treated econo-column (BioRad®). Alkali treatment of columns will be done by washing the columns with 0.2 M NaOH, then several volumes of binding buffer.
6. Wash oligo(dT) cellulose with 10 mL RNA preparation binding buffer followed by 2 mL RNA preparation washing buffer.
7. Elute RNA with 1 mL RNA preparation elution buffer. Ethanol precipitate RNA by adding 100 µL 5 M NaCl and 2.5 mL 100% ethanol. Freeze at −20°C.
8. Spin down RNA, wash twice with 70% ethanol, air dry the pellet and resuspend in an appropriate volume of RNase-free water or TE buffer. Use an aliquot for OD 260/280 measurement. Store at −80°C.
9. Regenerate the oligo(dT) cellulose in the following way:
10. Wash in several volumes of RNA preparation elution buffer, re-treat with 0.2 M NaOH, spin down and resuspend in 500 mM Tris-HCl (pH 7.4). Wash for 15 min, spin, check the pH of the supernatant and resuspend again in 500 mM Tris-HCl (pH 7.4) if it is still alkaline. Resuspend the neutralized oligo(dT) cellulose in sterile water, wash to remove the salt, spin, and resuspend in absolute ethanol. Store at −20°C protected from light.

Protocol 24   Kits and commercial reagents for the isolation of RNA

Several Kits are available to purify mRNA from total RNA and/or directly from cell and tissue samples. Some examples are listed in the following.

- The PolyATtract® mRNA Isolation Systems from Promega can be used to isolate mRNA from total RNA samples as well as directly from tissue or cultured cell samples (PolyATtract® System 1000). These systems utilize streptavidin paramagnetic particles which eliminates the need for oligo(dT) cellulose columns. Biotinylated oligo(dT) probe is annealed in solution to the poly(A)+ RNA in a total RNA sample. The hybrids are captured with the MagneSphere® Streptavidin Paramagnetic Particles in a Magnetic Separation Stand and the non-hybridized RNA is washed away. The mRNA can be isolated in about 45 min. The PolyATtract® Series 9600TM mRNA Isolation System provides reagents for the processing of mRNA from numerous small samples for RT-PCR or other applications in a 96 well plate format. It is available with or without reagents for performing first-strand cDNA synthesis.
- The MagNA Pure LC mRNA Isolation Kit from Boehringer is also based on magnetic bead technology. It is especially designed to isolate mRNA from blood, blood cells and cultured cells. Purified mRNA is suitable for RT-PCR reactions and other typical mRNA downstream applications such as Northern blotting, Northern ELISA, ribonuclease protection assay, and preparation of cDNA libraries.
- Stratagene's Poly(A) Quik® mRNA Isolation Kit is based on oligo(dT) cellulose chromatography and can be used to isolate mRNA from total RNA samples. Each column is designed to accommodate up to 500 µg of total RNA in volumes ranging from 200 to 1000 µL. It allows the isolation of poly(A)+ RNA from total RNA in about 15 min.
- The FastTrack™ Kit (Invitrogen) can be used to isolate mRNA directly from 1×10^7 to 3×10^8 cells, 0.4 gram of tissue, or 0.2–1 mg of total RNA in about 90 min. For smaller amounts of cells or tissues the Micro-FastTrack™ Kit is available (samples ranging in size from 1×10^2 to 5×10^6 cells, 10–200 mg of tissue, or 100 µg of total RNA). The procedure requires no ultracentrifugation or guanidinium lysis and is based on oligo(dT) cellulose binding. Obtained mRNA can be used for cDNA library construction, substracted probe generation, primer extension studies, Northern blot analysis, RNase protection assays, oocyte microinjection, and *in vitro* translation.

Commercial reagents for the isolation of total RNA similar to the protocol described in this section are available from a number of different vendors. Two examples are listed in the following:

- TRIzol (Gibco/BRL) is an acid phenol extraction reagent containing detergents and guandinium iso thiocyanate. The preparation procedure to isolate total RNA using TRIzol is pretty much similar to the GITC protocol described in this section (Protocol 22). The RNAClean™ systems from Hybaid are also based on the guanidinium thiocyanate and phenol extraction method and can be used for the isolation of total RNA from tissues and cells (RNAClean™) as well as from liquid samples (RNAClean™ LS). For the removal of polysaccharides from total RNA, which might be necessary in some cases for an exact quantification of the RNA (polysaccharides may absorb UV light at 260 nm), the "RNAClean™ Extension" system can be used.

Acknowledgements

Jennifer Ast, Jeff Groth, Scott Herke, Blair Hedges, Brian Jennings, Tod Reeder, and Christine Thacker, graciously shared their preferred DNA extraction methods.

References

1   Maniatis T, Fritsch EF, Sambrook J (1982) *Molecular cloning: A laboratory manual.* Cold Spring Harbor Laboratory, Cold Spring Harbor, NY

2   Hillis DM, Larsen A, Davis SK, Zimmer EA (1990) Nucleic Acids III: Sequencing. In: DM Hillis and C Moritz (eds): *Molecular Systematics.* Sinauer Associates, Inc., Sunderland, Massachusetts, 318–370

3   Hillis DM, Mable BK, Larson A, Davis SK et al. (1996) Nucleic Acids IV: Sequencing and Cloning. In: DM Hillis, C Moritz, BK Mable (eds): *Molecular Systematics,* second edition. Sinauer Associates, Inc., Sundland, Massachusetts, 321–381

4   Palumbi SR (1996) Nucleic Acids II: The Polymerase Chain Reaction. In: DM Hillis, C Moritz, BK Mable (eds): *Molecular Systematics,* second edition. Sinauer Associates, Inc., Sundland, Massachusetts, 205–247

5   Sambrook J, Fritsch EF, Maniatis T (1989) *Molecular Cloning: A laboratory manual,* 2nd ed. 3 volumes. Cold Springs Harbor Laboratory, Cold Springs Harbor, NY

6   Dowling TE, Moritz C, Palmer JD (1990) Nucleic acids II: Restriction site analysis. In DM Hillis, C Moritz (eds) *Molecular Systematics.* Sinauer Associates, Sunderland, Massachusetts, 250–317

7   Dowling TE, Moritz C, Palmer JD, Riesberg LH (1996) Nucleic acids III: Analysis of Fragments and Restriction sites. In: DM Hillis, C Moritz, BK Mable (eds): *Molecular Systematics,* second edition. Sinauer Associates, Sunderland, Massachusetts, 249–320

8   Miller SA, Dykes DD, Polesky GF (1988) A simple procedure for extracting DNA from human nucleated cells. *Nuc. Acids Res.* 16: 215

9 Coleman AW, Goff LJ (1991) DNA analysis of eukaryotic algal species. *J. Phycol.* 27: 463–73

10 Jorgensen RA, Cuellar RE, Thompson WF, Kavanagh TA (1987) Structure and variation in ribosomal RNA genes of pea. *Plant Mol. Biol.* 8: 3–12

11 Palmer JD (1991) Plastid chromosomes: structure and evolution. In: L Bogorad, IK Vasil (eds): *The molecular biology of plastids,* Vol. 7A. Academic Press, San Diego, 5–53

12 Birley AJ and Croft JH (1986) Mitochondrial DNAs and phylogenetic relationships. In: SK Dutta (ed): *DNA systematics,* Vol. I. CRC Press, Boca Raton, Florida, 107–37

13 Gray MW (1989) Origin and evolution of mitochondrial DNA. *Annu. Rev. Cell Biol.* 5: 25–50

14 Ward BL, Anderson RS, Bendich AJ (1981) The size of the mitochondrial genome is large and variable in a family of plants (Curcurbitaceae). *Cell* 25: 793–803

15 Harris EH (1989) *The Chlamydomonas Sourcebook.* Academic Press, San Diego, CA., 780 pp

16 Nishiguchi MK, Ruby EG, McFall-Ngai MJ (1998) Competitive dominance during colonization is an indicator of coevolution in an animal-bacterial symbiosis. *Appl. Environ. Microbiol.* 64 (9): 3209–13

17 Singer-Sam JR, Tanguay C, Riggs AD (1989) Use of Chelex to improve PCR signal from a small number of cells. *Amplifications* 3: 11

18 Walsh PS, Metzger DA, Higuchi R (1991) Chelex-100 as a medium for simple extraction of DNA for PCR-based typing from forensic material. *Biotechniques* 10: 506–513

19 Mundy NI, Unitt P, Woodruff DS (1997) Skin from feet of museum specimens as a non-destructive source of DNA for avian genotyping. *Auk* 114 (1): 126–129

20 Chirgwin JM, Przybyla AE, MacDonald RJ, Rutter WJ (1979) Isolation of biologically active ribonucleic acid from sources enriched in ribonuclease. *Biochemistry* 18: 5294–5299

21 Phillips AJ, Simon C (1995) Simple, efficient, and nondestructive DNA extraction protocol for arthropods. *Annl. Entomol. Soc. Amer.* 88(3): 281–283

22 Higuchi R (1989) Simple and rapid preparation of samples for PCR. In: HA Erlich (ed): *PCR technology, principles and applications for DNA amplification,* Stockton Press, New York, 31–38

23 Giribet G, Carranza S, Baguñà J, Riutort M et al. (1996) First molecular evidence for the existence of a Tardigrada + Arthropoda clade. *Mol. Biol. Evol.* 13: 76–84

24 Struwe LM, Thiv JW, Kadereit AS-R, Pepper TJ, Motley et al. (1998) *Saccifolium* an endemic of Sierra de La Neblina on the Brazillian-Venezuelan frontier is related to a temperate alpine lineage of Gentianaceae. *Harvard Papers in Botany* 3: 199–214

25 Ausubel FM, Brent R, Kingston RRE, Moore D et al. (1998) *Current Protocols in Molecular Biology,* John Wiley and Sons, New York

26 Goff LJ, Moon DA (1993) PCR amplification of nuclear and plastid genes from algal herbarium specimens and algal spores. *J. Phycol.* 29: 381–84

27 Chomczynski P, Sacchi N (1987) Single step method of RNA isolation by acid guanidinium iso thiocyanate-phenol-chloroform extraction. *Analytical Biochemistry* 162, 156–159

28 Chomczynski P (1993) A reagent for the single-step simultaneous isolation of RNA, DNA, and proteins from cell and tissue. *Biotechniques* 15,532–537

29 Chomczynski P (1992) Solubilization in formamide protects RNA from degradation. *Nucleic Acids Res* 20, 3791–2.

30 Sambrook J, Russell DW (2001) *Molecular cloning. A laboratory manual.* 3rd edition, Cold Spring Harbor Laboratory Press, Cold Spring Harbor, New York

Further Reading

Ferraris JD, Palumbi SR (1996) *Molecular Zoology.* Wiley-Liss Publications, New York.

Sambrook, J, Russell DW (2001) *Molecular Cloning. A laboratory manual.* Third Edition. Cold Spring Harbor Laboratory Press, Cold Spring Harbor, New York.

DNA isolation Protocols Websites
http://www.nwfsc.noaa.gov/protocols.html
http://bric.postech.ac.kr/resources/rprotocol/

RNA isolation Protocols Website
http://www.nwfsc.noaa.gov/protocols/methods/RNAMethodsMenu.html

13 Late Pleistocene DNA Extraction and Analysis

Alex D. Greenwood

Contents

1   Introduction

The study of DNA from ancient material adds molecular data from fossils to evolutionary biological research. Examples of successful retrieval and characterization of Pleistocene DNA include mammoths [1–8], mastodon [4], ground sloths [9–11], a sabre-tooth tiger [12], a cave bear [13], the Neanderthal type specimen [14], and modern human remains [15]. Although the fossils studied were from various locations, a few conditions are generally shared. Most samples are derived from permafrost, i.e., the mammoth samples, or cold caves, i.e., an extinct mylodontid ground sloth. Based on these studies and the fact that the only reproducible results date to the late Pleistocene, success is predicted to be confined to samples no older than 50,000 years of age and a

Methods and Tools in Biosciences and Medicine
Techniques in molecular systematics and evolution, ed. by Rob DeSalle et al.

higher probability of success is given to those samples retrieved from cold environments. In all cases, it is expected that the DNA retrieved will be of low molecular weight, low concentration, and often modified by hydrolytic and oxidative processes regardless of preservation conditions [16, 17]. In this context, ancient DNA (older than Holocene) can be seen as the study of samples from the late Pleistocene ranging from 10,000–50,000 years of age. Fortunately, many fossils exist from the late Pleistocene fossil record which are of interest for study.

The general approach to any ancient DNA study should be as follows:
- Is there a relevant question that requires ancient material to be answered?
- Do the samples necessary to answer the question exist?
- Do the samples exist in enough numbers or quantity to make sampling feasible?
- Are samples available for destructive sampling?

If any of the above questions can be answered in the negative, then a study involving ancient DNA is not worth pursuing. If all questions above can be answered affirmatively, then various protocols exist for retrieval of DNA from bone and other tissues. As bone is the most common Pleistocene fossil remaining, the protocol focuses on extracting DNA from this tissue type in detail.

Example
The following example is not meant to represent the only kind of question that can be addressed by using ancient DNA, but represents a recent example.
- Is there a relevant question whose answer requires ancient material?
- Can nuclear DNA be detected from mammoths and other extinct animals? Can these sequences be used for phylogenetic and population genetic studies of animals in the Pleistocene?
- Do the samples necessary to answer the question exist?
- Mammoths were abundant, large, and often lived in cold environments such that their fossils are frequently found in permafrost contexts.
- Do the samples exist in enough numbers or quantity to make sampling feasible?
- The size of mammoth bones makes repeated sampling possible without requiring complete destruction of the sample. There are thousands of mammoth samples in museum collections.
- Are samples available for destructive sampling?
- By following museum-specific destructive sampling protocols and clearly stating the objectives of a given project-many American, European and Russian institutes are willing to allow destructive sampling of mammoth specimens.

The end result of this study was to successfully demonstrate that nuclear DNA is preserved in some mammoth fossils and that it has the potential to allow for larger-scale phylogenetic and population genetic analysis of an extinct species [8]. Another recent high-profile case that fits the criteria above was the determination of sequences from the Neanderthal type specimen in order to evaluate the relationship between modern humans and Neanderthals [14].

2   Sample Collection, Lab Design and Other Special Considerations

2.1   Sample collection

Prior to initiating a project, the number of samples, locations, and availability should be determined, as collecting samples can be time-consuming and, in the case of rare samples, may take years to obtain. Although most museums have a destructive sampling protocol, not all samples will be readily available. If one must sample in the field, the time required and associated costs should be figured into the project.

2.2   Placement of lab

An obvious source of contamination is modern DNA and previously PCR-amplified DNA sequences. To avoid contamination of samples and reagents, the facility for storing samples, extracting DNA, and setting up the PCR reactions should be separated from the laboratory where the actual amplification occurs and where PCR products are analyzed. It is also helpful if the ancient DNA working area is in a facility that has not stored or analyzed animals from the same taxa as the fossils that will be analyzed, i. e., studying mammoths in an entomology department.

2.3   Airflow

Airflow to the room from outside should be restricted. Dust containing bacteria, skin, etc., could contaminate reagents and samples. If possible, positive pressure which would move air out of the room when doors are open, rather than in, is preferable. Another alternative is to use air filtration systems that keep particulate matter in the air to a minimum.

2.4   Ultraviolet light source

Ultraviolet (uv) light is highly destructive to DNA. Uv light sources that bathe all work surfaces prior to and after work has been done are very effective in preventing contamination or cross-contamination of samples or reagents. Another useful system is to use enclosed PCR workstations that incorporate a short-wave UV light source and an enclosed area for working. Such units are commercially available (i. e., PLAS LABS, MI).

2.5   Protective clothing

Clothing that prevents contamination of sample by the researcher should be worn at all times. Disposable lab coats, face masks, face shields and gloves will greatly reduce the amount of researcher-derived aerosol, sloughed cells and hair that might contaminate a sample while the researcher is performing extractions or making and handling reagents.

2.6   What should and should not be done in an ancient materials lab

Never bring in modern DNA, PCR-amplified products, or reagents generated in a general molecular biology laboratory to the dedicated ancient material facility. Any previously used equipment that one might wish to use in the ancient DNA facility must be bleached extensively and subjected to uv sterilization before using in the dedicated facility. If this is not possible then the equipment must be purchased new.

3   Materials

Equipment
The equipment in the ancient facility should be dedicated to ancient DNA work and not be used for modern DNA work.

Equipment used in a typical ancient DNA laboratory
- Centrifuge rated at 10,000 rpm that accepts 1.5–2.0 ml sample tubes
- Pipetting devices for 1–10 µl, 5–20 µl, 20–200 µl, and 200–1000 µl volumes
- Filtered pipette tips
- A freezer for DNA extract and PCR reagent storage
- A refrigerator for reagent storage
- 15 ml and 50 ml polypropylene tubes

- 0.5 ml, 1.5 ml, and 2.0 ml tubes (i. e., Eppendorf)
- Incubator
- Sample rotator for agitation of samples during demineralization and extraction
- Heat block that accepts 0.5–2 ml tubes
- Latex gloves
- Disposable surgical face mask
- Plexiglass face shield
- Disposable hair nets
- Disposable arm guards
- Disposable laboratory coats
- Plexiglass PCR workstations with UV light source, one dedicated to DNA extraction and one for PCR reaction setup, i. e., UV (PCR) Chamber, PLAS LABS, MI
- Wall or ceiling mounted uv light source
- Glass bottles 100 ml-1L volume
- Plastic pipettes (5 ml, 10 ml, and 25 ml)
- Pipette bulb
- Paraffin paper

Solutions
- 1M Tris-HCl, pH 8.8
- 1M Tris-HCl, pH 7.5
- 0.5 M EDTA, pH 8.0
- 5 M NaCl
- 1 M MgCl$_2$
- 5 M Guanidinium isothiocyanate (Sigma)
- 1 M (NH$_4$)$_2$SO$_4$
- 14 M 2-mercaptoethanol
- 10% Sodium Lauryl Sulfate (SDS)
- 10% N-Lauryl-sarcosine (sarkosyl, Sigma)
- 10 mg/ml proteinase K
- Silica
- 12N HCl
- double distilled water (i. e., Fluka)
- paraffin wax beads (i. e., Fluka)
- 25 µM each deoxynucleotide (dNTPs)
- Taq DNA polymerase
- Phenol
- Chloroform/isoamyl alcohol (24:1)
- Acetone
- 100% ethanol
- 10 mg/ml bovine serum albumin (BSA)
- pH paper indicators

Note: Reagents should be made in the ancient room. For most buffers, the solid powder can be weighed in 50 ml polypropylene tubes. If glass bottles are used, then they should be treated with acid, bleach, and then thoroughly rinsed with distilled water. Indicators to determine pH should not be placed in solutions but rather an aliquot pipetted onto a pH paper indicator and the pH adjusted until the desired pH is achieved. This system prevents contaminating reagents with the pH paper or with a pH electrode.

4   Methods

Protocol 1   Silica method: preparation

Note: The following protocol is a compilation of Boom et al. [19], Höss and Pääbo [20], Krings et al. [14], and Greenwood et al. [8].

All experiments should include a mock extraction without bone powder taken through all steps as samples are being extracted.

Buffers and reagents:

- 0.5 M EDTA, pH 8.0/10% sarkosyl (9:1) ratio
- 10 mg/ml Proteinase K
- Phenol, pH 8.0
- Phenol/(24:1)Chloroform:isoamyl, phenol and chloroform/isoamyl alchohol should be (1:1)
- 24:1 Chloroform:isoamyl alcohol
- 5 M Guanidinium Isothiocyanate, 0.1 M Tris-HCl, pH 7.5
- Silica solution

Silica preparation:

- Weigh out 4.8 grams silica dioxide into a 50 ml polypropylene tube
- Add 40 ml distilled water and agitate
- Let stand 24 hours
- Pipette 35 ml off
- Bring up to 40 ml with distilled water and agitate
- Let stand 5 hours
- Pipette 36 ml off
- Add 48 µl HCl
- Aliquot into 1.5 ml tubes and store in the dark

Protocol 2   Silica method: sample preparation and bone demineralization

The sample must be ground to fine powder, or in case of tissues, homogenized. For bone, if the sample is large enough, the surface of the sample can be decontaminated with 30% bleach and the samples can be drilled into with a fresh drill bit and a low-speed electric drill or hand-held drill. The collected shavings are fine enough to extract directly. Another protocol involves the use of a Spex Mill (Edison, NJ) and liquid nitrogen to grind the sample to generate sufficient surface area for proteinase K digestion of the sample [14].

Bone demineralization

- Weigh out approximately 0.2–0.5 g of bone powder directly into a 2 ml tube
- Add ~ 2 ml 0.5 M EDTA/Sodium Lauryl Sarcosine (9:1 ratio) solution.
- Wrap tubes in paraffin paper to prevent sample leakage and let incubate with agitation on a sample rotator in an incubator at 42°C for 48 hours.
- Centrifuge samples briefly
- Remove EDTA/Sarkosyl buffer
- Add fresh 1 ml EDTA/Sarkosyl buffer
- Add 10–20 µl of 10 mg/ml proteinase K solution
- Re-paraffin seal tubes and let digest 24–48 hours.
- If the sample has not digested completely add more proteinase K and continue the incubation.

Note: in some instances, samples will not completely digest. To scale up, samples should be weighed into 15 ml polypropylene tubes and correspondingly larger volumes of EDTA/sarkosyl and proteinase K used. However, this would also require a centrifuge that accepts 15 ml tubes and a 5 ml pipette device.
For tissues other than bone, a standard SDS buffer [21] can be used for tissue digestion.

Protocol 3   Silica method: extraction

- Centrifuge sample briefly
- Remove 1 ml supernatant
- Place in fresh tube
- Add 1 ml phenol and mix thoroughly
- Centrifuge 3–5 minutes
- Remove aqueous fraction (0.5M EDTA will cause the aqueous phase shift to the bottom rather than top so take bottom fraction)
- Place supernatant in fresh 2 ml tube
- Add 1 ml of phenol/chloroform/isoamyl alchohol and mix thoroughly
- Centrifuge 3–5 minutes
- Remove aqueous phase (this time aqueous phase is the top phase) and place in fresh 2 ml tube
- Add 1 ml chloroform/isoamyl alcohol and mix thoroughly
- Centrifuge 3–5 minutes
- Remove aqueous phase to fresh 2 ml tube

Protocol 4   Silica method silica purification

- Add 1.5 ml 5M Guanidinium isothiocyanate in 0.1 M Tris-HCl buffer, pH 7.5 and 40 µl of silica solution
- Paraffin seal tubes and let agitate at room temperature for 1–2 hours
- Centrifuge briefly
- Resuspend pellet in 1 ml cold 100% ethanol to remove the residual Guanidinium isothiocyanate buffer
- Centrifuge briefly and remove ethanol
- Resuspend pellet a second time in ethanol
- Briefly centrifuge and remove ethanol
- To dry pellet, add 1 ml acetone and re-suspend pellet
- Centrifuge briefly and completely remove acetone
- Place open tubes in heating block at 60°C for 5 minutes
- To elute DNA from silica, add 65 µl distilled water and re-suspend pellet completely
- Place tubes in heating block 15 minutes at 60°C
- Centrifuge samples 5 minutes
- Collect eluate
- Repeat elution step
- End result should be ~120–125 µl silica-free extract

Protocol 5   Hot start PCR

Hot stant is a method of PCR amplification that is commonly used. Amplification controls should include the mock extraction, and one or two water controls for all amplifications performed.

Step 1: *PCR setup*

Equipment:

DNA thermal cycler (located in laboratory separated from ancient DNA facility)

Primer design

If molecular sequence data exist for modern representatives of extinct taxa, they can be used to design primers specific to the taxa of interest. This helps to exclude contaminating DNA as well, which proved useful for both mammoth multi-copy and single-copy nuclear DNA and for independent reproduction of some of the Neanderthal type specimen DNA sequence [8, 14].
In addition, the use of short (< 200 bp) overlapping amplified fragments is necessary to exclude that one has detected a modern, or ancient, nuclear copy of mtDNA [14]. Short overlapping fragments also help avoid sequencing errors due to damage of templates, particularly for longer fragments, where longer fragments had many errors that did not appear in shorter overlapping fragments [22].

Setting up PCR reactions

Reactions are generally in a 20–30 µl volume. The reactions are split into two phases, one containing Tris-HCl buffer, MgCl$_2$, and primers, and a second phase containing Tris-HCl buffer, BSA, Taq DNA polymerase and template. Upon heating, the barrier to the two phases melts, allowing top and bottom phases to mix and PCR to begin. This prevents initiation of PCR at room temperature, which should reduce artifacts generated during PCR amplification and increase PCR specificity.

Bottom reaction

Final concentrations given:
  • 69.3 mM Tris-HCl, pH 8.8
  • 2 mM MgCl$_2$
  • 0.125 µM each dNTP
  • 0.33 µM mM each primer
Final volume: 10 µl

Add 1 paraffin wax bead to each tube and seal. Place reaction tubes in a heating block at 70°C until wax melts. Remove from block and let wax harden while top phase of the PCR is set up.

Top reaction

Final concentrations given:
  • 67 mM Tris-HCl, pH 8.8
  • 1.3 mg/ml BSA
  • 1 U Taq DNA polymerase
  • 0–5 µl distilled water per reaction depending on volume of extract used
Final volume without extract: 10–15 µl

Add 5–10 µl distilled water negative control, mock extract, or sample extract for a final volume of 20 µl for the top phase. The total PCR reaction volume is 30 µl.

Once reactions have been made, take them out of the ancient DNA facility to the laboratory housing the PCR thermocycler. At this point, it is advisable not to re-enter the ancient room until one has changed clothes. Upon exiting the ancient DNA facility, UV sterilize the room.

PCR cycling profile

  • 92°C 20 seconds
  • primer annealing temp 1 minute
  • 72°C 1 minute
  • 40 cycles

Step 2: *Product visualization and re-amplification of PCR products*

Equipment

- UV light table and camera for photographing agarose gels
- Glass Pasteur pipettes
- Pipette ball for Pasteur pipettes
- Mid-size or mini agarose gel casting tray, buffer chamber, and combs
- Pipette devices separate from those used in ancient DNA facility and aeroseal resistant tips
- Electrophoresis power supply
- 0.5 ml tubes

Reagents

- 1X Tris Acetate buffer
- Low melting temperature agarose
- 1X TBE
- agarose
- low molecular weight DNA marker
- 10 mg/ml ethidium bromide

Note: Particularly with single-copy nuclear DNA sequences, but often with mtDNA as well, the amplification product is very weak due to the low concentration of retrieved DNA from the sample [8, 14]. Therefore it is often necessary to re-amplify the initial product to generate sufficient amounts for downstream analysis such as sequencing.

- Prepare a 3% low melting point agarose gel with 1X Tris acetate buffer
- Load entire 30 µl reaction on gel
- Visualize gel on uv light table
- Photograph for records
- Isolate bands with glass Pasteur pipette and place in 1.5 ml tube
- Add 100 µl of distilled water and heat at 70°C until gel dissolves. Note: The negative controls should of course be negative. However, the same region isolated for the sample should be isolated from the mock extraction lane as well to serve as a control for the re-amplification step.

Isolated bands can be stored at –20°C indefinitely

Protocol 6 Re-amplification

Re-amlipfication is oftentimes required for obtaining enough PCR product in ancient studies.

Set up an identical PCR as for the first amplification but do not add extract. Bring tubes with no template to the laboratory housing the PCR thermal cycler. Add 5 µl of the isolated first amplification product to reactions. Perform water control for re-amplification and include reactions for the isolated mock extraction from first amplification. Cycle 20 cycles with at least a 60°C annealing temperature.

Visualize 5 µl of the 30 µl reaction on a 2% agarose gel in TBE buffer stained with ethidium bromide.

Assuming negative controls remain negative, the sample PCR product is ready for cloning into an appropriate vector.

Note: At any point that contamination is observed, one must remake all PCR reagents and repeat the procedure to see if the contamination persists. In the worst case, the sample and mock extraction are contaminated and the extraction procedure must be repeated with fresh reagents. The most likely sources of contamination are human DNA from the researcher, or cow DNA presumably deriving from the BSA used in the PCR reaction [10]

Cloning and sequencing

Reagents for cloning:

- Competent cells (either commercially available or see Sambrook et al. [21] pp 1.74–1.84 for preparation)
- Ampicillin/X-gal/IPTG bacterial culture plates (see Sambrook et al. [21] A. 4)
- Cloning kit for PCR products (i. e., pGem T vector, Promega Inc.)

Note: It is now clear that for analysis of mtDNA from ancient samples, cloning of the PCR products from overlapping fragments from several amplifications is required to distinguish damaged positions from bona fide *base changes and to exclude nuclear integrations of mtDNA genes as the source of the mtDNA sequences obtained [8, 14, 18, 23, 24]. When working on nuclear DNA, it is even more crucial to sequence clones from multiple PCRs to exclude damage due to low copy of DNA targets and to distinguish between damaged templates and alleles in diploid organisms [8]. Direct sequencing will not detect mixtures of organellar mtDNA and nuclear insertions or damage-induced variation and will obscure alleles. These problems could lead to production of misleading data.*

The easiest cloning method for PCR products is via a cloning vector with T overhangs (TA cloning). Kits are commercially available. Once the PCR product is ligated into the vector, it can be transformed into either heat shock or electrocompetent cells which may be purchased commercially or made from a stock ([21] pp. 1.74–1.84) and plated on standard agar plates with X-gal/IPTG/ampiciilin selection.

Sequencing template preparation by direct PCR from individual colonies:

Buffer conditions for PCR per reaction

- 67 mM Tris-Cl pH 8.9
- 4 mM MgCl$_2$
- 16 mM ammonium sulfate

- 10 mM 2-mercaptoethanol
- 0.125 µM each dNTP
- 25 pmol each primer
- 1 unit Taq DNA polymerase

Note: reaction volumes should not be less than 30 µl. At lower volumes, bacterial proteins may become too concentrated and interfere with PCR.

Cycling profile:

5 minutes 94°C to denature bacterial cell proteins followed by 25 cycles of:
- 94°C 30 seconds
- primer annealing temperature (1 minute)
- 72°C 1 minute

Pick a white bacterial colony after blue white selection on X-gal/IPTG/ampicillin plates and place in a 30 µl PCR reaction as described above with primers specific for the vector, i. e., M13 forward and M13 reverse primers. The PCR product can be cleaned on a column (Qiagen) to remove primers, Taq DNA polymerase, and dNTPs. The cleaned product can then be subjected to standard sequencing procedures (see chapter 15 this book).

Note: An aliquot of the column purified sample should be run on a 2% agarose gel stained with ethidium bromide to ensure that the colony contains the correct size insertion.

5   Remarks and Conclusions

Does the result make phylogenetic sense? If the results of sequencing are highly discordant with molecular studies from extant taxa or from morphology, one should strongly suspect contamination [18]. In addition, with samples that are over 50,000 years old or derive from conditions not conducive to DNA preservation, results should be verified by having a portion of the sequence obtained reproduced in another laboratory. A section of the bone or tissue should be sent to another lab and taken through the extraction steps and PCR to see if another laboratory can retrieve the same DNA sequence. This is the most convincing proof of authenticity available at present.

References

1   Hagelberg E, Thomas MG, Cook CEJ, Sher AV et al. (1994) DNA from ancient mammoth bones. *Nature* 370: 333–334

2   Höss M, Pääbo S and Vereshchagin NK (1994) Mammoth DNA sequences. *Nature* 370: 333

3   Hauf J, Baur A, Chalwatzis N, Zimmermann FK et al. (1995) Selective amplification of a mammoth mitochondrial cytochrome b fragment using an elephant specific primer. *Curr. Genet.* 27: 486–487

4   Yang, H., Golenberg EM and Shoshani J (1996) Phylogenetic resolution within the Elephantidae using fossil DNA sequences from the American mastodon (*Mammut americanum*). *Proc. Natl. Acad. Sci. USA* 93: 1190–1194

5   Ozawa T, Hayashi S, Mikhelson VM (1997) Phylogenetic position of the mammoth and Steller's sea cow within tethytheria demonstrated by mitochondrial DNA sequences. *J. Mol. Evol.* 44: 406–413

6   Noro M, Masuda R, Dubrovo IA, Yoshida MC et al. (1998) Molecular phylogenetic inference of the woolly mammoth „Mammuthus primigenius, based on the complete sequences of the mitochondrial cytochrome b and 12S ribosomal RNA genes. *J. Mol. Evol.* 46: 314–326

7   Derenko M, Malyarchuk, Shields GF (1997) Mitochondrial cytochrome b sequence from a 33,000 year-old woolly mammoth (*Mammuthus primigenius*). *Ancient Biomolecules* 1: 149–153

8   Greenwood AD, Capelli C, Possnert G, Pääbo S (1999) Nuclear DNA sequences from late Pleistocene megafauna. *Mol. Biol. Evol.* 16(11): 1466–1473

9   Höss M, Dilling A, Currant A, Pääbo S (1996) Molecular phylogeny of the extinct ground sloth *Mylodon darwinii*. *Proc. Natl. Acad. Sci. USA* 93: 181–185

10 Taylor P (1996) Reproducibility of ancient DNA sequences from extinct Pleistocene fauna. Mol. Biol. Evol. 13: 2839

11 Poinar H, Hofreiter M, Spaulding WG, Martin PS, et al. (1998) Molecular coproscopy: dung and diet of the extinct ground sloth *Nothrotheriops shastensis*. *Science* 281: 402–406

12 Janczewski DN, Yuhki N, Gilbert DA, Jefferson GT et al. (1992) Molecular phylogenetic inference from saber-tooth cat fossils of Rancho La Brea. *Proc. Natl. Acad. Sci. USA* 89: 9769–9773

13 Hanni C, Laudet V, Stehelin D and Taberlet P (1994) Tracking the origins of the cave bear (*Ursus spelaeus*) by mitochondrial DNA sequencing. *Proc. Natl. Acad. Sci. USA* 91: 12336–12340

14 Krings M, Stone A, Schmitz RW, Krainitzki H, Stoneking M et al. (1997) Neanderthal DNA sequences and the origin of modern humans. *Cell* 90: 19–30

15 Béraud-Colomb E, Roubin R, Martin J, Maroc N et al. (1995) Human beta-globin gene polymorphisms characterized in DNA extracted from ancient bones 12,000 years old. *Am. J. Hum. Genet.* 57: 1267–1274

16 Lindahl T (1993) Instability and decay of the primary structure of DNA. *Nature* 362: 709–715

17 Höss M, Jaruga P, Zastawny TH, Dizdaroglu M et al. (1996) DNA damage and DNA sequence retrieval from ancient tissues. *Nucleic Acids Res.* 24: 1304–1307

18 Zischler H, Höss M, Handt O, von Haesler A et al. (1995) Detecting dinosaur DNA. *Science* 268: 1192–1193

19 Boom R, Sol CJ, Salimans MM, Janson CL et al. (1990) Rapid and simple method for purification of necleic acids. *J. Clin. Microbiol* 28(3): 495–503

20 Höss M, Pääbo S (1993) DNA extraction from Pleistocene bones by a silica-based purification method. *Nucleic Acids. Res.* 21: 3913–3914

21 Sambrook J, Fritsch EF, Maniatis T (1989) Molecular cloning: A laboratory manual. In: N Ford, C Nolan and M Ferguson (eds): Cold Spring Harbor Laboratory Press, New York

22 Handt O, Richards M, Trommsdorff M, Kilger C et al. (1994) Molecular genetic

analysis of the Tyrolean ice man. *Science* 264: 1775–1778

23 Greenwood AD, Pääbo S (1999) Nuclear insertion sequences of mitochondrial DNA predominate in hair but not in blood of elephants. *Mol. Ecol.* 8: 133–137

24 Zhang D, Hewitt GM (1996) Nuclear integrations: challenges for mitochondrial DNA markers. *Trends Ecol. Evol.* 11: 247–251.

14 PCR Methods and Approaches

James Bonacum, Julian Stark and Elizabeth Bonwich

Contents

Methods and Tools in Biosciences and Medicine
Techniques in molecular systematics and evolution, ed. by Rob DeSalle et al.
© 2002 Birkhäuser Verlag Basel/Switzerland

1   Introduction

While it is possible to perform molecular systematic studies without the Poly-merase Chain reaction (PCR; [1]), there is no doubt that PCR is responsible for the tremendous explosion of the field, both in terms of the number of sequences generated and the number of taxa which have been sampled. This technique, which at its most basic can be accomplished with three water baths, a pair of hands and the necessary reagents, has made it possible to quickly and easily search entire genomes for particular sequences. Researchers using PCR have extracted genetic information from ancient samples (reviewed in [2]; see also "Enjoy 577 Ancient DNA references (or even more)" http//www.comic.sbg.ac.at/staff/jan/ancient/references.htm) and as minimal a starting material as a single cell [3]. Its effect has been felt in all fields of biology ranging from medicine and forensic science to ecology and evolutionary biology.

2   The Basics

A PCR reaction is comprised of a combination of chemical components and a series of temperature manipulations, all of which can affect the outcome. The chemical components are template DNA from the organism of interest (this can consist of either genomic or mitochondrial DNA, or in most cases a combination of both), a pair of primers which flank the region of interest, a mixture of the four individual DNA bases in dinucleotide triphosphate form(dNTPs), DNA polymerase, and a buffer which provides the optimal salt concentration and pH conditions for the reaction to proceed. The one biochemical component that varies from reaction to reaction in PCR is the primer and we discuss their design and use below. Combinations of primer pairs which flank a chromosomal or organellar DNA region of interest are designed so that they will hybridize to opposite strands and provide the reaction with its ability to find a specific sequence within a sample of total cellular DNA.

The reaction is driven by controlled changes in temperature that accomplish the following three steps: denaturation, annealing and extension. During the denaturation step the reaction cocktail is heated to 95°C. At this temperature, the hydrogen bonds which link the strands of the double helix break apart, separating the two strands. The reaction is then quickly cooled to a temperature ranging anywhere from 40° to 60°C. It is important at this point to realize that a typical DNA extraction contains numerous copies of any particular gene region. As the reaction cools, the double helix will be reconstituted in many cases, but if the two primers have been added to the reaction mixture in sufficient quantity, they will frequently anneal to the sites they were designed to complement. Next the reaction is heated to a temperature that is based on the efficiency of the particular DNA polymerase that is being used (usually 72°C for reactions using *Taq* DNA polymerase). During this stage, the DNA polymerase extends the strands to which the primers have attached, using the opposite strand as a template, and using the dNTPs as the building material. If there were a single copy of the region of interest in the starting template DNA, after one cycle there would be two copies. Following a second cycle there would be four copies, and after a third, eight copies and so on. This means that if one started with a single copy, after 40 cycles there would be 1.1×10^{12} copies of the region of interest.

While PCR provides a quick and easy method for locating and amplifying specific regions of a particular organism's genome, it is obvious that there are numerous parameters that can be varied, all of which can affect the results of the reaction. In general, the parameters which are manipulated in the PCR reaction are the concentration of the chemical components, and the temperature and duration of time of the various steps in the reaction. In the following section we will address these individual parameters and the effect that varying them can have on the success of the reaction.

2.1 Chemistry

The success or failure of a PCR reaction is generally judged by visualizing the reaction on an agarose gel. Generally as little as 5 µl of the reaction product is sufficient to determine if a reaction has been successful. A successful PCR reaction will appear as a single bright band on an agarose assay gel. Reaction failures are indicated by smears, multiple bands, or no product at all. Assuming that the primers have been properly designed, and that the template DNA contains the sequence of interest, some amplification should be visible. Smeared multi-banded amplifications indicate that too much amplification is taking place, and that the primers, in addition to binding to their intended targets, are also binding to other sites in the template. In general these types of problems can be fixed more easily than cases where no amplification occurs. Reaction conditions must be optimized for every primer pair and sometimes for individual templates. Cases where no amplification is observed can also be optimized, but these tend to be more difficult as there are many potential reasons why this can occur. Due to the ability of this technique to amplify even minute amounts of template DNA, great care must be taken to prevent contamination when setting up PCR reactions. When performing PCR amplifications it is especially important to include a negative control, i. e., a reaction which contains all of the reagents used in the experimental amplifications except the DNA target template. If amplification is observed in the negative control it indicates that a source of DNA contamination has been introduced into the reaction. This contamination can come from the chemical reagents used in the reaction, the tubes, tips, or pipettors used to prepare the reaction, or as a result of human error when setting up the reactions. A number of simple precautions can help to minimize contamination (Protocol 1).

Protocol 1   PCR contamination and controls

1. Individual reagents should be stored in individual aliquots so that if the source of contamination can be traced to a particular reagent it will only be necessary to discard a small amount of it.
2. Exposing reaction tubes, tips and other disposable items to intense UV light will crosslink any contaminating DNA, thereby preventing it from amplifying.
3. Human error is the most difficult area to control. Researchers performing PCR reactions should always wear rubber gloves to prevent contamination when handling reagents or reaction tubes.
4. Distractions should be minimized when preparing PCR reactions. It is a very simple matter to forget to change a tip, thereby contaminating an entire tube of reagents. We also recommend that protocol checksheets with detailed steps listed be used to set up and execute PCR experiments. These sheets can have a list of the reactants on them that can be checked off as they are added to the PCR. The sheets can also keep track of the primers used, the cycling conditions used and the result of the experiment.
5. It is also a good idea to include a positive control, a reaction which uses a template which has successfully amplified in the past under the same conditions. If the positive control template amplifies while others do not, it indicates that amplification problems may be template-specific.

Templates which fail to amplify often contain impurities which have been carried over from the extraction process which can "poison" the PCR reaction. The best course of action in such cases is to perform a new extraction on the original tissue sample but in many cases this may not be practical. Various methods may be used to purify these templates. These include commercially available kits, spin columns, or re-extracting the contaminated template using phenol and/or chloroform. Protocol 2 lists several precautions and guidelines for the preparation of template DNA and for optimizing PCR reactions.

Protocol 2   Template purity and optimizing PCR reactions

1. When extracting DNA for use in PCR reactions, it is important to remove all traces of proteins and especially any phenol or chloroform residue, as these will have an adverse affect on the reaction. Urea, ethanol and sodium dodecyl sulfate (SDS), reagents commonly used in many extraction protocols, have been shown to inhibit the amplification process [31].
2. It is recommended that DNA extractions which prove difficult to amplify be diluted as this will also dilute the impurity and allow the amplification to proceed. It is important to bear in mind that PCR is an amplification process, and even minuscule amounts of template can yield good results. We recommend making a series of template dilutions ranging from 1:10 to as much as 1:10,000. In cases where the template is severely contaminated, however, this strategy may not be successful.

3. A simple experiment known as a "back-poison PCR" can be performed to determine if template impurities are inhibiting the reaction. Prepare four PCR reactions. One reaction should contain a positive control template. The second reaction should contain the template which has failed to amplify. The third reaction should contain a combination of both the positive control and the template which has failed to amplify, and the fourth reaction should be the negative control. If the positive control amplifies, but the reaction with the combined templates does not, it indicates that there is something in the suspect template which is inhibiting the amplification process.

4. In general, increasing the concentration of $MgCl_2$ in PCR reactions tends to promote binding of primers, and adding $MgCl_2$ can prove useful in cases where no amplification products are observed. If the concentration is too high, however, nonspecific primer binding can occur, resulting in amplifications that are multi-banded or smeared when visualized on an agarose gel. If this is the case, decreasing the amount of $MgCl_2$ may solve the problem.

5. Several manufacturers offer PCR optimizer kits which consist of a series of buffers which contain a graded series of different $MgCl_2$ concentrations. These buffers make it possible to quickly and easily assay the effect of a range of $MgCl_2$ concentrations while keeping all other reaction parameters constant.

There are a number of commercially manufactured buffers available for PCR which are usually shipped along with the DNA polymerase. These buffers contain 10 to 20 mM Tris-CL and 10–50 mM of either NaCl or KCl. pH values for PCR buffers vary between 7.5–8.8. Almost all of the commercial PCR buffers contain $MgCl_2$ and the concentration of this particular salt affects the ability of the primer to bind to the template. The concentration of the dNTPs does not appear to have much effect on the reaction as long as they are added in sufficient quantity. All four nucleotides must be added in equal amounts, and the final concentration generally varies between 0.2mM and 0.8mM. Aside from the DNA polymerase, dNTPs are the most expensive component of the reaction, and using the minimum amount is a good strategy for extending the lab budget. In addition to these components, it has been shown that a number of chemical additives can enhance both the specificity and the yield of PCR reactions, particularly in cases of GC rich templates. In Protocol 3 we list several of these additives and describe their use.

Protocol 3   Trouble-shooting and additives to PCR reactions

1. A number of additives have been found which enhance the amplification process. These additives usually function by either increasing the efficiency of the denaturation step, or by stabilizing the structure of the polymerase, but they should be used cautiously, however, as beyond a certain point they tend to inhibit rather than enhance the reaction process.

2. Certain templates can prove difficult to amplify due to their high G+C content [33]. It has been shown that the addition of the amino acid analogue betaine (N,N,N-trimethylglycine) [32] or tetramethylammonium chloride (TMAC) [35] can reduce the effect of base composition on the melting temperature of DNA, resulting in more efficient denaturation. A number of reports indicate that these additives can improve the results of PCR reactions when templates have a high G+C content. Henke et al. [34] showed that while betaine improved the amplification of GC-rich templates, the optimal concentration appears to be dependent on the GC concentration of the template. The optimal concentration varied between 1M for a template with a 66% GC concentration to 2.5M for a template with a 72% GC concentration. TMAC has been shown to improve amplification of GC-rich templates when added in concentrations in excess of 100mM [35]. It has also been demonstrated to increase both the specificity and yield of reactions using AT-rich templates when added in amounts between 15 and 120 mM [36]. Adding TMAC in excess of this amount, however, was shown to inhibit amplification.

3. Another approach to solving the problem of incomplete denaturation in GC-rich templates is to replace 75% of the dGTP in the reaction mixture with 7-deaza-2'-deoxyguanosine5'triphosphate (dc^{7}GTP) [37]. dc^{7}GTP forms a weaker GC base pair, facilitating denaturation.

4. Dimethylsulfoxide (DMSO) is also often used to improve the efficiency of the denaturing step. It achieves this effect by disrupting base pairing [35]. The concentration of DMSO, when used in conjunction with *Taq* polymerase, should not exceed 5%, as it has been shown to decrease amplification above this level [31]. Formamide is sometimes used to improve the specificity of PCR reactions in cases where the addition of DMSO has proven ineffective [38]. It is added in concentrations between 1.25–10%.

5. Another approach to enhancing the yield and specificity of PCR reactions is to stabilize the structure of the polymerase, thereby improving its efficiency. Gelatin, Bovine Serum Albumin (BSA) and a number of nonionic detergents, such as Tween-20 or NP-40, are common additives which have this effect [39]. In addition to stabilizing the polymerase, nonionic detergents can also block the inhibitory effect of SDS. BSA is added in concentrations ranging from 0.001–0.1%. Gelatin was a component of the first buffer used with *Taq* polymerase [4]. These reactions used 0.02% gelatin although Stommel et al. [39] have shown that gelatin derived from different animal sources has different effects on the ability to improve reaction specificity and yield, and the actual concentration may vary depending on the type of gelatin used. Glycerol may also be added in amounts between 2.5–20% in order to increase specificity [31].

The final component of the reaction is the DNA polymerase which performs the actual sequence replication. In 1988 the DNA polymerase of *Thermus aquaticus*, a species of Archaebacteria, was successfully isolated [4]. As this enzyme functions effectively in nature at temperatures in excess of 95°C, it is capable of withstanding the extreme heat of the denaturing step. This enzyme, commonly referred to as *Taq,* has made it possible to automate most of the PCR process, as it can be added to the individual reactions before any cycling has begun. The reactions can then be placed in a PCR machine which raises and lowers the temperature of the reaction according to the instructions of the user. There are a wide variety of kinds of PCR machines commercially available (Appendix, commercially availably PCR machines). Since the introduction of *Taq* polymerase, additional thermostable DNA polymerases have been isolated, notably Vent®, from New England Biosystems, which in addition to providing polymerase activity, also provides a proofreading function by removing erroneous base incorporations which can occur with *Taq.*

2.2 Temperature and cycling conditions

Although the denaturing and extension portions of PCR always occur at 95° and 72°C respectively, the annealing temperature varies for each individual primer pair and often for individual templates. Annealing temperatures usually range from 40° to 60°C. A simple rule of thumb is that higher annealing temperatures promote more specific binding while lower annealing temperatures provide less stringent conditions. As we noted earlier, successful PCR reactions appear as a single bright band when visualized on an agarose gel. If the annealing temperature is too low, the primers will bind to additional sites on the template, resulting in multi-banded or smeared amplifications. If the annealing temperature is too high, on the other hand, the primers will not be able to bind to the target sequences and the gel will appear blank as no amplification will have taken place. In general, if the results are smeared or multi-banded, the annealing temperature should be raised, while if no reaction product is seen, the annealing temperature should be lowered. Changes in the annealing temperature should be made in small increments, as a difference of one or two degrees can have a dramatic effect on the success of the reaction. Protocol 4 discusses several methods that we have used that alter temperature or cycling parameters to increase the efficiency of PCR reactions.

Protocol 4 Altering temperature of cycling parameters

1. The starting point in temperature optimization is important; the melting temperature (T_m) of a primer is the temperature at which 50% of the copies of a particular primer will anneal to its target sequence. This temperature can be approximated by adding 4°C for every G or C the sequence contains and 2°C for every A or T in the sequence. After calculating the T_m for each primer, subtracting 10°C from the lower temperature in the pair often gives an accurate estimate of the appropriate annealing temperature. While this estimate may not yield perfect amplifications on the first attempt, it can usually provide a place to begin the optimization process.

2. The length of time and the number of cycles can also be varied during the optimization process, and it is a good idea to minimize the length of time spent at each step in the process as well as the number of cycles. This is because longer annealing times can lead to non-specific primer binding. Along with the target sequence there are almost always additional sequences which are amplified in very low concentrations due to the fact that the primers do not properly match the sequences. By increasing the number of cycles, the amplification and concentration of these products can increase to the point where they compromise the specificity of the reaction.

3. Primer pairs which prove difficult to optimize can sometimes benefit from a technique known as ramping. This is particularly helpful in cases where no amplification occurs above a certain temperature, but smeared amplifications occur below this threshold. The first three or four cycles of the reaction are performed at a low annealing temperature, making it possible for some amplification to take place. Although this makes it possible for non-specific amplification to take place, and a certain amount of "junk" sequences are amplified along with the target sequence the concentration of the target sequence, will be increased. When the remaining cycles are performed at the higher annealing temperature the specificity of the reaction will increase and the primers will preferentially amplify the target sequence .

4. In addition to the standard reaction profile of denature/annealing/extension, most workers add two additional steps to the profile, one at the beginning and one at the end. In the beginning it is advisable to heat the reaction to 95°C and hold it there for 2–5 minutes to ensure complete denaturing of the template. Following this long denaturing step the standard PCR cycling begins and continues until the optimal number of cycles has been reached. After the last round of extension has been performed, a final hold of 10 minutes at the extension temperature is performed in order to both maximize the yield of the extension products and to use up any leftover polymerase.

As we have pointed out, non-specific amplification is one of the most common causes of PCR reaction failure. When visualized on an agarose gel, these types of reaction failures usually appear as smears or multiple bands. This is due to the fact that while there is an optimal temperature for the polymerase to operate, it actually functions over a much broader range of temperatures, and non-specific amplification can begin to occur as soon as the polymerase is added to the reaction. For this reason many workers prepare a reaction cocktail containing all of the necessary reagents, adding the template during the final step before placing the reactions in the PCR machine. Another procedure frequently used to combat this problem is to keep the individual PCR reactions on ice until they can be placed in the PCR machine. Even these techniques may not be sufficient to prevent this problem, however, and it may be necessary to perform a hot start (see Protocol 5).

Protocol 5   Hot starting PCR reactions

1. A hot start prevents non-specific amplification by heating the PCR cocktail to 95°C before allowing the polymerase to come into contact with the rest of the reagents in the reaction. There are several ways to accomplish this. The simplest is to simply place the reactions into the PCR machine and allow the heating block to reach 95°C, and then manually add the polymerase to the individual reaction tubes. Mineral oil can then be added to the tube if needed and the tubes are capped. The machine is then programmed and the regular PCR run can be started.
2. A second method is to place a small bead of paraffin into the tube containing the template, dNTPs and water. The tubes are then heated to melt the paraffin and then cooled. The paraffin, being lighter than water, will then form a cap on the top of the reaction mix. The buffer, primers and polymerase can then be placed on top of this cap and the reactions are then returned to the machine. The cap keeps the individual reagents separate until the temperature is high enough to melt the cap, at which point the reagents mix and the reaction can proceed.
3. *Amplitaq* Gold achieves a similar effect by biochemical means. This version of *Taq* polymerase is bound to antibodies which render it incapable of performing its function. These antibodies disassociate when the critical temperature is reached, allowing the reaction to proceed.
4. Commercially available beads pre-soaked with Taq polymerase, nucleotides and buffer are also available for hot-start reactions.

3   Cloning and Sequencing PCR Products

As we noted earlier, the specificity of PCR makes it possible to isolate a particular genomic region without resorting to the laborious process of cloning, but there are situations in which the cloning of PCR products is required. Often PCR products are used as templates to synthesize oligonucleotide probes. In cases where these probes are made from RNA, it is necessary to insert the PCR product in a plasmid vector in order to accomplish this. Although it is very common to directly sequence PCR products using the same primers that were used for the amplification process, there are occasions when this is not possible, for example when using highly degenerate primers. In other cases it may be difficult to optimize the conditions of the sequencing reaction for a particular primer. Fortunately, due to the high concentration of the target seqeunce in a PCR reaction, they are usually quite easy to clone.

The simplest method is to use one of the commercially available kits which are designed specifically for this purpose. These kits take advantage of a unique side-effect of *Taq* polymerase. *Taq* adds an additional adenine nucleotide to the 3' end of each amplified strand. This makes it possible for specially designed plasmid vectors, which in their linearized state have an extra thymine residue on their 3' ends, to readily incorporate PCR products. An alternative method is to include a restriction site on the 5' end when designing primers. As long as the primers provide a good match on the 3', end they should still bind to the target sequence. It is also necessary to add three or four additonal bases on the 5' end as it is necessary to digest the PCR product in order to create a "sticky end" for the ligation reaction. By designing each primer with a different restriction site, it is possible to control the orientation of the insert in the vector. PCR can also be used to simplify the screening of clones (see Protocol 6). There are several good websites that discuss the nuances of PCR product cloning (http://www.the-scientist.com; http://www.invitrogen.com/catalog_project/cat_topohtp.html.)

Protocol 6   Colony PCR

A quick method that works quite well for rapid isolation and screening of plasmas is the colony PCR technique. This method uses the PCR reaction to amplify recombinant plasmid sequences straight from a picked colony. The solution needed for the reaction is a regular 10X PCR buffer. The reactions are accomplished in a final volume of 30 µl by preparing enough 1X PCR buffer with primers, TAQ polymerase, nucleotides and dH$_2$O for the number of clones to be screened. We routinely use microtitre dishes to screen 96 clones or 384 clones at a time. .For example, for 96 well dishes we prepare 3 ml of reaction mixture (300 µl of 10X, buffer, 300 µl of 8mM DNTP's, 10 µl of forward universal primer (100 µM concentration), 10 µl of reverse universal primer (100 µM concentration) and 10 µl of TAQ polymerase)

1. Dispense 30 µl of the 10X PCR solution into each well of the microtiter dish.
2. Pick colonies that appear to be recombinant (either through color selection or drug resistance) with a pipette tip or a toothpick.
3. If desired before placing the picked colony in the reaction well or tube, streak a master plate so that the colony can be recovered later in larger amounts, then place the picked colony in the well or tube.
4. We have found that the trick to this reaction is not to overpick the recombinant colonies. Slightly touching the colonies is usually sufficient to obtain enough cells to drive the reaction.

5. Also it only takes a very brief period of sitting in the reaction buffer to drive the PCR reaction. 10 to 15 seconds is sufficient.
6. The microtiter dish or tubes are placed in a thermal cycler and processed 25 cycles for 30 seconds at 96°C, 15 seconds at 50°C and 15 seconds at 72°C.
7. A final cycle of 72°C for ten minutes can be added but we usually omit this step.
8. PCR products can be assayed via 1X TBE, 1% agarose gels using a 100 bp ladder as a marker.

4   Troubleshooting

The World Wide Web contains a vast number of sites devoted to PCR, many of which are maintained by manufacturers of PCR supplies and reagents. Often these sites contain troubleshooting guides which generally appear in the form of a table with a particular PCR problem listed in one column and possible causes and solutions listed in adjacent columns. These sorts of sites are useful for quickly reviewing potential causes of problems (see Appendix, troubleshooting web sites).

5   Primer Design

Primer design is perhaps the most critical step in successful PCR reactions. Degeneracy of the primer is a major concern in design [5]. Degeneracy is dependent on the amino acid: one codon (methionine and tryptophan) to six (arginine, leucine, serine) codons (Fig. 1). The more degenerate the amino acid sequences used in designing a primer, the less specific it will be with respect to the sequences it anneals to. Depending on the gene region and objective, this

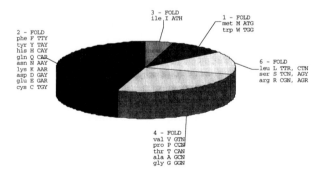

Figure 1 Diagram showing the degeneracy of the genetic code. The diagram indicates the frequency of the various degeneracy classes as well as the single letter and three letter shorthand notations for the twenty amino acids. The genetic code for the various amino acids is also indicated. IUPAC Codes, H = A,C, or T; N = G,A,C or T; R = G or A; Y = C or T.

might be more or less desirable. If the primer is designed from a region that is highly conserved over a wide range of taxa or genes, it may amplify too many sequences to be of much use, unless the goal was to look specifically for as many regions possessing a particular motif, such as a homeobox region, as possible.

The total degeneracy value is the product of the individual degeneracies of each of the amino acids in the primer sequence. For example, the primer based on the amino acid sequence MCIAL would be 1*2*3*4*6 = 144-fold degenerate. This translates practically into avoiding the use of four- or six-fold degenerate amino acids wherever possible when designing primers. Typically, degeneracy is limited to a value at or below 516, which is considered a practical upper limit [5]. If doing systematic work where the goal is the amplification of a particular gene region, rather than looking for several similar regions in several genes, degeneracy should probably be as low as practically possible: in practice, we find that the upper limit is 32 to 64-fold. Degeneracy at the 3' end of the primer should be avoided, because a mismatch in this region might prevent further extension [5]. Degeneracy is also affected by codon bias, or how amino acids are coded for in particular taxa or groups of taxa. This can also be considered in primer design. In this case, only the codons used in this taxon or group could be used, further increasing the stringency of the primers. Stop codons should not be ignored if they happen to be in an appropriate region. Another "trick" that may be used, especially with short primers, is to add a 6 to 9 base 5' extension to the primer, which will amplify after the first cycle and increase annealing stability [5]. There is no better way to find out how primers might function than by trying them out, however: primers thought to be suboptimal might turn out to be useful (or vice-versa).

Primers can be designed using manual inspection of aligned sequences (see Protocol 7). A search of the literature is an easy way to obtain sequence information for primer design. Articles dealing with any manner of character-ization of a particular gene or genes very often will either give sequence information on the primers used in the study, and/or provide aligned amino acid sequences in which a gene sequence is compared with similar sequences in other organisms. If the sequence is long enough, several forward and reverse primers can be designed to amplify the desired gene region. The targeted amino acid sequences can be manually translated into degenerate nucelotide se-quences or the short amino acid stretches can be fed into an oligo-determining program such as CODEHOP (see below). Primers for structural RNAs are designed by visual inspection of aligned RNA sequences. The conservation of sites determined visually can be enhanced by using a sequence alignment program that allows for the viewing of sequences in a matrix that shows only variable positions (the matchchar=. option in PAUP or the match first character = . option in Macclade).

Protocol 7   Some common-sense guidelines to designing primers using visual inspection

1. Published amino acid alignments are visually inspected for regions of similarity between two or more distantly related taxa.
2. Once a region of five or more exactly similar amino acids is determined, the region is inspected for the occurrence of six-fold degenerate amino acids. If one occurs in the region, we generally accept such an amino acid stretch as a potential primer site only as a last resort.
3. Acceptable conserved amino acid stretches are back-translated using a genetic code table and a codon bias table for the species of interest.
4. Only the first four amino acids on the 3' end of the designed primer are degenerated in the third position, leaving the 5'-most amino acids designed with codon bias in mind.
5. Generally, we design oligonucleotides of between 20 and 25 nucleotides and with a GC to AT ratio of about 1:1.
6. The 3'-most nucleotide should never reside in the third position of a codon unless that triplet codes for M or W.
7. For primers designed for RNA genes such as rDNA or tRNA genes, the 3'-most nucleotides in the primer should be designed to be a G or a C if at all possible.
8. Positions that vary among taxa in RNA gene alignments should be designed to be degenerate in those positions.

Primer Design Example: Primers from published alignments

Primer design programs are widely available on the World Wide Web and easily accessible by use of search engines. Specifically, we show primer design procedures using the CODEHOP (COnsensus-DEgenerate Hybrid Oligonucleotide Primers) program (for website see Appendix, websites with primer design programs). Below we show a region from the tgo (tango, a bHLH-PAS protein) locus in *Drosophila* that is orthologous to the human Arnt locus (a human bHLH-PAS protein) with highly similar amino acid sequences from Sonnenfeld et al. [6]:

FLY tgo	EIERRRRNKMTAYITELSDM VPTCSALARKPDKLTILRMAV
HUM arnt	EIERRRRNKMTAYITELSDM VPTCSALARKPDKLTILRMAV

CODEHOP will provide the user with a wide range of parameters about the sequence and its possible performance after the desired primer sequence has been input. The primer annealing may be optimized using an extensive codon usage table built into the program. The information required for the use of this program can be found in the CODEHOP documentation. CODEHOP, as well as several other primer design programs, will degenerate the first four amino acids in a potentially good amino acid stretch. The last three or four amino acids are not degenerated and are called a CLAMP in CODEHOP. The clamp will oftentimes need to be extended in order for the designed primer to be useful. Pressing the "Look for Primers" key on the CODEHOP page tells the program to design primers for the block above. The program then displays a page with the designed primers as shown in Figure 2. As can be seen, several primers are designed in both directions. The annealing temperature and degeneracy of the designed primer are given in this format. In some cases the CLAMP region needs to be extended (primers above where "Extend clamp" appears). The extension of clamps can be accomplished manually or by making a new set of blocks for CODEHOP.

Primer Design Example 2: Automated primer design using GADFLY and CODEHOP.

Design of primers can be accomplished more readily and easily using existing databases. As an example, we illustrate a primer design procedure using GadFly as accessed in Flybase (flybase.bio.indiana.edu:82/) and the CODEHOP program. By searching GadFly, the new annotated *Drosophila* genome database, one can focus on any one of the 14,000 or so *Drosophila* genes in the database. Part of the annotation of genes in the GadFly database includes a list

Block tango__

EIERR
oligo:5'-GAGathgarmgnmg-3' degen=96 temp=-62.1 Extend clamp
RRRRNKMT
oligo:5'-CGGCGGCGGCGnaayaaratga-3' degen=16 temp=63.4
RRRNKMT
oligo:5'-GCGGCGGCGGaayaaratgac-3' degen=4 temp=61.3
RRRNKMTA
oligo:5'-GCGGCGGCGGAAyaaratgacng-3' degen=16 temp=61.3
RRRNKMTA
oligo:5'-CGGCGGCGGAACaaratgacngc-3' degen=8 temp=62.7
RRNKMTAY
oligo:5'-GCGGCGGAACAAGatgacngcnta-3' degen=16 temp=62.8
AYITELSDMV
oligo:5'-CGCCTACATCACCGAGCTGwsngayatggt-3' degen=32 temp=62.9
AYITELSDMVP
oligo:5'-GCCTACATCACCGAGCTGTCngayatggtnc-3' degen=32 temp=60.8
ITELSDMVP
oligo:5'-CATCACCGAGCTGTCCgayatggtncc-3' degen=8 temp=62.1
ITELSDMVPT
oligo:5'-CATCACCGAGCTGTCCGAyatggtnccna-3' degen=32 temp=62.1
ELSDMVPT
oligo:5'-CGAGCTGTCCGACatggtnccnac-3' degen=16 temp=60.9
ALARKPDK
oligo:5'-GCCCTGGCCCCGGaarccngayaa-3' degen=16 temp=63.0
LARKPDKL
oligo:5'-CCTGGCCCCGGAAGccngayaaryt-3' degen=32 temp=63.9
ARKPDKLT
oligo:5'-GCCCGGAAGCCCgayaarytnac-3' degen=32 temp=63.1
PDKLTILRMA
oligo:5'-CCCGACAAGCTGACCATCytnmgnatggc-3' degen=64 temp=60.2
DKLTILRMAV
oligo:5'-CCGACAAGCTGACCATCCTnmgnatggcng-3' degen=128 temp=61.2
DKLTILRMAV
oligo:5'-CGACAAGCTGACCATCCTGmgnatggcngt-3' degen=32 temp=60.4

Complement of Block tango__

EIERRRR
ctytadctykcCGCCGCCGCC oligo:5'-CCGCCGCCGCckytcdatytc-3' degen=24 temp=61.9
IERRRRNKM
tadctykcnkcCGCCGCCTTGTTCT oligo:5'-TCTTGTTCCGCCGCcknckytcdat-3' degen=96 temp=60.5
NKMTAYITE
ttrttytactgGCGGATGTAGTGGC oligo:5'-CGGTGATGTAGGCGgtcatyttrtt-3' degen=4 temp=61.9
KMTAYITELS
ttytactgncgGATGTAGTGGCTCGACAG oligo:5'-GACAGCTCGGTGATGTAGgcngtcatytt-3' degen=8 temp=60.8
DMVPTCSA
ctrtaccanggGTGGACGAGGCGG oligo:5'-GGCGGAGCAGGTGggnaccatrtc-3' degen=8 temp=62.6
KPDKLTILR
ttyggnctrttCGACTGGTAGGACGCC oligo:5'-CCGCAGGATGGTCAGCttrtcnggytt-3' degen=16 temp=62.4
PDKLTILRM
ggnctrttyraCTGGTAGGACGCCTAC oligo:5'-CATCCGCAGGATGGTCaryttrtcngg-3' degen=32 temp=60.2
DKLTILRMA
ctrttyrantgGTAGGACGCCTACCG oligo:5'-GCCATCCGCAGGATGgtnaryttrtc-3' degen=32 temp=63.1
RMAV
ankcntaccgncAC oligo:5'-CAcngccatnckna-3' degen=128 temp=-135.3 Extend clamp

Figure 2 List of primers designed by CODEHOP for the tango region for *Drosophila* and human orthologs.

of genes in other databases that show similarity to the *Drosophila* entries. Figure 3 shows an annotation page from GadFly for the *Drosophila* signal transduction protein *wingless*.

The sequences at the bottom of this page can be retrieved from the relevant databases and used as input into the CODEHOP program, by simply clicking on the *Drosophila* sequence (accession # – JO3650) and on the human sequence (accession # 4885655). To use these sequences in CODEHOP, blocks must first be made that correspond to regions of the sequences that are of high similarity. The blocks are made with the "blockmaker" function in CODEHOP. The best way to accomplish the construction of the blocks in blockmaker is to keep the sequences in FASTA format. An example of the blockmaker format for the human and *Drosophila* sequences mentioned above for *wingless* is given in Figure 4. Note that it doesn't matter if the sequences are formatted with sequence position numbers or not.

Blockmaker will align the two sequences and choose areas (termed blocks) of maximal similarity for further consideration for primer design. For the *wingless* example, the blocks determined by the blockmaker program are shown in the reproduced webpage in Figure 5. The blocks from this page can then be used in CODEHOP to design primers for these blocks by utilizing the [CODEHOP] button in the "Primers" line in Figure 5. Doing so will automatically take the blocks from this page and put them directly into the CODEHOP page. Once the blocks

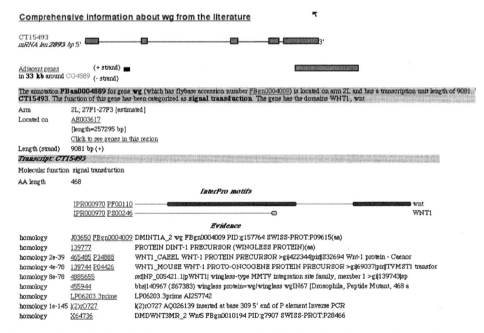

Figure 3 Part of the wingless annotation from GadFly. The list of entries below "Evidence" are the genes with similarity to *Drosophila* wingless. Clicking on the accession numbers of these genes retrieves them from the database.

>flywingless

MDISYIFVICLMALCSGGSSLSQVEGKQKSGRGRGSMWWGIAKVGEPNNITPIMYMDPAIHSTLRRKQRRLVRDNP

GVLGALVKGANLAISECQHQFRNRRWNCSTRNFSRGKNLFGKIVDRGCRETSFIYAITSAAVTHSIARACSEGTIESCT

CDYSHQSRSPQANHQAGSVAGVRDWECGGCSDNIGFGFKFSREFVDTGDRGRNLREKMNLHNNEAGRAHVQAEM

RQECKCHGMSGSCTVKTCWMRLANFRVIGDNLKARFDGATRVQVTTSLRATNALAPVSPNAAGSNSVASNGLIIPQ

SGLVYGEEEEERMLNDHMPDILLENSHPISKIHHPNMPSPNSLPQAGQRGGRNGRRQGRNHNRYHFQLNPHNPEHKP

PGSKDLVYLDPSPSFCEKNLRQGILGTHGRQCNETSLGVDGCGLMCCGRGYRRDEVVVVDRCACTFHWCCEVKCK

LCRTKKVIYTCL

>humanwingless

  1 mglwallpgw vsatlllala alpaalaans sgrwwgivnv asstnlltds kslqlvleps

 61 lqllsrkqrr lirqnpgilh svsgglqsav reckwqfrnr rwncptapgp hlfgkivnrg

121 cretafifai tsagvthsva rscsegsies ctcdyrrrgp ggpdwhwggc sdnidfgrlf

181 grefvdsgek grdlrflmnl hnneagrttv fsemrqeckc hgmsgsctvr tcwmrlptlr

241 avgdvlrdrf dgasrvlygn rgsnrasrae llrlepedpa hkppsphdlv yfekspnfct

301 ysgrlgtagt agracnsssp aldgcellcc grghrtrtqr vtercnctfh wcchvscrnc

361 thtrvlhecl

Figure 4 Example of formatting for blockmaker in CODEHOP for human and *Drosophila* wingless sequences.

```
BLOCKS from MOTIF

Logos:    [Postscript][PDF][GIF]  [About Logos]  [Map]
Tree:     [Data] [XBitmap] [Postscript] [PDF] [GIF] [Newick]   [About Trees]
Structures: [3D Blocks]   [About 3D Blocks]
Search: [LAMA]  [About LAMA]   [MAST]   [About MAST]
Primers:[CODEHOP]  [About CODEHOP]


            **BLOCKS from MOTIF**

>wingles family
2 sequences are included in 6 blocks

        winglesA, width = 43
   flywingless     64 LRRKQRRLVRDNPGVLGALVKGANLAISECQHQFRNRRWNCST
   humanwingless   64 LSRKQRRLIRQNPGILHSVSGGLQSAVRECKWQFRNRRWNCPT

        winglesB, width = 41
   flywingless  (  11)  118 KIVDRGCRETSFIYAITSAAVTHSIARACSEGTIESCTCDY
   humanwingless (  8)  115 KIVNRGCRETAFIFAITSAGVTHSVARSCSEGSIESCTCDY

        winglesC, width = 38
   flywingless  (  20)  179 DWECGGCSDNIGFGFKFSREFVDTGDRGRNLREKMNLH
   humanwingless (  8)  164 DWHWGGCSDNIDFGRLFGREFVDSGEKGRDLRFLMNLH

        winglesD, width = 34
   flywingless  (   4)  221 GRAHVQAEMRQECKCHGMSGSCTVKTCWMRLANF
   humanwingless (  4)  206 GRTTVFSEMRQECKCHGMSGSCTVRTCWMRLPTL

        winglesE, width = 22
   flywingless  (  13)  268 ATRVQVTTSLRATNALAPVSPN
   humanwingless ( 26)  266 ASRAELLRLEPEDPAHKPPSPH

        winglesF, width = 48
   flywingless  ( 132)  422 DGCGLMCCGRGYRRDEVVVVDRCACTFHWCCEVKCKLCRTKKVIYTCL
   humanwingless ( 35)  323 DGCELLCCGRGHRTRTQRVTERCNCTFHWCCHVSCRNCTHTRVLHECL
```

Figure 5 Blocks generated by blockmaker for the wingless sequences from human and *Drosophila*. On the Primers line is the [CODEHOP] button that will automatically deliver these blocks to the codehop program.

are in the CODEHOP page, one needs only to press the "Look for Primers" button and the program will design primers for each of the blocks shown in Figure 5. The format output of the designed primers is identical to that shown in the example in Figure 2. Primers can then be chosen on the basis of their location, degeneracy and annealing temperature.

6   Nested Primers

Primers may be designed as *nested,* in which the priming sites of an inner pair are contained within the nucleotide sequence of a fragment produced by another pair of primers. Nested primers can function as a check on the specificity of the fragment amplified, in the case of multiple genes with identical motifs, repeats, or other forms of common sequence [1]. In this application, the amplification is done as a two-stage process, first with the more inclusive outer primer pair. The PCR product(s) of this reaction, perhaps representing amplifications from several genes, are then used as templates for the second reaction with the internal nested set, specific to the gene of interest. In this way, greater yields are often obtained than if the internal set had been used exclusively.

7   Inverse PCR

This is a method for amplifying gene regions flanking primer sequences. Restriction digestion of source DNA results in some fragments containing the PCR core region and flanking sequences. The fragments are then ligated. A PCR reaction with primers directed outward from the core and annealing to the circularized fragment containing the core and flanking regions will result in amplification of the flanking sequences. The size of this fragment is limited by the maximum practicable amplification size, about 3 kb.

Several practical considerations must be accounted for in setting up the inverse PCR reaction [5]. If the core region is uncut by the restriction enzyme(s), both flanking sequences are obtained. However, if there is a cut within the core region, upstream or downstream sequences can be selectively obtained. Size-fractionation of putative core-region-containing fragments can be used to enrich yields.

Applications for inverse PCR include short genome walking for the purpose of orienting lambda, cosmid, YAC, or restriction fragment libraries. It may also be used for site-directed mutagenesis, by incorporating a base change within one of the primers [7], by having the primer 5' ends hybridize to adjacent bases on the circular DS DNA containing the region of interest, and/or by having the 3' ends prime synthesis in the opposite direction. After amplification, the 5' end is phosphorylated (T4 polynucleotide kinase) to facilitate host transformation.

8   "Jumping" PCR

This method of using PCR to construct longer-length fragments from highly degraded DNA was developed by Nasidze and Stoneking [8]. In this approach, a target region of interest is first specified. Nasidze and Stoneking chose the Hypervariable region 1 of human mtDNA, the DNA source material being a Paleolithic tooth. Primers were designed spanning this region in short (the longest being 159 bp) fragments. Results of these initial PCR reactions are then gel-purified to remove excess primer. Rather than then having to sequence each of the fragments, forward and reverse primers from any pair, or the pair spanning the entire region, can be used to generate a recombinant product which matches the original full-length sequence, because some of the products from each primer will extend long enough to be primed in the other direction. By limiting the PCR reactions to 15–20 cycles, *Taq* polymerase errors are minimized. In addition, by designing the primer sequences with identifiable insertions deviating from the natural sequences, "authentic" template sequences can be distinguished from overlapping product sequences containing primer sequences internally.

9   "PCRASS" (PCR Assays of Variable Nucleotide Sites for Identification of Conservation Units)

This technique [9], is a method for assaying the uniqueness of given populations in order to establish parameters for conservation decisions. Advantages of the approach are the ability to use relatively easily obtainable and preservable DNA sampling methods from live animals (e. g., hair, whole blood) or museum specimens in a quick and highly discriminatory assay. Polymorphic sites of interest are first determined by the sequencing of preliminary samples from various populations, which may or may not be preceded by a RFLP technique (PG-RFLP). Restriction sites are incorporated within primer sequences in the variable region, although this will not unambiguously identify specific base changes [9]). Non-degenerate primers are designed for each of the variable sequences, such that a mismatch will result in non-amplification of the complementary sequence. Each primer can then be used as both a positive and a negative control on the amplification itself, and any ambiguous amplification product, perhaps resulting from a previously unencountered polymorphism, can be sequenced for identification. The presence of one or more fixed base differences between populations, unique to each population, is the criterion for the conservation unit.

10 Asymmetric PCR SSCP (Single-strand Conformation Polymorphism)

This technique is useful in assaying for single base allelic differences in the screening of samples from different individuals. A symmetric PCR reaction is first performed with sense and antisense primers for the region of interest, and the product purified. A second PCR is performed on the products of the first in the sense or antisense direction only (determined empirically). Homozygotes and heterozygotes for each allele are clearly identified using this technique, because the purification step and asymmetric PCR eliminate extraneous sense-antisense combinations, which usually show up as a smaller fragment, and thus reduce the asymmetric signal when sense and antisense primers are used simultaneously [10].

11 RT-PCR (Reverse Transcription PCR)

Reverse transcription PCR is a method of using RNA transcripts as the source material for a PCR reaction. The source material may be from total RNA, a fraction thereof, or an RNA virus. Several types of RNA virus-derived reverse transcriptases, such as Avian myeloblastosis virus-RT (AMV-RT) or Moloney Murine Leukemia Virus-RT (M-MLV-RT), or DNA polymerases which show reverse transcriptase activity, such as *rTth*, have been used for reverse transcription [13, 14]. The use of RNA as the source material allows the use of RT-PCR as a gene expression assay, and also allows sequencing of cDNAs derived from mRNA transcripts. There are several variations on the technique, mainly differing in the separation of the reverse transcription and amplification reactions and the enzymes used [13, 14]. In the "Uncoupled Reaction" [13], the reverse transcription from total RNA is done separately, and then PCR is performed in a separate reaction using the cDNA as the template. In the "Continuous Two Enzyme method" [13], both the reverse transcription and amplification occur in the same reaction volume. Finally, using the dual properties of the *rTth* enzyme, the "Continuous One Enzyme method" also occurs in the same reaction volume [13]. The "Continuous Two Enzyme" method was found to be the most sensitive, and the "uncoupled" method the least [13]. The "Two Enzyme" method also uses the least reagents, the "Single Enzyme" method the most polymerase. Of the various other advantages ascribed to the continous methods, primary is the decrease in the possibility of contamination due to less handling. The "Uncoupled" method has the advantage of the superior storage properties of cDNA *versus* RNA. It also has a lower error rate than methods using the *rTth* polymerase, which additionally suffers from the inability to distinguish amplification from RNA or contaminating DNA pseudogenes. [13, 14]. A further method, nucleic acid sequence-based amplification (NASBA), relying on RNA T7 Polymerase to generate single-stranded RNA from genomic

DNA, which is then reverse-transcribed, was found to have a high error rate if this was done with *rTth* [12].

12   Gene Walking with PCR

This is accomplished with a simple strategy, using a 14–16 base primer consisting of a 5' 10 base degenerate portion and a 4–6 base invariant portion. The idea is that the 4–6 base section will find an exact match in a region of unknown sequence, and the degenerate portion will allow annealing to the complement. If the random primer is combined with primers designed from a region of known sequence, an amplification product may result. If the reaction is repeated using a series of nested primers, with each round using a primer closer to the unknown region, more and more of the unknown sequence can be reliably amplified, and then sequenced. The process can then be repeated using primers designed from the newly sequenced region, and gradually walked further and further along the chromosome.

13   "Long" PCR

This gene-walking technique has been demonstrated to generate much longer amplification products than most other PCR methods [13–15]. In a three-step reaction, an initial primer specific to the 3' end, but not at the end of the 3' sequence, generates ssDNA product which includes the unknown region. This product is tailed with poly (dC) using *Tdt*, and then used as template for another PCR reaction using a primer designed closer to the 3' end of the known sequence, and an adapter primer with guanines and an *Eco*RI site at the 5' end, which will anneal to the poly dC tail. In this way, fragments up to 35kb have been generated [13]. Using a second primer spaced closer to the 3' end of the known sequence, but not at its end, also provides a control to establish the identity of resulting products: the initial sequence 5' to the second primer must match the known sequence in a non-artifact [15].

14   DOP-PCR (Degenerate-Oligonucleotide-Primed-PCR)

This technique makes use of degenerate primers to increase the amount of DNA template available for analysis [3, 16]. By using highly degenerate primers, large tracts of genomic DNA are made available for amplification and further analysis in the form of short (~500bp) fragments [16]. This is particularly useful in situations where small quantities of template are expected, such as in the screening of neoplasms for mutant sequences [16].

15 AFLP – PCR (Amplification Fragment-Length Polymorphism – PCR)

This technique, a PCR-based successor to RFLP (Restriction-Fragment-Length Polymorphism), is used to determine polymorphic loci and allow the separation of discrete phylogenetic units, primarily at low levels. Genomic DNA is restriction-digested with one [17] or two different restriction enzymes (frequent and infrequent cutter), and the cut fragments are "pre-amplified" with primers specific to the restriction site [18, 19]. The PCR products of this reaction are then amplified with so-called adapter primers, which have an additional 1–3 nucleotides at the 3' end. This reduces the number of fragments to a manageable number for investigation, and the radiograph banding patterns of different samples are compared to reveal polymorphisms. The selection of adapter extensions is determined empirically [19]. Visualization of the banding patterns can be done by end labeling of one of the primers with γ-[^{32}P]-dATP or γ-[^{33}P]-dATP, or more efficiently by using α-labeled nucleotides in the reaction mixture [19]. The one enzyme method reveals less polymorphism than the two enzyme method, but has the advantage of simplicity, in that the restriction / ligation and PCR reactions can be combined into one reaction each [19]. An interesting variation on this technique is Ligation-Mediated-Suppression PCR (LMS-PCR; [22]). Using different adapter primers which are added to the original restriction fragments in a ligation step, locus-specific amplification in an upstream or downstream direction can be performed with locus-specific primers [20]. Chemiluminescent detection, using labeled primers, can also be performed, eliminating the need to use radioactivity [21]. This method requires the stringent maintenance of reaction conditions to ensure reproducibility [19], and also reveal primarily dominant markers [17].

16 RAPD-PCR (Random-amplified Polymorphic DNA – PCR), AP-PCR (Arbitrarily-primed – PCR)

This technique can be applied to similar study objectives as AFLP – PCR, but dispenses with the restriction step. Simply put, short random oligonucleotide primers, typically decamers [22–27] are used in PCR reactions that produce polymorphic PCR products in different individuals. Fragments can be separated by agarose or polyacrylamide electrophoresis, the bands visualized by ethidium bromide or silver staining [27]. If fluorescently labeled primers are used, automated sequencing machines can be used to distinguish RAPD fragments, as in the GENESCAN program used with ABI Prism machines [24, 28]. A subset of the RAPD technique, Microsatellite – Primed PCR (MP-PCR), uses primers designed to anneal to microsatellite repeats, i. e., simple sequence repeats variously present in genomes [27]. RAPD-PCR is unfortunately prone to reaction condition changes, such as different Taq polymerases, cycle settings, cycling machines, reaction constituent concentrations, and DNA target [24–26], which must be carefully controlled, in addition to primer-derived, non-specific amplification [24, 25, 29], and difficulty in recognizing heterozygotes [25, 28]. Additionally, co-migrating, same-sized bands may not in fact be homologous between species or populations [27, 30], resulting in ambiguity that can only be resolved by further investigation using different methodologies. The same proviso could also pertain to AFLP – PCR, as it does to RFLPs.

Appendix

Commercially available PCR machines

1) Heatblock thermal cyclers: These machines have been the standard for PCR. Since they take standard 0.5μl and 0.2μl microtubes, no special equipment is needed for the machines. However, due to the time it takes for the block and the tubes to equilibrate, an average PCR run takes about three hours to complete. A heatblock thermal cycler with an unheated lid also requires the use of mineral oil to prevent sample evaporation. To address the issue of sample evaporation while eliminating the need for mineral oil, many heatblock cyclers come with heated lids. By keeping the lid of the tube at a significantly higher temperature than the area where the reaction is taking place, evaporation is controlled.

2) Gradient Cyclers: There are two kinds of heating mechanisms available in heatblock cyclers. The first has a uniform temperature across the block and the second allows for a temperature gradient across the block. This gradient can assist in optimizing annealing and/or denaturation temperatures during the same PCR run, thus eliminating the extra time spent performing multiple runs.

3) Airflow thermal cyclers: These machines use high-velocity heated air circulating as a vortex to control the temperature changes for PCR. Samples are contained in capillary tubes to optimize the amount of surface area of the sample that is in contact with the heated air. Because of the speed with which the sample can equilibrate with the air, cycling times are drastically shorter than they are with the heatblock thermal cycler. For example, the LightCycler (Roche Molecular Biochemicals) can complete a 30 cycle PCR in less than 10 minutes, whereas a heatblock thermal cycler would take 2–3 hours to complete the same task.

4) Water bath; The era of the Human Genome Project has brought the need and the technology for large-scale automated PCR. By using robotic plate handlers such as Incyte's sequencing robot and microtiter liquid dispensers, a laboratory can quickly and efficiently process large numbers of reactions in preparation for sequencing. In brief this technology uses waterbaths and robotic arms that move the reaction plates between the different waterbaths.

5) Portable systems: PCR technology is being used in the field for pathogen identification and other applications. Field units are available which include all equipment needed for PCR analysis, including agarose gel electrophoresis and photo documentation (Ruggedized Advanced Pathogen Identification Device (RAPID), Roche Molecular Biochemicals and the Mobile Molecular Laboratory, MJ Research). Weighing around 50 pounds, these systems are designed and packaged to be compact enough to be taken as luggage and carried into the field.

Troubleshooting web sites

1. Two sites which offer these sorts of guides can be found at the Promega UK web site at http://www.promega.com/uk/ukAccessRT.htm, and the Boehringer -Mannheim web site at http://www.biochem.boehringer-mannheim.com/techserv/pcrts.htm. Many researchers and academic institutions also offer troubleshooting advice on their web sites, for example the PCR decision-tree-based troubleshooting guide provided by the Biological Institute of the University of Oslo which can be accessed at http://www.biologi.uio.no/ascomycete/PCR.troubleshooting.html.
2. One of the most elaborate troubleshooting guides around is the Promega Amplification Assistant (http://www.promega.com/amplification/assistant/). This site functions as an interactive decision tree, making it possible to examine numerous potential causes for various problems that might arise.
3. An excellent PCR product cloning website is http://www.the-scientist.com.

Websites with primer design programs

http://bioinformatics.weizmann.ac.il/mb/bioguide/pcr/software.html:

What kinds of programs are available for designing PCR primers?
- PrimerGen
- Primer (Stanford) Sun Sparcstations
- Primer (Whitehead) Unix
- Amplify
- PrimerDesign 1.04
- PC-Rare
- Primer Design
- CODEHOP
- Primer3
- NetPrimer (PREMIER Biosoft International).

http://www.chemie.uni-marburg.de/~becker/prim-gen.html:

Online Primer/Oligo Design
- Primer v2.2: World Wide Web Front End
- Primer Selection (Text)

- Oligonucleotide Vital Statistics
- POLAND REQUEST FORM: (ss and ds DNA/RNA)
- Find repeats within a sequence and much more
- Genefisher: Degenerative primer (german version)
- DNA Sequence Translation Primer Selection (Text)
- STS Pipeline v1.2

http://www.hgmp.mrc.ac.uk/GenomeWeb/nuc-primer.html:

Primer Prediction and Analysis programs
The PCR Jump Station
- GeneFisher
- GeneWalker
- CyberGene
- Web Primer
- Primer Design
- Primer3
- Web Primers
- POLAND – melting profiles of double+stranded DNA
- CODEHOP – PCR primers designed from protein multiple sequence alignments
- NetPrimer
- Cassandra – Recognition of protein-coding segments for PCR
- GenePrimer – Computational support of gene identification experiments
- rawprimer – a tool for selection of PCR primers

http://www.cbi.pku.edu.cn/genome/nuc-primer.html

The Genome Web
With links to CODEHOP
www.blocks.fhcrc.org/codehop.html

References

1 Mullis K, Faloona F, Scharf S, Saiki R et al. (1986) *Specific enzymatic amplification of DNA in vitro: the polymerase chain reaction.* Cold Spring Harbor Symposia on Quantitative Biology 51: 263–273

2 DeSalle R, Bonwich E (1996) DNA isolation, manipulation and characterization from old tissues. *Genetic Engineering* 18: 13–32

3 Zhang L, Cui X, Schmitt K, Hubert R et al. (1992) Whole genome amplification from a single cell: implications for genetic analysis. *Proc. Natl. Acad. Sci. USA* 89: 5847–5851

4 Saiki RK, Gelfand DH, Stoeffel S, Scharf S et al. (1988) Primer directed amplification of DNA with a thermostable DNA polymerase. *Science* 239: 487–491

5 Innis MA, Gelfand DH, Sninsky JJ (eds.) (1990) *PCR Protocols: A Guide to Meth-*

ods and Applications. Academic Press, San Diego

6   Sonnenfeld M, Ward, Nystrom, Mosher G, Stahl J et al. (1997) The *Drosophila* tango gene encodes a bHLH-PAS protein that is orthologous to mammalian Arnt and contols CNS midline and tracheal development. *Development* 124: 4571–4582

7   Helmsley A, Arnheim N, Toney MD, Cortopassi G et al. (1989) A simple method for site-directed mutagenesis using the polymerase chain reaction. *Nucleic Acid Res.* 17: 6545–51

8   Nasidze I, M. Stoneking. 1999. Short Technical Reports: Construction of larger-size sequencing templates from degraded DNA. *Biotechniques* 27 (September): 480–88

9   Amato G, Gatesy J (1994) PCR assays of variable nucleotide sites for identification of conservation units. In: B Schierwater, B Streit, GP Wagner and R DeSalle (eds) *Molecular Ecology and Evolution: Approaches and Applications,* Birkhauser Verlag, Basel/Switzerland

10   Selvakumar N, Ding B, Wilson SM (1997) Improved resolution of asymmetric-PCR SSCP Products. *Biotechniques* 22: 606–608

11   Sellner LN, Turbett GR (1998) Comparison of three RT-PCR methods. *Biotechniques* 25 (August): 230–234

12   Chadwick N, Wakefield J, Pounder RE, Bruce IJ (1998) Comparison of Three RNA Amplification Methods as Sources of DNA for Sequencing. *Biotechniques* 25 (November): 818–822

13   Barnes WM (1994) PCR amplification of up to 35-kb DNA with high fidelity and high yield from lambda bacteriophage templates. *Proc. Natl. Acad. Sci. USA* 91: 2216–2220

14   Cheng S, Fockler C, Barnes WM and Higuchi R (1997) Effective amplification of long targets from cloned inserts and human genomic DNA. *Proc. Natl. Acad. Sci. USA* 91: 5695–5699

15   Min G, Powell JR (1998) Long-Distance Genome Walking Using the Long and Accurate Polymerase Chain Reaction. *Biotechniques* 24 (March): 398–400

16   Sanchez-Cespedes M, Cairns P, Jen J, Sidransky D (1998) Degenerate Oligonucleotide-Primed PCR (DOP-PCR): evaluation of its reliability for screening of genetic alterations in neoplasia. *Biotechniques* 25 (December): 1036–1038

17   Suazo A, Hall HG (1999) Modification of the AFLP Protocol Applied to Honey Bee (*Apis mellifera* L.) DNA. *Biotechniques* 26 (April): 704–709

18   Vos P, Hogers R, Bleeker M, Reijans M et al. (1995) AFLP: A new technique for DNA fingerprinting. *Nucleic Acids Res.* 21: 4407–4414

19   Reineke A, Karlovsky P (2000) Simplified AFLP Protocol: Replacement of primer labeling by the incorporation of α-labeled nucleotides during PCR. *Biotechniques* 28 (April): 622–623

20   Schupp JM, Price LB, Klevytska A, Keim P (1999) Internal and flanking sequence from AFLP Fragments Using Ligation – Mediated Suppression PCR. *Biotechniques* 26 (May): 905–912

21   Lin J-J, Ma J, Kuo J (1999) Chemiluminescent detection of AFLP markers. *Biotechniques* 26 (February): 344–348

22   Atienzar F, Evenden A, Jha A, Savva D et al. (2000) Optimized RAPD analysis generates high – quality genomic DNA profiles at high annealing temperature. *Biotechniques* 28 (January): 52–54

23   Ellinghaus P, Badehorn D, Blümer R, Becker K et al. (1999) Increased efficiency of arbitrarily primed PCR by prolonged ramp times. *Biotechniques* 26 (April): 626–630

24   Corley-Smith GE, Lim CJ, Kalmar GB, Brandhorst B (1997) Efficient Detection of DNA Polymorphisms by Fluorescent RAPD Analysis. *Biotechniques* 22 (April): 690–699

25   Pan Y-B, Burner DM, Ehrlich KC, Grisham MP et al. (1997) Analysis of primer – derived, nonspecific Amplification products in RAPD – PCR. *Biotechniques* 22 (June): 1071–1077

26   Gallego FJ and Martinez I (1997) Method to improve reliability of random – amplified polymorphic DNA markers. *Biotechniques* 23 (October): 663–664

27 Ramser J, Weising K, Chikaleke V, Kahl G (1997) Increased informativeness of RAPD analysis by detection of microsatellite Motifs. *Biotechniques* 23 (August): 285–290

28 Leamon JH, Moiseff A, Crivello JF (2000) Development of a high – throughput process for detection and screening of genetic polymorphisms. *Biotechniques* 28 (May): 994–1005

29 Williams JGK, Kubelik AR, Livak KJ, Rafalski JA (1990) DNA polymorphisms amplified by arbitrary primers are useful as genetic markers. *Nucleic Acids Res*. 25: 6531–6535

30 Rieseberg LH (1996) Homology among RAPD fragments in interspecific comparisons. *Mol. Ecol*. 5: 99–105

31 Varadaraj K, Skinner DM (1994) Denaturants or cosolvents improve the specificity of PCR amplification of a G+C rich DNA using genetically engineered DNA polymerases. *Gene* 140: 1–5

32 Rees WA, Yager TD, Korte J, von Hippel PH (1993) Betaine can eliminate the base pair composition dependence of DNA melting. *Biochemistry* 32: 137–144

33 Melchior WB and von Hippel PH (1973) Alteration of the relative stability of dA•dT and dG•dC base pairs in DNA. *Proc. Nat. Acad. Sci. USA*. 70: 298–302

34 Henke W, Herdel K, Jung K, Schnorr D et al. (1997) Betaine improves the PCR amplification of GC-rich DNA sequences. *Nucleic Acids Research*. 25: 3957–3958

35 Baskaran N, Kandpal RP, Bhargava AK, Glynn MW et al. (1996) Uniform amplification of a mixture of deoxyribonucleic acids with varying GC content. *Genome Research* 6: 633–638

36 Chevet E, Lemaitre G, Katinka D (1995) Low concentrations of tetramethylammoniumchloride increase yield and specificity of PCR. *Nucleic Acids Research* 23: 3343–3344

37 McConlogue L, Brow MA, Innis MA (1988) Structure independent DNA amplification of PCR using 7-deaza-2'-deoxyguanosine. *Nucleic Acids Research* 16: 9869

38 Sarkar G, Kapelner S, Sommer SS (1990) Formamide can dramatically improve the specificity of PCR. *Nucleic Acids Research* 18: 7465

39 Stommel JR, Panta GR, Levi A, Rowland LJ (1997) Effects of gelatin and BSA on the amplification reaction for generating RAPD. *Biotechniques* 22: 1064–1066

Michael F. Whiting

Contents

1 Introduction

Nucleotide sequence data are currently the predominant type of molecular data used in phylogenetic inference, and the trend does not appear to be dissipating. Coupled with the ability to easily amplify specific gene regions via PCR, automated sequencing has produced an explosion in the amount of data used for phylogenetic inference simply because the time and expense required to generate sequence data for a particular group of organisms is dramatically less than it was even a few years ago. Workers can routinely go from organism to primary sequence data in about two days, and if PCR has been carefully optimized, then samples can be processed in batches, allowing for high

Methods and Tools in Biosciences and Medicine
Techniques in molecular systematics and evolution, ed. by Rob DeSalle et al.
© 2002 Birkhäuser Verlag Basel/Switzerland

throughput. Current instrumentation can handle such high volumes of sequencing template, and many institutions have gone towards centralized core sequencing facilities. With an increase in throughput comes a corresponding increase in sequencing accuracy; the polymerases and dye markers used in automated sequencing are more sophisticated than their manual counterparts, and assure greater accuracy in the primary data [1, 2, 3]. Errors associated with manual sequencing, and the potential for embarrassing retractions of research results (e. g., [4]) can be reduced. The ease and simplicity of sequencing chemistry reduces the amount of time required to optimize sequencing conditions and reduces the number of failed reactions.

Nucleotide by nucleotide, automated sequencing is less expensive than manual sequencing, and depending on how a core facility is operated, it can be *much* less expensive. While the initial outlay of resources to purchase a sequencer and support equipment is rather high, the cost per reliable nucleotide in terms of chemical expense is lower than with manual methods. Current instrumentation serves a large arena of users, and the expense to operate an automated sequencing facility can be spread among multiple investigators. If a core facility builds into its price structure the expense of technicians, service contracts, replacement costs and chemical expenses, then the actual charge may be a bit pricy. However, if the institution will support instrumentation costs and technician salaries, the actual chemical expenses can be less than $1/sample with appropriate cost-saving measures (see below). Perhaps more importantly, since this method does not employ radioisotopes, the cost to the environment by dumping radioactive by-products and the potential human health risks associated with heavy radioisotope usage is eliminated.

Perhaps the greatest advantage of automated sequencing is that the primary data can be captured, stored and edited entirely in digital format. This allows the user to always have direct access to the chromatogram data, so that discrepancies are resolved by referring directly to the raw data. The ability to directly compare hundreds of chromatograms easily and swiftly is a major practical advance over manual sequencing methods, which require the user to dig through a pile of gel autoradiographs, find the exact band on a set of gels, and then make some sort of interpretation to resolve a discrepancy in the sequence data. For automated sequencing, any potential synapomorphy or diagnostic feature for a set of taxa can be quickly visualized on the chromatograms for those taxa.

The centralization of equipment and expertise reduces the overall cost of sequencing, increases accuracy and throughput, and makes automated sequencing available to the masses. However, this may also cause systematists to become less coupled with the data they are generating, and many researchers use automated sequencing without an understanding of the underlying limitations of the chemistry and instrumentation, and hence the limitations of the accuracy of the data produced. The chapter will outline the basics in chemistry and instrumentation used in automated sequencing, provide tips and protocols for processing samples in large volumes, provide suggestions on troubleshooting, offer caution on raw data interpretation, and provide a list of considerations in selecting a core facility.

2   How It Works (in Theory)

Automated sequencing is a modification of the Sanger dideoxynucleotide terminator sequencing [5], where labeled DNA fragments are produced from an initial template in a reaction to which a polymerase, deoxynucleotides (dNTP), dideoxynucleotides (ddNTP) and some sort of label has been added. The incorporation of a ddNTP inhibits the extension of the sequence fragment, leading to a production of a large number of fragments of different lengths, all terminated on known nucleotides. In manual sequencing, the DNA template is split into four reactions corresponding to one of four ddNTPs, labeled with a radioisotope, and placed on adjacent lanes on an acrylamide gel. Each reaction is fractionated via electrophoresis and visualized by exposure to radiography film. The banding pattern is then scanned into a computer for band interpretation or (more commonly) nucleotide sequence is read manually off the film and directly entered into a computer. Automated sequencing modifies the Sanger method by using fluorescent markers to label DNA fragments, thermocycling to increase the yield of labeled templates, and a laser excitation and detection system to view the fragments as they migrate through the electrophoretic gel. These data are then fed directly into a computer and the information is collected, interpreted and archived. It is the modification and automation of these three processes that results in the high throughput and simplified chemistry of automated sequencing.

2.1   Fluourescent tags

Current chemistries are based upon two classes of dyes: near-infrared (NIR) fluorescent dyes and rhodamine or dichlororhodamine dyes (dRhodamine). The utility of dyes in flourescent sequencing depends on two fundamental characteristics of the dye: wavelength spectra and excitation levels. It is essential in multiple dye systems that the spectrum of each dye show as little overlap as possible, since overlap of spectral frequencies reduces the accuracy of discrimination among dyes. The dyes must also fluoresce under the excitation of the laser such that they are brighter than the surrounding background fluorescence. Instrumentation and chemistries have been developed to take advantage of the characteristics of each class of dyes.

NIR dyes form the basis of the LI-COR IR^2 System. The NIR dyes have a spectral frequency which is separated by 100 nm, which virtually eliminates signal cross-over in data interpretation [6]. NIR dyes are excited by a diode laser; infrared wavelengths are excellent for fluorescence detection since biomolecules, gels and gel plates have little inherent infrared fluorescence [7]. The disadvantage of NIR fluorescence is that there are only two dyes developed for automated sequencing (IRD700 and IRD800), and thus chemistry is restricted exclusively to dye primer chemistry, which may limit the throughput of a sequencer (described below) [8].

Rhodamine and dRhodamine dyes form the basis of the Applied Biosystems (ABI) sequencing instrumentation. The advantage of dRhodamine dyes over the earlier rhodamines is that their emission spectra are narrower, thus giving less spectral overlap and thus less noise (though some overlap is still present, unlike NIR dyes) [9]. Rhodamines use an argon ion laser for excitation, and because their excitation level is lower than is ideal for sequencing, acceptor-donor dye chemistries have been developed (e. g., ABI Prism® BigDye™ Terminators). These tags consist of two dyes: one which accepts the excitation energy of the argon ion laser and then passes it to a second dye which fluoresces at its own spectral frequency, but at a higher excitation level. These tags take advantage of the excitation sensitivity of a donor dye (e. g., 6-carboxyfluorescein) and then transfer that excitation level to the acceptor dye (e.g, dRhodamine), producing the spectral differentiation of a dRhodamine dye at the excitation level of a fluorescein dye. This causes the dRhodamine to fluoresce brighter than straight rhodamines resulting in brighter, more distinct banding patterns, and hence more reliable sequence data. There are multiple kinds of dRhodamine tags available (e. g., dR6G, dROX, dR110, dTAMRA, etc.), allowing for more flexible chemistries, including dye terminator chemistry, than NRI dyes.

2.2 Dye primer chemistry

In dye primer chemistry, a fluorescent tag is added to the primer during synthesis. If a single dye is added to the primer, then the reactions must be subdivided and run in four separate lanes, one for each nucleotide (as in manual sequencing). The LI-COR IR² System allows a second dye to be applied to another primer (e. g., the primer for the complementary strand), and then two samples can be loaded in the same wells (duplexing), though each sample still requires four lanes. In multi-dye primer chemistry, a primer is synthesized four times, each with a different fluorescent tag, and each tag is then associated with a particular ddNTP in setting up the sequencing reactions. For example, if the primer has the dROX tag, then ddCTP is used as the terminator in cycle sequencing. After thermocycling, the reactions are pooled together and the sample can be loaded on a single lane as with dye terminator chemistry.

The main advantage of dye primer chemistry is that longer sequence reads are possible than with dye terminator chemistry, due in part to a bias against the incorporation of tagged ddNTPs by the polymerase in dye terminator reactions. This bias leads to shorter reads and unevenness in peak height (though new polymerases have been developed to address this problem). If the samples are separated into four lanes, then there is additional gain in sequence length since there are fewer problems with the intensity of one band masking the intensity of another. A second advantage is that if PCR is carefully optimized, then there is no need to clean the PCR product prior to sequencing. A small fraction of the original PCR volume can serve as the template for dye primer sequencing reactions, because only fragments that have incorporated a labeled primer can be visualized on the gel; therefore there is no need to remove excess primer from the PCR reaction. Full automation of the sequencing process, from PCR amplification through the sequencing process, is possible with this chemistry.

The disadvantages of this chemistry are obvious. First, subdividing samples into four reactions is laborious and introduces a greater possibility of human error; additionally, the reactions occupy four times the space on a thermocycler. In single dye chemistry, the increase in read length comes at the sacrifice of the total number of samples that can be processed per day. Every sample requires four lanes, and currently the most advanced instrument which takes advantage of this chemistry (LI-COR IR² System) can run only 32 samples per gel (or 64 with duplexing), and since the run times are for 8 hours, it is hard to process more than two gels per day. Second, the cost of tagging every primer with four dyes can be prohibitively expensive, especially when one considers the number of internal primers required to sequence a lengthy gene for complementary strands. This cost can be reduced by adding universal primer tails (such as M13 primer sites) to the 5' end of the forward and reverse primers, thus using one set of universal primers for sequencing. However, the addition of these bases may cause problems in optimizing PCR, particularly with degenerate primers that are commonly used in molecular systematics. Third, as with radioisotope chemistry, false terminations are possible which will lead to greater background noise in the sequence data. Fourth, when specific primers are designed and tagged with a fluorescent label, care must be taken in primer design because the tag will cause a mobility shift in the fragments by the interaction of the first 5 bases of the primer with the dye tag. While mobility shifts can be corrected for in the software, the investigator must be aware of this and design the primers according to specific instructions of a particular instrument.

2.3 Dye terminator chemistry

In dye terminator chemistry, each ddNTP is tagged with a different rhodamine or dRhodamine dye. As the ddNTP terminates the extension of product during cycle sequencing, it also labels the fragment with the specific tag. The polymerases used in this chemistry are modified from *Taq* (e. g., Taq FS), reducing the bias towards incorporating one terminator over another and resulting in more even chromatogram peaks. The length of read from this chemistry can be maximized to 800 bp and current instruments (e. g., ABI 377) can process 96 samples per gel and up to 4 gels per day.

The advantages of dye terminator chemistry are in its flexibility and ease of use. There is no need to subdivide a template into four reactions prior to thermocycling, reducing the total number of tubes and tips used. The reagents are purchased in a single master mix, so the investigator only has to add purified template and primer to the reaction. There is no need to purchase individual tags for the plethora of primers used commonly in systematics, at a tremendous savings to the investigator. Only those fragments that are terminated by a ddNTP are visualized by the sequencer, thus reducing background noise. With the continuing innovation of polymerases and dyes, newer chemistries are approaching the sequence lengths obtained by dye primer chemistry. The efficiency and accuracy of donor-acceptor dye terminator chemistry is such that the investigator can use much less sequencing reagent than is recommended by the manufacturer, resulting in a tremendous financial savings (see Protocols). The universal nature of dye terminator chemistry makes it by far the most widely used in core facilities [10], and most molecular systematics laboratories have made this the chemistry of choice.

2.4 Thermocycling

Thermocycling allows a single sequence template to produce multiple fragments, thus increasing the overall yield of fragments for detection via electrophoresis. The quantity of starting template can be lower for automated sequencing than manual methods, but because the template is cycled, in general it must be of higher quality and purity than for most radioisotope methods. The cycling conditions for most chemistries have been meticulously optimized, so that altering temperatures will generally result in lower yield, though if you are sequencing under 600 bp, there is no need to have the extension cycle run more than 3 minutes. As with PCR, temperature ramping speeds vary among brands of thermocyclers, so care must be taken in optimizing the cycling conditions for new machines.

2.5 Electrophoresis

Instrumentation for electrophoresis consists of two major types: gel-slab and capillary systems. A gel-slab system requires the pouring of an acrylamide gel which is allowed to polymerize before a comb is inserted and samples are loaded in adjacent lanes. In 64- and 96- well systems, samples can be loaded individually with a standard pipetter, or in sets using a multichannel pipetter or multichannel syringe. Capillary systems use specifically designed, proprietary polymers which are injected into a silicone capillary, a sample is automatically loaded into the capillary, and the template is separated via electrophoresis. A disadvantage of gel-slab systems is the potential to contaminate adjacent lanes with spilled sample as the template is loaded, or as it leaks from lane to lane. One way to alleviate this problem is through the use of disposable combs which absorb the sample directly into the teeth of the comb, the comb is placed in the gel, and the samples are electrophoresed from the comb teeth into the gel matrix. While this eliminates the problem of spillage between lanes, extra care must be taken in positioning the delicate comb so that samples are not lost. Another disadvantage of gel-slab systems is that they cannot be easily automated; thus the gel must be poured and prepared, samples loaded manually, and plates cleaned. Despite these problems, gel-slab systems have been the most widely used in core sequencing facilities because they offer a much higher throughput than the standard capillary systems, with a 96-lane gel-slab (e. g., ABI 377; 288 samples/day) having 24 times the capacity of a single capillary instrument (e. g., ABI 310; 12 samples/day). Moreover, slab systems have a much smaller margin of error in correctly discriminating among nucleotides for difficult templates [11]. Capillary systems are much more sensitive to template quality and concentration than gel-slab systems, making it a poor instrument of choice for core facilities that process a wide range of DNA from many users, or for users who are trying to cut costs by using smaller sequencing volumes or generic chemicals. In addition, capillary systems are notorious for accumulating proteins, detergents and unlabeled DNA templates which cause fouling and clogging of the capillaries, signal reduction and sample failure. Capillary systems are used mainly by researchers who are willing to sacrifice throughput and cost for the ease of sample manipulation, or by core facilities who would use them as make up instruments to process "drop outs", the few samples that might fail on a gel-slab system. The increased cost and reduced capacity of single capillary systems make them ill-suited for most molecular systematic applications. However, recent developments in instrumentation have resulted in multiple capillary systems that offer tremendous throughput, while claiming to have solved the problems with previous capillary systems (e. g., ABI 3700 and 3100; discussed below).

2.6   Data collection

In gel-slab systems, purified, lyophilized templates are re-suspended in a buffer and a loading dye, denatured, and loaded into lanes on a vertical acrylamide gel. The samples migrate towards the bottom of the gel, separating according to size. On a lower portion of the gel, the samples pass through a region where a laser scans continuously across the gel and excites the fluorescent dyes attached to the fragments. Each dye emits light at a specific wavelength which is filtered to remove background noise, then split by a spectrograph and detected with a cooled, charged couple device (CCD) camera (ABI Systems). Alternatively, the fluorescence is detected by two thermally controlled, solid-state avalanche photodiodes (Li-Cor Systems). Multiple dyes can be detected by a single pass of the laser and then the data collection and analysis software processes, analyzes, and translates the information into nucleotide sequence. In capillary systems, the fragments migrate through the capillary where they are excited by a stationary laser, and then interpreted in the same way as the gel-slab system. Every sequencing system has an accuracy rate related to its chemistry, the quality of sequence template provided, instrumentation, and data processing algorithms [3]. No system is fool-proof, so the conscientious systematist will carefully check the sequence data for errors.

3 How It Works (in Reality)

3.1 Quality and quantity of DNA template

The following is a description of protocols for dealing with the most common sequencing chemistry used in systematics: dye terminator chemistry. The lack of properly purifying and quantifying template prior to cycle sequencing is by far the most common cause of failed sequencing reactions. PCR product is the most common template used in systematic applications, though clones may be used for genes that are difficult to amplify. Any method that removes excess primer and dNTP should be adequate to purify PCR product. Micro-filtration columns (e. g., Centricon®-100 Micro-concentrator; PE Applied Biosystems) provide excellent quality PCR template, but are expensive and time consuming to use. Columns which use a matrix to bind the PCR product (e. g., Quiagen spin columns) work very well. Other brands of columns abound at various prices and qualities. I have found GeneClean (Bio 101, Vista Ca.) to provide consistently good quality at a very economical price. Concentration of DNA is dependent on particular chemistries used, but a few guidelines are helpful. Too much template causes top-heavy reactions, i. e., the polymerase generates too many short fragments so the sequence read terminates at about 250 bp. Too little template causes very faint bands, and since the CCD camera has a hard time distinguishing signal from background noise, the chromatogram will have many secondary peaks. As a general rule of thumb, 5ng/100 bases should be sufficient for PCR product, but exact concentration should be determined empirically by titrating template and primer concentrations. Cloned products need to be carefully purified using either commercially available kits (e. g., Quiagen spin columns) or Alkali lysis/PEG precipitation. Alternatively, for some applications (e. g., TA cloning), it is easier to screen positive colonies via PCR, and then sequence the PCR product rather than growing the colony and performing miniprep purification. This is done by picking a colony, placing it in 20 µl of dH$_2$0, then using 2 µl of this as template for PCR. The PCR must have a hot start

(95°C for 8 minutes), can use either the original PCR primers or the vector universal primers, and should only require 25 cycles. Sequencing primers should be high purity, no secondary structure present, especially at the 3' end, 18 bases minimum, with a Tm of around 55-60°C. Avoid runs of more than 4 of the same nucleotide and not more than 32-fold degenerate, with no degeneracy on the 3' end. Taq FS is relatively robust to variations in template/primer concentration; however, long reads require a fine tuning of primer to template concentration, and clean reads (i. e., little background) require very specific, well-purified template.

3.2   Reaction setup and thermocycling

Setting up sequencing reactions is somewhat trivial, but a few hints are worth mentioning here. The majority of manufacturers suggest running large sequencing reactions (e. g., 20 µl) which are then purified, lyophilized, and suspended in some volume prior to loading. More often than not, the manufacturer's protocol suggests loading only 1/4th to 1/8th of the total volume of the sequencing reaction, leaving the remainder as "backup" in case the gel fails. Since gels rarely fail in labs run by experienced workers, this is a tremendous waste of resources for the systematist. Moreover, full sized reactions contain nearly twice as many sequencing errors as half-sized reactions [12], and I have observed less background noise with 1/4 reactions than with 1/2. My core facility routinely runs 1/8 reactions using ABI Prism® BigDye™ terminator chemistry, by simply cutting the total volume down to 2.5 µl (protocol below). These reactions are amplified (with no loss of volume), purified, and half the entire reaction is placed on the sequencer (ABI 377). Because we are working with extremely small volumes, we use electronic pipetters to reduce pipetting error; however, the savings in sequencing costs more than make up for the price of a good pipetter. We consistently get high-quality sequences (~800 bp) for less than a dollar per reaction.

3.3   Sequencing reaction purification

After thermocycling, the unincorporated dyes must be removed from the reaction or else they will fluoresce so brightly that they saturate the laser, causing large dye blobs that obscure the sequence. There are two major protocols for purifying sequencing reactions. The first uses a hydrated matrix to which the unincorporated dyes bind while passing through a spin column. Though spin column kits are expensive, once purchased they can be re-packed with a hydrated medium and used about 300 times (see Protocols). While spin columns are very efficient at removing unincorporated dyes with a minimum of template loss, they are labor-intensive and are inefficient for large-scale sequence production. New products are being introduced to the market that allow one to pack a 96-well plate with the hydrated matrix, and then treat the entire plate as a spin column. Alternatively, ethanol precipitation methods can process 96-well microtiter plates (or PCR racks with micro tubes) simultaneously with only minor expense and effort (see Protocols). While this method does not remove all the unincorporated dyes, you lose only about 10–20 bases from the 5' end of the sequence, which is considered an acceptable loss in large-scale sequencing operations.

3.4   Raw data interpretation and analysis

The primary data collected by automated sequencing are an electronic gel file, which is produced by combining the data from each scan of the laser across the gel. This produces an image of all the bands that have migrated through the gel. A lane-tracking algorithm tracks the series of flourescent peaks found in each vertical lane of the gel file, and then converts each lane into a chromatogram. The chromatogram, which displays the distribution and peak height of the fragments marked by each fluorescent dye (as well as any background noise or secondary signal) may be considered secondary data. These peaks are then interpreted as nucleotide bases using a base-calling algorithm, generating the actual nucleotide sequence (tertiary data). *Never accept sequence data without first checking the gel file and the chromatograms.* Mistakes can and do occur in any of these steps, so it is imperative that sequence accuracy be assessed in every stage of the data generation process. Moreover, a large class of mistakes cannot be detected without sequencing complementary strands, regardless of how carefully you check the chromatogram. Therefore, sequencing complementary strands is essential to collect accurate data for molecular systematic studies.

From an algorithmic standpoint, correctly tracking lanes is a very difficult task, and recent algorithms have included neural net subroutines which allow them to learn from previous tracking failures [9]. Even so, tracking errors are still present on the majority of gels and must be manually corrected. A commonly encountered error is that the tracker will miscount lanes and assign the wrong identity to each sample. The algorithm may also begin tracking in one lane and end in an adjacent lane, creating a hybrid sequence. Also, if a particularly bright sample is loaded adjacent to a faint sample, and there was minor spillage in loading the samples on the gel, then the tracker may pick up the bright signal rather than the faint signal. Failure to detect these errors will have obvious devastating consequences on a phylogenetic study. These kinds of sequence contamination can be reduced by loading templates with different primers in adjacent lanes, thus making it easier to distinguish among samples and correctly counting and tracking lanes on the gel file. It is extremely useful to view a picture of the gel image, and a carefully run core facility should provide access to the gel image (such as on a web site), complete with tracking lanes, for every sample processed. If there is a question as to a sequence's identity, the user can consult the gel file to verify tracking results.

The base caller attempts to assign nucleotides to each peak in the chromatogram; the accuracy of the algorithm depends on the sequence quality, sequence length, background noise, and peculiarities of fragment migration on a particular gel. Base callers commonly make mistakes, but these mistakes are predictable and the experienced systematist will proof-read all sequences for base-calling errors. Errors commonly occur at the beginning and end of sequences, and tend to be more frequent towards the end of the sequence. Sequence processing software (e.g., Sequencher© GeneCodes Co.) allow you to trim the ends of the sequences by setting the degree of tolerance for ambiguous data. The first proof is performed by comparing complementary strands of sequence data and then resolving any discrepancies. The second is an alignment proof, where all the taxa in an analysis are aligned and any unusual features (i.e., single insertion/deletion events, bases that are called as ambiguities in some taxa and one state in another) are screened and back-checked with the chromatogram to ensure that the differences are not due to errors in base calling. For large-scale sequencing projects, sequence correction software should be considered a necessary expense required for accurate data analysis. A hard copy of a chromatogram does not allow you to effectively compare ambiguities that may be present in your data, and for any large-scale work it becomes too cumbersome to compare. Some workers prefer to keep hard copies of chromatograms as backup in case of data loss. However, it is much more effective to keep electronic copies of the raw data as backups. I routinely make permanent backups of sequence data on CD, and since the CD is not rewriteable, it serves as a permanent record that is more useful than hard copies.

3.5 Protocols

Protocol 1 Spin column purification

1. Completely hydrate Sephadex (G-50, medium) with water by placing 3 g in 50 ml dH$_2$0 for 30 minutes. Solution can be stored for up to 6 months at 4°C.
2. Remove the hydrated Sephadex from the fridge and vortex. Aliquot out 750 µl to each spin column (I recommend Centri-Sep™ from Princeton Separations).
3. Place each column in a fixed-angle micro centrifuge (with their collection tubes) and spin for 2 minutes at 4000 rpm (Eppendorf 5415C Centrifuge). Mark the tops of the columns with a dot and keep the dot up during all subsequent spins to help orient your samples. Do not over-spin the columns or your sample will be absorbed when it is added to the column.
4. Remove the column, decant the water from the collection vial, replace, and spin again for 2 minutes.
5. Remove columns and carefully load your sample on top of the packed Sephadex. (With 1/8th reactions, add 10 µl of dH$_2$0 to each reaction before loading on the columns). Place the columns on the appropriate collection tubes. Do not allow any of the sample to adhere to the walls of the columns or it will run past the Sephadex and remain unpurified.
6. Place the column with your loaded sample inside the corresponding 1.5mL tube and spin at 4000 rpm (Eppendorf 5415C Centrifuge) for 4 minutes. If no sample comes through the columns, add 20 µl of dH$_2$0 to the column and spin again.
7. Lyophilize samples in a speed vac and store at –20°C for up to 2 weeks. *Spin columns can be reused about 300 times. Discard the old Sephadex and leave the column in the light to help degrade dyes. When new Sephadex is added and the columns are pre-spun, any remaining template is washed off the filter. Commercial companies now offer 96- well plates that can be used for loading the Sephadex and processing the samples in batches.*

Protocol 2   ETOH precipitation of sequence reactions

Reagents Needed:

95% ETOH (HPLC grade) with 120 mM NaAc, pH 4.5 and 5.0 mM $MgCl_2$
70% ETOH (HPLC grade)

Procedure:

1. Add 30 µl of the prepared 95% ETOH solution to each reaction in the 96-well plate (or micro tubes in a rack). A multi-channel pipetter is most effective. Mix well by pipetting up and down three times. Place contact film on the top of the tray/rack. (Alternatively, if you are using single tubes, add the 95%ETOH solution and transfer to micro-centrifuge tubes).
2. Place the samples at 4°C for 15 minutes.
3. Spin at 2500g for 30 minutes. This works best with a microtiter plate rotor on a desktop centrifuge.
4. Remove contact film and carefully shake out the supernatant. Do not shake too vigorously as this may cause loss of the precipitant.
5. Wash the precipitant by adding 80 µl of 70% ETOH.
6. Let sit at room temperature for two minutes.
7. Spin at 2500g for 10 minutes.
8. Carefully shake out the 70% ETOH.
9. Repeat steps 5–7.
10. Place a paper towel on top of the tray, invert, and spin at 200g for 30 seconds to remove the excess ETOH.
11. Seal, and store at –20°C.

    Be careful to remove all the ETOH after precipitation or else you will get large dye blobs in your lanes which will obscure the first 100 bp of your sequence. It is normal to get slight dye blobs with this protocol which will obscure only about 10–20 bp of sequence (see Fig. 1).

4   Troubleshooting

No chemistry will sequence all DNA perfectly. Some sequences (e. g., GC-rich, TA repeats, strong secondary structure) and some kinds of clones (Cosmid and P1 clones) are more difficult to sequence than others. Troubleshooting depends on the chemistry and instrumentation used, but a few general points are worth considering.

- *High levels of background signal:* This is the most common complaint with automated sequencing, and may be difficult to troubleshoot because it can be attributed to multiple causes.

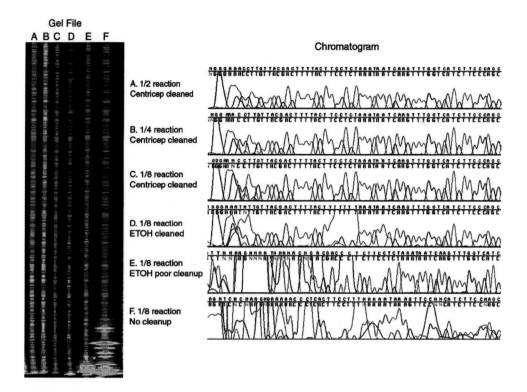

Figure 1 Comparison of reaction size and method of template purification in automated sequencing for a 28S rDNA flea template. Lane A is a ½ sequencing reaction (10 µl), lane B is 1/4th reaction (5 µl) and lane C is 1/8th reaction (2.5 µl; see boxes). Lanes A-C were purified via Centricep columns. The signal strength observed in the gel file is similar for lanes A and B, and slightly weaker for lane C. The background noise and accuracy of base calling for the first 61 bases in the chromatogram is nearly identical for lanes A-C. For the first 800 bp length of sequence, the ½ reaction miscalled 6 nucleotides, 1/4th reaction miscalled 5 nucleotides, and the 1/8th reaction miscalled 7 nucleotides, indicating that the reactions are nearly indistinguishable in accuracy; the length of read was also identical (data not shown). Lane D is a 1/8th reaction purified via ETOH precipitation (see boxes). The unincorporated dye blob present at the base of lane D in the gel file is normal for ETOH precipitation, and obscures the calling of 6 nucleotides (bases 26–31) in the chromatogram. If the ETOH is not entirely removed after precipitation (lane E), dye blobs will obscure a greater number of nucleotides (~20 bases). For lane F the sequencing reaction was lyophilized without purification and resulted in larger dye blobs that obscured the first 160 nucleotides. All reactions were run with the ABI Prism® BigDye™ Terminator Chemistry on an ABI 377 DNA Sequencer using a 64-lane comb, nucleotides were called with the ABI 200 base caller, and chromatograms were analyzed via Sequencher© 3.1.1 (GeneCodes Co.).

Cause 1: Amplification of non-target bands during PCR and/or cycle sequencing. During electrophoresis, multiple bands will be superimposed and the laser will detect multiple signals. The chromatogram may look like a primary sequence (high peaked bands) and a secondary sequence below it, but because multiple bands will cause spectral interference, you should not trust the primary sequence.

Solution: Increase stringency of PCR by increasing temperature and/or primer specificity, or remove the target fragment by "plugging" (i. e., excising the target fragment from an agarose gel) or cloning. Always run your sample on an agarose gel in order to detect multiple bands, increasing the agarose percentage in the gel to provide better resolution among fragments.

Cause 2: PCR template has not been properly purified, so residual primers are amplifying complementary strands during cycle sequencing.

Solution: Be certain that you are following your PCR purification protocol correctly, or try using one of the many commercial kits that are available.

Cause 3: The primer is old and/or of poor quality, such that the primer is missing a nucleotide. While the primer may still amplify well, it will generate n-1 fragments as well as n-sized fragments, which become superimposed when sequencing. Likewise, if a primer has a homopolymer sequence (strings of a single base, typically more than 4 in a row), then it may produce n+1 and/or n-1 fragments, depending on the interaction of the template and primer, which become superimposed when sequencing. Note that agarose gels do not have the resolution power to detect these problems.

Solution: Redesign primers and/or order new aliquots of higher quality primer.

Cause 4: Your reaction was very faint and the laser could not distinguish between the normal background noise due to gel fluorescence and your sample. If a sample is too faint, the base caller will not call the bases, but if it has a slight amount of signal, the base caller will attempt to assign bases, but because of the background noise, there will be many artifacts. When you see the gel file, the sample will look very faint; alternatively, view the chromatogram and it should give you a low signal level for each nucleotide.

Solution: Faint reactions have many causes, including improper chemistry setup, incomplete thermocycling, dirty template, loss of most of the sample during purification, resuspending the sample in too high a volume of loading buffer, and/or not adjusting the laser sensitivity to the proper level. The most common cause is an improper balance between primer and template concentrations. Try titrating the reaction by increasing the template concentration by 2-, 4-, and 8- fold.

Cause 5: The acrylamide gel contained high levels of contaminants that fluoresce, causing all samples to have high background noise. On the gel file, this is often observed as a milky green or blue haze, or it may be caused by bubbles migrating through the gel during electrophoresis ("red rain"). In the chromatogram you will see background of a single base, i. e., insertions of A in random positions throughout the chromatogram.

Solution: Carefully clean the plates and rerun the samples. If this is a recurrent problem, the plates may need an acid wash or may need to be replaced. If "red rain" is a problem, try adding a moist paper towel to the bottom of the gel immediately after pouring it, seal the bottom of the gel in plastic-wrap, and allow the gel to polymerize. In a properly run facility, only one out of a hundred gels has this level of fluorescence, and when it is a problem, the facility should rerun the samples gratis.

- *Sequence length shorter than expected:* Overall length of sequence depends on chemistry used, instrumentation, and the how the instrument was operated (long plates *versus* short plates; fast run versus slow run); the user may have little control over these variables. However, a short sequence is commonly due to the lack of optimization of sequencing components.

Cause 1: Primer and template concentrations are unbalanced. As more template is added to a reaction (while holding the primer concentration constant), too many short fragments are produced during thermocycling, which leads to top-heavy data: the chromatogram peaks are very high for the first 150 bases, then they quickly decrease. Likewise, adding too much primer, while holding the template concentration constant, has a similar effect.

Solution: The exact concentrations of primer and template are dependent on the chemistry used. As a rule of thumb, start with 100 ng/100 bases of PCR product and 3–10 pmol of primer. If the results are less than optimal, then titrate.

Cause 2: Primer degradation. Degraded primers are less effective at generating long fragments than are non-degraded primers.

Solution: Order new aliquots of high-quality primer.

- *A gene region that is difficult to sequence:* The chromatogram will show high peaks and low background, and suddenly high background appears and the sequence cannot be read downstream. Manufacturers are beginning to produce chemistries specifically for hard-to-sequence templates (e. g., ABI Prism® dGTP BigDye™ Terminator Kit designed for GC-rich templates). Before purchasing additional reagents, the following tips may be helpful.

Cause 1: GC-rich regions.

Solution: The easiest modification is to add 5% DMSO final concentration to the reaction mix (10% DMSO is too high and tends to kill the reaction) [1]. Some labs use 5% DMSO routinely in all chemistry, but you may see poor results with AT-rich regions. Alternatively, try a combination of any of the following: Increase cycles to 30 for dye terminators, try a pre-denaturation step at 96°C for 4-5 minutes, or use 2 × dNTPs in the sequence reaction to balance out top-heavy reactions [13]. The ABI Prism® dGTP BigDye™ Terminator Kit has better performance in sequencing GC-rich regions than other chemistries [12], but it is not as robust a chemistry as standard BigDye™ and should only be used for GC-rich regions and not routine sequencing.

Cause 2: AT-rich Regions

Solution: These templates are generally not as problematic as GC-rich templates. If a primer is AT-rich, increase its length to about 24-26 bases to increase the melting temperature closer to 55°C. The other option is to try dye primer chemistry since it tends to sequence more evenly through AT-rich regions.

Cause 3: Repeat Regions. Repeat sequences such as di-, tri-, and tetra-nucleotide repeats are problematic in PCR products because enzyme slippage occurs and sequence cannot be obtained through the repeat region.

Solution: Longer repeat sequences such as variable tandem repeats of 30 or more bases are difficult to deal with. If possible, run the gels on 48 cm plates to increase read length. AG repeat sequences can be problematic because Taq FS produces a weak G signal after A in terminator data. Try using more template or sequencing the opposite strand.

Cause 4: Homopolymer Region. Problems can arise when more than 10 identical bases occur consecutively (sometimes fewer than 10 for G and C). Generally, one can sequence through the shorter runs of poly Ts, but enzyme slippage can occur when the number of bases is greater than 25. Surrounding sequence may also play a role in the success of sequencing through homopolymer regions.

Solution: For poly T regions use an anchored primer of 25 T residues with a degenerate 3' base of G, C, and A. This is a useful tool to help sequence through cDNA clones. If you are trying to get through Poly G and Poly C sequences, design a primer that will anneal about 20 bases upstream from the homopolymer to force the polymerase through the problem. It is important to use an exonuclease-deficient enzyme for sequencing through homopolymer regions such as Taq FS.

Cause 5: Palindromes/ secondary structure

Solution: Try the techniques described for GC-rich regions.

- *Large peaks occur on the chromatogram that obscure 5–20 bases:*

Solution: This is due to unincorporated dyes that have not been removed during template purification. If you are using a spin column, then be certain that the samples are being loaded in the middle of the matrix and not running down the sides of the tube. If you are doing ETOH precipitation, dye blobs will obscure the first 10 bases. If they are extending further, then you are not effectively removing the ETOH with each wash. Try an additional 70% ETOH wash to remove the residual dyes.

5   Shopping for a Core Sequencing facility

If resources are not in place for generating sequence data in a lab or an institution, there are many avenues a researcher can explore to obtain automated sequencing. Many commercial companies will provide high-quality automated sequencing for a fee, although this tends to be the most expensive option available. Core facilities at some universities are happy to accept outside work since they can charge a higher rate than for in-house users, helping drive down costs in their own facility. These costs are generally lower than with commercial sequencing because the institution often provides some support for the facility. The key is to shop around; sequencing rates will vary by more than four- fold among institutions. All core facilities will do the sequencing chemistry for you, if you provide templates and primers. Some will allow you to perform the sequencing chemistry and submit lyophilized template which is re-suspended and run on the sequencer, billing only for lane charges. The advantage of this is that the user can cut down on sequencing volumes to save on chemical expenses, while paying as little as 1/10th of the price that the facility might charge for performing the sequencing chemistry. The disadvantage is that the inexperienced user may perform the chemistry improperly, making it difficult for the facility to troubleshoot. For the experienced user on a budget, this is an excellent option to pursue. Charges among core facilities vary considerably, and in many cases it may be less expensive to be an outside user with a different facility, than an in-house user at your own facility. Do not feel that you are stuck with the core facility at your institution. It is a minor burden to drop your samples in the mail and send them to a remote sequencing facility, in order to obtain better or more economical service.

When deciding on a facility you should consider the following:

1. How much will it cost? In 1996, the average core facility in-house charge for sequencing a template was $30 [14]. In 1998, the average charge for in-house work was $15 and outside work was $24 [15]. Current data for 1999 and 2000 are not available, though prices should be going down. If you are paying significantly more than this, you should consider shopping around.

2. If I am a new user, will the facility give me a chance to optimize my chemistry before it begins assessing a charge? It sometimes takes a few tries for the facility and the user to make sure the chemistry and protocols are optimal.

3. Does the facility give me the option to submit sequence reactions and pay only lane charges? If they do not, ask why, since this is a very easy request for a facility to accommodate. Likewise, does the facility provide discounts for large-volume work? There is flexibility in the price structures of most facilities, and they should provide discounts to heavy users.

4. How do I receive my data, and in what form does it come back? A good facility will provide an interface for submitting sample information prior to sequencing and web access to data for instant downloading (see for example the core facility web site at my institution, http://dnasc.byu.edu/index.htm). Some facilities provide only printouts of the chromatograms, some provide the chromatogram data file, and some provide both. If you do not have access to sequence correction software, then you will need a printout. However, printouts are difficult to correct, and for anyone doing large-scale sequencing, they soon become burdensome. A good facility will provide access to sequence correction software in a centralized computer lab if the user cannot afford it.

5. Do I have access to a copy of the gel file? You need to be able to check the gel files to determine if the anomalies in any particular sequence may be due to errors in tracking and base calling lanes.

6. Does the facility provide any sort of primary data backup in the event my data are lost or erased? While backing up of electronic data is always the responsibility of the user, it is a minor burden for the facility to archive gel files on a CD. ($15,000 of data can be archived on a $2 CD. Seems worth it.)

7. What is the turn-around time? In 1996, the average time was 3 days, in 1998 it was 2 days, and the trend should be towards faster turn-around time [14, 15].

6 Which System is Best for my Lab?

In general, the most economical way for an individual or institution to perform sequencing is through a core sequencing facility. A core facility centralizes equipment and expertise, reducing the amount of redundant equipment in labs throughout the institution while providing expertise in sequencing chemistry. The fact that most institutions have moved towards core facilities is evidence of the economic and practical success of this approach. Current automated sequencers, if properly operated, can have such a high throughput that only an exceptionally prolific lab can keep it busy night and day on a continual basis. Automated sequencers are relatively easy to use, but also easy to misuse, and inexperienced operators can unwittingly force a machine out of calibration, or ruin expensive parts (plates, combs, etc.) by carelessness.

In many cases, systematists may wish to generate sequence data in their own labs. The chief advantage of owning and operating a DNA sequencer within your own lab is the opportunity to develop cost-saving measures which may not be available in a core facility, as well as controlling the schedule and use of the instrument. While the choice of instruments and chemistry may be a matter of taste, finances and circumstance, a 1999 survey of 74 DNA sequencing core facilities is illuminating. This survey revealed that 98% of sequencing reactions were performed with PE Applied Biosystems instrumentation and 2% performed with Li-Cor instrumentation. For the PE Applied Biosystems instruments, 75% of all work was performed on the ABI 377, 23% on older models (ABI 373), and only 2% on the single capillary system (ABI 310) [11]. Of chemistries, the BigDye™ chemistry was found to have the greatest accuracy, was most readily diluted into smaller-sized reactions, and was the most common chemistry used in these labs [11]. The fact that there are more core facilities using ABI instrumentation and chemistry than any other system may be sufficient reason to go that route.

It appears that the PE Applied Biosystems instrumentation will continue to retain dominance in the marketplace, at least for the immediate future, with the introduction of a production-grade multiple capillary system. In late 1999, PE Applied Biosystems began marketing a 96-capillary system (ABI 3700) which boasts the ability to sequence up to 1,536 samples/day. This system has combined the best of the slab systems (high throughput) with the best of the capillary systems (increased automation), while correcting many of the problems associated with earlier capillary systems (e. g., clogging and degradation of the capillaries). The ABI 3700 has a fixed laser to excite templates in each of the 96 capillaries and is entirely automated, such that the investigator need only place a microtiter plate of sequencing reactions within the instrument and walk away. While this instrument seems ideal for high-production core facilities, it comes at a hefty price in terms of initial cost, ongoing chemical expenses to keep the instrument operational (you don't just turn it off), and it is not yet clear how effective it is at processing sub-optimal or partial reaction volumes. This instrument is probably best suited for a core facility, as it is hard to imagine a molecular systematics lab which is sufficiently productive to keep it continually in operation. At time of writing, PE Applied Biosystems announced the release of a 16 capillary version of the same instrument (ABI 3100) which may prove more suitable for the production level of the molecular systematist. If nothing else, the marketing of these new instruments is causing previously owned ABI 377, the current workhorse of sequencing instruments, to appear on the market at a much reduced price, as much as 1/6th of the cost of the ABI 3700.

7   Data for Sale: Molecular Systematics in the 21st Century

The future of data generation in molecular systematics is closely tied to the development of novel chemistries and instrumentation. While it is hard to predict future innovations in instrumentation and chemistries, some challenges still elude the best of sequencing systems. For instance, there has been relatively little progress made in improving chemistry for sequencing through difficult regions, and there is still a great amount of trial and error required to get through the GC or homopolymer templates. While the length of sequencing reads are now beginning to approach 1 kb on a consistent basis, getting reads over 2 kb is still not possible, even though polymerases have been developed which can amplify DNA product of that magnitude and greater. One of the major limitations of automated sequencing has been the reliance on acrylamide gels which can only withstand a certain amount of heat and electrical current before they begin to degrade. This limits the length of time the gel can effectively resolve nucleotide differences among sequencing fragments, and hence the length of a sequence read. While the new capillary polymers present an alternative to acrylamide, they too are limited by the effects of heat and current before degradation occurs. Perhaps the greatest innovation in instrumentation would be the development of a solid-state gel matrix which does not degrade and can be decontaminated between sequencing runs. This would allow longer sequencing runs while relieving the user of pouring gels, or the instrument of flushing capillaries with fresh polymer. In terms of chemistry, the development of dyes with less spectral overlap will lead to greater accuracy, while additional dyes would open the door to large-scale multiplexing of DNA sequence products. Wherever the future innovations in sequencing technology may lie, what seems clear is that molecular systematics is no longer an enterprise that is easily pursued by a single researcher with a manual gel rig working isolated in a lab. Molecular systematics, for good or bad, is becoming a group enterprise that relies on expensive instruments and chemicals to gain new insights into the most basic questions of phylogeny and evolution.

Acknowledgments

I thank Paige Humphreys, Helaman Escobar, and Alison Whiting for assistance in optimizing the protocols and Matthew Terry, Taylor Maxwell, and Jason Cryan for useful comments on the manuscript. This work was supported by NSF grants DEB-9983195, DEB-9806349, and DEB-9615269.

References

1   Adams PS, Dolejsi MK, Grills G, Hardin S et al. (1997) Effects of DMSO, Thermocycling and Editing on a Template with a 72% GC Rich Area: Results from the 2nd Annual ABRF Sequencing Survey Demonstrate that Editing is the Major Factor for Improving Sequencing Accuracy. In: Ninth International Genome Sequencing and Analysis Conference. *Microbial and Comparative Genomics* 2: 198

2   Adams PS, Dolejsi MK, Grills G, McMinimy D et al. (1999) An Analysis of Techniques Used to Improve the Accuracy of Automated DNA Sequencing of a GC-Rich Template. *Journal of Biomolecular Techniques* 9: 9–18

3   Naeve CW, Buck GA, Niece RL, Pon RT et al. (1995) Accuracy of automated DNA sequencing: a multi-laboratory comparison of sequencing results. *BioTechniques* 19: 448–453

4   Baker RJ, Bussche RAVD, Wright AJ, Wiggins LE et al. (1996) High Levels of genetic change in rodent of Chernobyl: retraction. *Nature* 380: 707–708

5   Sanger F, Nicklen S and Coulson AR (1977) DNA sequencing with chain-terminating inhibitors. *Proc. Natl. Acad. Sci. USA* 74: 5463–5467.[1]

6   Narayanan N (1998) New NIR Dyes: Synthesis, Spectral Properties and Applications in DNA Analysis. In: S Daehne (ed): *Near-Infrared Dyes for High Technology Applications.* Kluwer Academic Publishers, The Netherlands, 141–158

7   Middendorf L (1998) Near-Infrared Fluorescence Instrumentation for DNA Analysis. In: e. a. S Daehne (ed): *Near-Infrared Dyes for High Technology Applications.* Kluwer Academic Publishers, The Netherlands, 21–54

8   Wiemann S, Stegmann J, Grothues D, Bosch A et al. (1995) Simultaneous on-line sequencing on both strands with two fluorescent dyes. *Anal. Biochem.* 224: 117–121

9   Biosystems PA (1998) *ABI Prism Big-DyeTM Terminator Cycle Sequencing Ready Reaction Kit Protocol.* PE Applied Biosystems, Foster City, CA

11 Thannhauser T, Adams PS, Dolejsi MK, Grills G et al. (1999) Analysis of the effects of different DNA sequencing methods on accuracy, quality, and expansion of a web-based sequencing resource: Results of the ABRF DNA Sequencing Group 1999 Study *in* American Society for Biochemistry and Molecular Biology DNA Sequencing Research Committee 1999 Web Poster. *http: //www.abrf.org/ABRF/Research-Committees/dsrcreports/dsrg99ss/ dsrg99ss.pdf.*

12 Grills G, Dolejsi MK, Hardin S, McMinimy D, et al. (1998) *Assessing the Current State of the Art in DNA Sequencing and Creating a Quality Control Resource.* American Society for Biochemistry and Molecular Biology. DNA Sequencing Research Committee 1998 Web Poster. http://www.abrf.org/ABRF/Research-Committees/dscreports/DNASEQ98.htm#ABSTRACT

13 Adams PS, Dolejsi MK, Hardin S, Mische SB et al. (1996) DNA Sequencing of a Moderately Difficult Template: Evaluation of the Results from a Thermus thermophilus Unknown Test Sample. *Biotechniques* 21: 678

14 Grills G, Adams PS, Dolejsi MK, Hardin S et al. (1996) Evaluation of the Effects of Different DNA Sequencing Methods on

---

[1] The authors describe the technique of using dideoxynucleotides to terminate DNA sequences during extension for DNA sequencing applications. This technique forms the basis of all current automated sequencing technology, and this is considered one of the most influential papers in DNA technology.

Sequencing Standard and Difficult Templates, Expansion of a Web Based Quality Control Resource, and Establishing a Test Array of Sequencing Templates. ABRF DNA Sequencing Research Committee 1996 Web Poster

15 Grills G, Dolejsi MK, Hardin S, McMinimy D et al. (1998b) Assessing the Current State of the Art in DNA Sequencing and Creating a Quality Control Resource. American Society for Biochemistry and Molecular Biology DNA Sequencing Research Committee 1998 Web Poster

16 A Practical Guide for Microsatellite Analysis

Kenneth D. Birnbaum and Howard C. Rosenbaum

Contents

1   Introduction

This chapter's primary emphasis is on providing readers with a practical understanding of microsatellites, so that the technical aspects of obtaining them, using them for genotyping, and analyzing data become straightforward. In order to do so, we first provide some general background on microsatellites, which leads us into more practical protocols and suggestions. The majority of the techniques are centered around fluorescent dye labeling and automated DNA sequencer technology, since many laboratories are moving in this direction. However, we do provide some brief tips and discussion on the use of radioisotopes because they are still used and provide high-quality data. Towards the end of this chapter, we review some specific concerns and troubleshooting ideas when calling microsatellite alleles (or lack thereof). For problems that occur at the early stage of PCR amplification, we offer a troubleshooting flow chart for thermocycler reactions. Although we refer to other reports for details on technical aspects of microsatellites, it is our hope that this chapter

Methods and Tools in Biosciences and Medicine
Techniques in molecular systematics and evolution, ed. by Rob DeSalle et al.
© 2002 Birkhäuser Verlag Basel/Switzerland

provides readers with a logical framework for initial data collection and evaluation of microsatellite data quality. Chapter 10, Analyzing Data at the Population Level, provides an overview of analysis methods and statistical techniques that can be used to address a variety of ecological and evolutionary questions with microsatellite allelic data.

Microsatellites are short, tandemly repeated arrays with a core pattern of 1–5 base pairs that may be entirely consistent or may deviate somewhere in the repeat (see [1, 2]). Perfect (or pure) repeats maintain the core repetitive unit throughout the region, while imperfect (or interrupted) repeats have internal segments that do not follow the core repeat pattern. Interrupted microsatellites are believed to express different levels of evolutionary stability and polymorphism when compared to perfect microsatellites (e. g., [3]). Finally, compound microsatellites consist of different types of tandemly repeated sequences adjacent to one another (Appendix, types of microsatellites).

2   Considerations for obtaining microsatellite loci

The most common sources for identifying microsatellites are the numerous DNA databases such as GENBANK, GDB, or EMBL. In addition, the published literature may contain microsatellites that have not yet been listed in public databases. The greatest advantage to using known microsatellite markers is that they have often been fully characterized: the PCR primer sequences and PCR conditions are both known. Consequently, one does not have to spend time optimizing conditions. Moreover, it may be possible to genotype microsatellite loci originally cloned in one species in other closely related species, even though sequence differences may exist in priming sites. Numerous studies have successfully demonstrated that such cross-species use of primers is feasible for closely related taxa (e. g. [4–12]).

Although Primmer et al. [9] attempt to establish a correlation between microsatellite utility and evolutionary distance from the species from which the loci were originally cloned, there is still no documented benchmark as to the extent to which cross-species microsatellite analysis may be conducted. The nature and dynamics of these loci may vary considerably with increasing evolutionary divergence from the species for which the loci were originally cloned [3, 10, 11, 13]. Although the conservation of these loci (or the surrounding flanking/priming sites) across taxa has been a contributing factor to their popularity, it is only recently that empirical studies have been conducted which illustrate the complexity of these loci. A discussion of these topics is covered in chapter 10.

If previously identified microsatellites are not available, then screening for new ones becomes necessary. The process involves the construction and screening of a genomic library and subsequent sub-cloning of the desired DNA region (e. g., [14, 15]). Newer methods of library production and subsequent screening have greatly facilitated isolation of microsatellites [16]. Discussion of such techniques is summarized in Hammond et al. [17], and will not be discussed further in this chapter. Several companies will screen for microsatellites in a given species for a few thousand dollars providing 10 to 30 microsatellite sequences (see Genetic Identification Services, http://www.genetic-id-services.com/ and Research Genetics, http://www.resgen.com/resources/xref.php3). However, newly identified microsatellite loci may not necessarily be polymorphic or informative.

3   Microsatellites and PCR

Amplification of microsatellites has traditionally been done with radioactively labeled DNA fragments (procedures for which are discussed at the end of this chapter). The use of fluorescent labeling along with automated sequencers has not only eliminated the need for radioactivity but also offers advantages in speed and consistency in calling alleles. When using fluorescently labeled PCR products, one of the PCR primers is typically synthesized with a 5' fluorescent label. Alternatively, some laboratories have used direct incorporation of fluorescent dNTPs [18] to reduce the cost of custom-labeled primers.

Fluorescent labeling enables microsatellites with overlapping allele patterns to be multiplexed on the same gel if they are labeled with different dye colors (multiplexing PCR reactions is not recommended at this time). With one of the fluorescent dyes reserved for the internal lane standard, three (or four) separate colors are available and multiplexing loci with non-overlapping allele patterns is also possible. Thus, careful planning when designing and choosing dye colors for labeled primers can enable up to as many as eight or ten loci to be run on a single gel. Checking the strength of PCR reactions by running a small amount on an agarose gel will help determine optimal amount of each microsatellite for a multiplexed run.

Optimization of the PCR requires some testing to find the reaction conditions which produce the best results. The goal is to amplify only the target sequence, eliminating or at least minimizing amplification of non-target bands, while producing strong, bright bands with a minimum of PCR artifacts or stutter bands. The flow chart outlines (Troubleshooting: microsatelite PCR reactions) a step-by-step process for adjusting conditions, such as annealing temperature, template and $MgCl_2$ concentration, and cycle duration and number, to help optimize PCR with microsatellites.

4   Genescan and Genotyper

If PCR amplification is successful, detection and sizing of the alleles is done using Genescan software. Sizes of the alleles are determined based on a comparison with standard markers, which are loaded into each lane. The internal lane standard eliminates lane to lane variation in sizing alleles. After processing your data, it can be imported into ABI's Genotyper software which aids in "binning" or grouping together alleles that fall within a particular size range; this process assists with calling alleles where there is slight variation in band migration within and between gels. The results from Genotyper can be exported to a database, spreadsheet or statistical packages.

In this section, we address some features of Genescan and Genotyper that can be used to detect and then solve problems that often lead to inaccurate calling of alleles. The emphasis is on problems that could easily go undetected because we regard these as potentially the most serious for subsequent analyses. For instruction on learning Genescan or Genotyper, we refer to extensive documentation and tutorials that accompany the two programs. We focus on three critical points that should be considered when analyzing a microsatellite gel: a) tracking gels b) spotting and correcting false peaks due to seepage of one sample into an adjoining lane and c) setting up categories for automated allele calling.

4.1   Tracking gels and assigning standards

Once gels are tracked, extracted and analyzed, they should be checked to ensure that the standard peaks, which are used to determine the size of all other peaks, are sized correctly by Genescan. This can be done in Genescan or Genotyper by aligning standard peaks by size. For instance, the red signal for each lane can be imported into a Genotyper file and all lanes superimposed in the main window. Any peaks that fall outside the expected standard peaks can be identified. Check all regions of the lanes since size-calling problems may be limited to certain parts of the gel (Troubleshooting: solving tracking problems).

4.2   Checking for lane seepage

Small amounts of PCR product can seep from one lane to the next even when loading appears to have gone smoothly. False peaks from even minor seepage can be particularly misleading when a lane with a strong signal leaks into a lane with a weaker signal. Genotyper can be used to systematically check for seepage as follows: 1) import the dye color of interest 2) pull down the "Views" menu on the toolbar and select "Display by Scan" 3) adjacent lanes can then be checked for peaks that appear at the same scan number, which are potential leakage artifacts in gels where odd and even lanes are loaded at different times. This step is especially important when making categories for automated allele calling since creating a category for a seepage artifact would lead to obvious errors. It is worth noting that seepage artifacts can appear as consistent microsatellite peaks in many lanes, even on gels where odd and even lanes are loaded in a staggered fashion.

4.3 Setting up categories and macros

Categories, which define alleles, can be created either by using the main window to get a peak average or by using the plot window for a representative peak. The automatic category setting is a convenient feature, which can be used by first drawing a "box" to define a peak and executing [Shift]+[Apple]+[L]. This will define a category at the center of the peak with a range of ±0.5 bases. At this point, it is also convenient to set up macros for labeling peaks, creating a table for editing, and creating a final table (see Protocol 1).

If the same loci are being repeated for a moderately large number of samples, it is worth creating a Genotyper "template" file that contains specific allele definitions for certain dyes with custom macros/windows. Once analyzed, be sure to use FILE, "Save as....", so that the template can used again for subsequent analyses.

Protocol 1   Convenient macros for Genotyper

Labeling Peaks

Once categories are designated, this macro will label peaks using an automated routine that can be used in subsequent steps to analyze the assignment of allelic identities:
   1 define color category
   2 in the "Analysis" menu, choose "label peaks" (select "size in base pairs" and "categorys name")
   3 in the "analysis" menu, choose "filter peaks." While default settings work well on most microsatellites, filter settings should be tailored to each microsatellite depending on its banding pattern. Note that default settings are designed for dinucleotide repeats.
   4 in the main window, select the plot window.
   5 end macro.

Editing Peaks

This macro first labels peaks and then sets up a table with allele warnings that aid in correcting allele calls. Using the table and its warnings as a guide, allele calls can be checked and corrected by deleting or adding labels in the plot window.

   1. run the labeling macro

2. in the "Table" menu, choose "set up table." Select desired lane, sample, and category information, also select "low signal" warning and "text if > N labels" warning. For example, it can be set to warn if more than two alleles are found in a diploid organism
3. in the "Table" menu, choose "append rows to table"
4. in the main window, select the table window icon
5. end macro

Creating a Final Table

This macro will set up a data file based on the corrections made to the allele calls in the previous step.

1. in the "Analysis" menu, choose "clear table"
2. in the "Table" menu, choose "setup table" again. This time include all fields that are relevant for analysis or appending records to a database
3. end macro

The table just created can be exported as an Excel file.

5   Specific cautions on calling alleles

The seemingly complex patterns generated by microsatellites may appear to be a problem at first, but these characteristic signatures can actually help distinguish microsatellites from artifacts on a gel. Genescan and Genotyper have made allele calling more consistent and rapid but they have not eliminated the need to inspect and interpret allele patterns. This section will briefly cover the phenomena that give rise to microsatellite banding patterns as an aid to identifying and calling alleles once you have analyzed data with Genescan and Genotyper.

The banding patterns on a microsatellite gel are generally the result of two phenomena that occur during PCR amplification: frameshift products and the addition of an extra base by *Taq* polymerase. Several mechanisms have been proposed to explain microsatellite frameshifts in PCR reactions; the most prominent is slipped-strand mispairing [19] (Other mechanisms proposed to explain frameshift include recombination [20] and template switching [21]). Slipped-strand mispairing occurs when a part of the microsatellite repeat denatures during synthesis and anneals to a new but complementary position on the opposite strand. A frameshift results from the formation of a loop in the DNA, which typically forms on the template strand [22]. The frameshift product is, therefore, usually shorter than the original templated version and the new product is amplified in subsequent rounds of PCR.

The frameshift pattern is complicated by the frequent addition of a non-templated 3' nucleotide by *Taq* polymerase [23], which is not usually complete and results in a mixture of products. The strength of the +1 reaction appears to depend largely on the base at the 5' end of the template [24]. Thus, a microsatellite allele usually has a series of peaks, which typically start with the +1 band as the peak with highest molecular weight (peak a in Figure 1) followed by the template-length band (peak b in Figure 1). Each of the first two bands (a and b) will usually have one or more sets of frameshift artifacts, causing "shadow" or "stutter" bands one, two and possibly three repeat lengths shorter. They have the same relative amplitudes of the first two peaks but usually diminish in height (c and d). Locus-specific differences in the strength of the +1

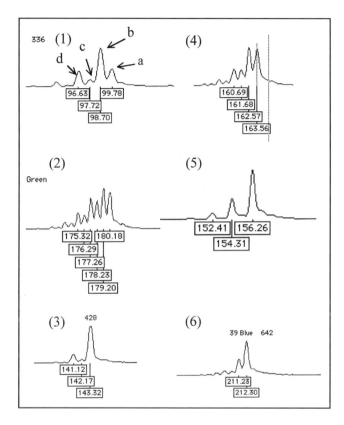

Figure 1 Six different microsatellite patterns all derived from frameshift products and non-templated additions. (1) Peak b represents the PCR product as it was encoded on the template. In this case it is the dominant peak. Peak a represents the products derived from peak b on which *taq* polymerase added a non-templated nucleotide, which is a minor product at this locus. Peaks c and d represent shadow bands of peaks a and b, respectively. Panels 2–6 are patterns generated by the same processes but with differing proportions of +1 products and shadow bands. Panel 2 has a +1 band of almost equal amplitude as the the template band and two sets of stutter bands. Panel 3 has one dominant band, probably the +1 product, and a low amplitude stutter band. Panel 4 has template and +1 bands of about equal amplitude and two sets of rapidly diminishing stutter bands. Panel 5 has no visible +1 bands and two stutter bands. Panel 6 has a high amplitude +1 band and no discernable stutter bands.

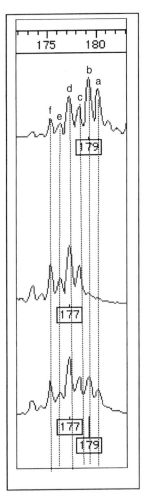

Figure 2 A dinucleotide microsatellite repeat showing two alleles that are two base pairs apart and a heterozygote between the two. In the top panel, the +1 band (top panel), which is lower in amplitude than the template-length band, and two sets of stutter bands, first, c and d, and then, e and f, are shown. The second allele, in the middle panel, shows a similar pattern but is two base pairs shorter. In the bottom panel, the heterozygote is a superimposition of the two single allele patterns although it contains all the same peaks as the allele in the top panel. Note that the 177 allele buried in the heterozygote pattern is higher in amplitude than the 179 allele, probably because it is the result of both the primary peaks of the 177 allele and the stutter bands of the 179 allele.

reaction and the number of stutter bands allow for many variations of the pattern (panels 2–6 in Figure 1 and also see Appendix, aids in interpreting allele patterns).

The pattern becomes more complex when peaks from two alleles overlap, such as in the case of a dinucleotide repeat with alleles that are one repeat apart. In Figure 2, the top panel shows a single allele, which has a stronger template-length product (peak b is typically higher than peak (a) and has two sets of shadow bands (c and d are shadows of a and b, respectively; e and f are the second set of shadows). This allele gets labeled at the templated length (peak b), which is 179 base pairs. (Genotyper tends to work better when when the larger of the templated or +1 product gets the label). By the same reasoning, the peak in the middle panel is labeled at 177.

A signature of the heterozygote is that the highest peak is buried within the shadow bands as in the bottom panel of Figure 2. The first set of peaks (a and b) corresponds to the template length and +1 bands of the 179 allele. The next set of peaks (c and d) represents both the first set of shadow peaks from the 179 allele and the template length and +1 band of the 177 allele. Thus, the shorter length allele (177 in this case) is typically larger in amplitude for a heterozygote that has two alleles separated by two base pairs. The third set of bands in the heterozygote is the first set of shadow bands from the 177 allele and the second set of shadow bands from the 179 allele. Note that the heterozygote has all the same peaks as the single 179 allele but the relative peak amplitudes are different. Genotyper can recognize this pattern, but it is important to check that it does so properly and adjust settings if necessary. In addition, it can be very difficult to determine heterozygotes for certain loci that have alleles with a high repeat number (Figure 3).

Some techniques have been developed to quantitatively dissect overlapping allele patterns and reconstruct the signals for each allele [25]. However, such methods will be difficult to apply when the relative amplitude of each allele is unpredictable and/or the banding patterns for each allele are different. This means that heterozygotes with long-repeat alleles are often difficult or impossible to distinguish and any analysis with these loci should take this into account (Figure 3).

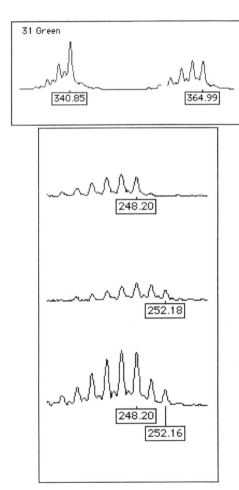

Figure 3 Long repeats are a problem when calling alleles. The top panel shows an allele with a high repeat number that is about 24 base pairs longer than the shorter allele and shows the shift from a single dominant peak to several peaks with a similar amplitude. In the bottom panel, the heterozygote pattern is difficult to assess. The two alleles shown on the top of the box appear to be different sizes at 248.2 and 252.18, respectively, although the gradual fading of the longest peaks makes even these calls dubious. The pattern at the bottom of the box may represent a heterozygote between the two alleles or a single 252 allele. In this case, the putative heterozygote at the bottom panel and the 252.18 allele in the middle of the box have no distinguishing features. All these alleles could be binned into one category, that is, labeled as a single allele, but the locus will have limited use.

6   Detecting alleles with radioisotopes

The use of radioisotopes to detect alleles has become less common in recent years as automated sequencers have become widespread. However, "manual" generation of microsatellite alleles remains a reliable way to analyze microsatellite alleles. In radioisotope analysis, PCR reactions are often scaled down to 10 µl to conserve costly radioactive nucelotides. For incorporation reactions, alpha-labeled nucleotides are used (e. g., α-^{33}P dCTP (10mCi/ml, 2,000–4,000Ci/mmol) using 0.1 µl per 10 µl PCR reaction and a lower concentration of non-labeled dCTP so all dNTPs are at the same concentration). The most common two choices for radionucleotides are ^{33}P, which is more costly, emits less powerful energy and has a half-life of 25 days, or ^{32}P, which is less costly, emits a more energetic particle and has a half-life of 14 days (DuPont NEN). The use of ^{35}S is not recommended as it volatizes more readily and can contaminate PCR machines. Of course a geiger counter, appropriate shielding, disposal facilities, lab certification, and training are also required for work with radioisotopes.

Analysis of microsatellites can be done on any manual sequencing apparatus. Narrow sequencing gels are easier to pour than wide gels and most power supplies allow two gels to be run at once. A wide gel with a water-cooled system that is run at a constant temperature can greatly increase throughput and keep gels easy to score. We have used a 38×50 cm plate SequiGen apparatus (BioRad) and a power source with a temperature sensor to automatically adjust voltage (e. g., see products from Stratagene, Pharmacia, or BioRad). The wide gels can run up to about sixty samples since the water-cooling prevents distortions that can make scoring difficult. Any sequencing reaction (ladder) can be run on one or both ends of the gel and in the middle to help size microsatellite alleles on different parts of the gel. Gels are then dried (for 2–3 hours with heat) and exposed overnight. For increased sensitivity, especially when using ^{33}P, BioMax film (Kodak) can be used. For ^{32}P, non-coated films are cheaper and usually effective.

7   Troubleshooting

The increasing availability of microsatellite primer sequences allows high-resolution population-level studies on a wide variety of species. Automated technology has sped up the process but understanding complex microsatellite patterns is still important for interpreting raw data. We have presented an overview of microsatellites and a practical guide to working with them in the hope that a researcher new to microsatellite markers will be able to start generating results relatively easily and rapidly. The following section provides more detail in addressing some common problems.

7.1 Microsatellite PCR reactions

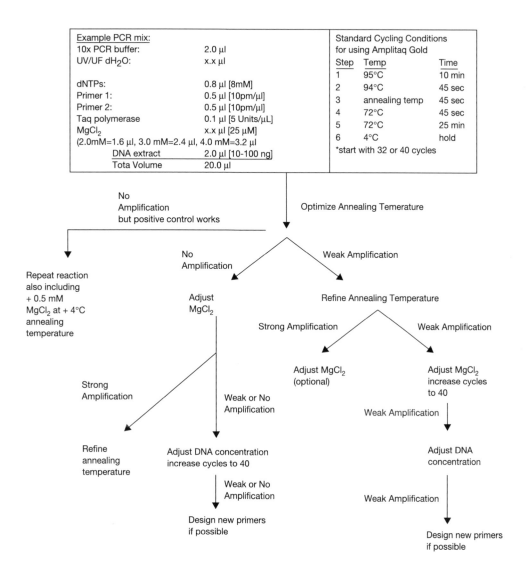

Suggested Adjustments:
- Optimize annealing temperature = 50°–60°C at 5°C intervals
- Refine annealing temperature = ± 4°C at 2°C intervals starting with annealing temperature of best reaction
- Adjust $MgCl_2$ = 0.5–4.0 mM $MgCl_2$ at 0.5 mM intervals
- Adjust DNA concentration = 50 ng genomic DNA (1X), try 10X,0.1X
- Design new primers = preferably non overlapping, follow published guidelines for primers = design.

7.2   Solving tracking problems

- *Bubbles and distortions:* Although Genescan attempts to correct gel distortions, the final gel image may contain distorted or sloping bands. If such a sloping band is tracked at the high end of the slope in one part of the gel and the low end of the slope in another, standards will be off in certain regions. The problem is usually corrected by consistently tracking the same end of the sloping band in all parts of the gel.

- *False standard peaks:* We have also noticed that problems with standards often arise in the regions of low molecular weight standards (e. g., 50, 75 and 100 bp). For instance, when primer dimers in these regions are intense, they can create artifacts that mislead Genescan's automated assignment of standard peaks. The solution is to manually create a standard for each lane where such problems exist, since even a good generic standard may fail with erroneous peaks present.

Appendix

Types of microsatellites

```
PURE        CAGCAGCAGCAGCAGCAGCAGCAGCAGCAG
COMPOUND    CAGCAGCAGCTGCTGCTGCTGCTGCTGCTG
IMPERFECT   CAGCAGCAGTTCAGCAGCAGCAGTTCAGCAG
FLANKING    ACTGTATTTATGCAGCAGCAGCAGCAGCAGCAGCAGCAGCAGTTACAGT-
TACCC
SEQUENCE
```

This schematic is an example of the different types and compositions that make up a repetitive array of a microsatellite. This particular array consists of ten CAG tri-nucleotides. The underlined sequence is an example of the non-repetitive DNA sequence external to or flanking the repetitive array. These regions may also contain base substitutions, insertions and deletions that influence allelic size variance at microsatellite loci (from [1]).

Aids in interpreting allele patterns

- *Run a known cross*

Many microsatellites do not exhibit Mendelian patterns of inheritance (e. g., [26]). Running microsatellite products of several individuals and their known parents is the best way to check for Mendelian inheritance. If available, we recommend running a test cross the first time a microsatellite is used, even if other labs have used the same primer pair in previous studies. These will also provide a check to ensure allele calling is accurate.

• *Use a standard sample*

Many evolutionary studies, especially at the population level, will involve the analysis of hundreds of individuals requiring comparison of alleles run on different gels. We have found as much as a one base pair shift between different gels (also see [27]). It is good practice to run one or two samples from the same individuals (using new PCR reactions) on each gel. This provides a tool to determine the extent of band migration shift on a gel. It also provides a control for PCR reactions that can alter banding patterns, for instance, by much more efficient addition of the +1 products in a given set of reactions.

• *Sequence alleles*

It is not possible to determine the exact size of an allele using Genescan and Genotyper [27]. If exact sizes are needed, at least one allele from the microsatellite should be sequenced. Since size homoplasy has also been shown to occur, spot check sequencing of alleles that are binned together is also a good idea if possible, especially when necessary to ensure identity by descent relationships.

References

1   Rosenbaum HC, Deinard AS (1998) Caution before claim: An overview of microsatellite analysis in ecology and evolutionary biology. In: R DeSalle and B Schierwater (eds): *Individuals, Populations and Species: Molecular Approaches and Perspectives*. Birkhäuser, Basel, Switzerland

2   Jarne P, Lagoda PJL (1996) Microsatellites, from molecules to populations and back. *Trends in Ecology and Evolution* 11: 424–429

3   Estoup A, Garnery L, Solignac M, Cornuet J-M (1995b) Size homoplasy and mutational processes of interrupted microsatellites in two bee species, *Apis mellifera* and *Bombus terrestris* (Apidae). *Molecular Biology and Evolution* 12(6) 1074–1085

4   Schlötterer C, Amos B, Tautz D (1991) Conservation of polymorphic simple sequence loci in cetacean species. *Nature* 354: 63–65

5   Morin PA, Moore JJ, Chakraborty R, Jin L, Goodall J, Woodruff DS (1994a) Kin selection, social structure, gene flow, and the evolution of chimpanzees. *Science* 265: 1193–1201

6   Meyer E, Wiegand P, Rand SP, Kuhlmann D, Brack M, Brinkmann B (1995) Microsatellite polymorphisms reveal phylogenetic relationships in primates. *J. Mol. Evol.* 41: 10–14

7   Pépin L, Amigues Y, Lépingle A, Berthier J-L, Bensaid A and Vaiman D (1995) Sequence conservation of microsatellites between *Bos taurus* (cattle), *Capra hircus* (goat) and related species. Examples of use in parentage testing and phylogeny analysis. *Heredity* 74: 53–61

8   Valsecchi E, Amos W (1996) Microsatellite markers for the study of cetacean populations. *Molecular Ecology* 5: 151–156

9   Strassman JE, Solis CR, Barefield K, Queller DC (1996) Trinucleotide microsatellite loci in a swarm-founding neotropical wasp, *Parachartergus colobopterus* and their usefulness in other social wasps. *Molecular Ecology* 5: 459–461

10 Primmer CR, Møller AP, Ellegren H (1997) A wide-range survey of cross-

species microsatellite amplification in birds. *Molecular Ecology* 5: 365–378

11 Ellegren H, Moore S, Robinson N, Byrne K, Ward W, Sheldon BC (1997) Microsatellite evolution-a reciprocal study of repeat lengths at homologous loci in cattle and sheep. *Molecular Biology and Evolution* 14(8): 854–860

12 Palsbøll PJ, Bérubé M, Larsen AH, Jørgensen H (1997) Primers for the amplification of tri- and tetramer microsatellite loci in baleen whales. *Molecular Ecolgy* 6: 893–895

13 Angers B, Bernatchez L (1997) Complex evolution of a Salmonid microsatellite locus and its consequences in inferring allelic divergence from size information. *Molecular Biology and Evolution* 14(3) 230–238

14 Rassman K, Schlötterer C, Tautz D (1991) Isolation of simple-sequence loci for use in polymerase chain reaction-based DNA fingerprinting. *Electrophoresis* 12: 113–118

15 Edwards et al. (1996)

16 Rico C, Rico I, Hewitt G (1994) An optimized method for isolating and sequencing large (CA/GT)n (n>40) microsatellites from genomic DNA. *Molecular Ecology* 3: 181–182

17 Hammond RL, Saccheri IJ, Ciofi C, Coote T, et al. (1998) Isolation of microsatellite markers in animals. In: *Molecular Tools for Screening Biodiversity*: A Karp, PG Isaac, DS Ingram (eds): Chapman & Hall, London

18 Rhodes M, Darelove A, Straw R, Fernando S, et al. (1997) High-throughput microsatellite analysis using fluorescent dUTPs for high-resolution genetic mapping of the mouse genome. *Genome Research* 7: 81–86

19 Levinson G, Gutman GA (1987) Slipped-strand mispairing: a major mechanism for DNA sequence evolution. *Molecular Biology and Evolution* 4(3): 203–221

20 Meyerhans A, Vartanian J-P, Wain-Hobson S (1990) DNA recombination during PCR. *Nucleic Acids Research* 18(7): 1687–1691

21 Odelberg SJ, Weiss RB, Hata A, White R (1995) Template-Switching during DNA synethesis by *Thermus aquaticus* DNA polymerase I. *Nucleic Acids Research* 23(11): 2049–2057

22 Hite JM, Eckert KA, Cheng KC (1996) Factors affecting fidelity of DNA synthesis during PCR amplification of D(C-A)•d(G-T)$_n$ microsatellite repeats. *Nucleic Acids Reseach* 24(12): 2429–2434

23 Clark JM (1988) Novel non-templated nucleotide addtion reactions catalyzed by procaryotic and eucaryotic DNA polymerases. *Nucleic Acids Research.* 16: 9677–9686

24 Magnuson VL, Ally DS, Nylund SJ, Karanjalwala ZE, et al. (1996) Substrate nucelotide-determined non-templated addition of Adenine by Taq DNA polymerase: Implications for PCR-based genotyping and cloning. *BioTechniques.* 21: 700–709

25 Miller MJ, Yuan B-Z (1997) Semi-automated resolution of overlapping stutter patterns in genomic microsatellite analyis. *Analytical Biochemistry* 251: 50–56

26 Sharon D, Cregan PB, Mhammed S, Kusharska M, et al. (1997) An integrated genetic linkage map of avocado, *Theoretical and Applied Genetics* 95: 911–921

27 Haberl M, Tautz D (1999) Comparative allele sizing can produce inaccurate allele size differences for microsatellites. *Molecular Ecology.* 8: 1347–1350

17 Determining the Spatial and Temporal Patterns of Developmental Gene Expression in Vertebrates and Invertebrates Using *in situ* Hybridization Techniques

Ruth D. Gates, Thorsten Hadrys, Cesar Arenas-Mena and David K. Jacobs

Contents

Methods and Tools in Biosciences and Medicine
Techniques in molecular systematics and evolution, ed. by Rob DeSalle et al.
© 2002 Birkhäuser Verlag Basel/Switzerland

1   Introduction

Over the past century, histologists have utilized specific stains and microscopy to resolve cellular components, tissue structure and classes of molecules. In addition, immuno-histochemistry continues to be widely used to localize proteins, a technique that has been employed extensively in comparative studies of both structural proteins and regulatory genes important in development. More recently, *in situ* hybridization protocols have been developed which allow for the precise localization of specific nucleic acid sequences in embryos and tissue sections. The advantage of this approach over antibody studies is that a species-specific probe can be generated from a cloned gene product produced by PCR, RT-PCR or cDNA library screening, a methodology that eliminates both the intermediate step of developing an antibody and the concern regarding cross-species reactivity of the antibody. Thus, in the absence of an effective polyclonal or cross-reactive monoclonal antibody for a protein known to function across the taxa, *in situ* hybridization is more time- and cost-effective.

In comparative developmental studies, antibody studies address the location of proteins and *in situ* hybridization is most often used to locate messenger RNA (mRNA). Depending on the function of a gene and its protein product, the mRNA and protein may be localized in different cellular compartments. In addition, there are post-transcriptional regulatory inputs that can prevent the translation of a protein, and a protein can be very stable and remain long after transcription ceases. Thus, the information from antibody studies and *in situ* hybridization does not overlap completely and the inclusion of both techniques in a study can be informative. In this chapter we focus on *in situ* hybridization.

In situ hybridization is a technique that has broad applications in biology. These include: 1) The resolution of spatial and temporal patterns of gene expression 2) genetic mapping and the localization of specific DNA sequences on chromosomes 3) confirming the source of sequences generated by PCR to resolve whether the sequences originate from the organism of interest, or from an ingested contaminant [1] 4) the isolation of non-cultured micro-organisms from mixed samples using specific probes generated against one of multiple sequences (such as 18SrDNA) produced using environmental PCR and 5) the localization of viruses and other disease agents in tissues.

In this chapter we concentrate on *in situ* hybridization as it is applied to resolve the spatial and temporal patterns of developmental gene expression across diverse taxa. The utility of this technique in addressing comparative aspects of gene expression across a wide range of organisms has already been demonstrated and includes studies of butterfly wing pattern formation [2], Hox gene expression in heteronomous polychaete worms [3], molluscs [4], echinoderms [5, 6] and hydrozoans [7, 8], to mention just a few. Here we discuss key aspects of the technique of *in situ* hybridization that may be pertinent to a range of organisms which have not yet been the focus of detailed investigations.

2　An Overview of *in situ* Hybridization

In situ hybridization was first developed in 1969 by Pardue and Gall [9], and by John and his colleagues [10]. The technique exploits the property of complementary nucleic acid sequences to anneal to one another under specific experimental conditions. A radioactively or non-radioactively labeled antisense RNA or DNA probe is generated using a cloned fragment of the gene of interest, and hybridized to the target RNA or DNA in sectioned tissues or whole mounts of the organism under investigation. The label attached to the antisense probe during synthesis is either directly visualized using fluorescent microscopy or auto-radiography, or indirectly localized using an antibody conjugated to an enzyme. The labeled antibody conjugate is visualized by providing a substrate for the enzyme that yields a colored insoluble precipitate which can then be observed by microscopy.

2.1　Probe type and length

Single stranded RNA, single-stranded DNA, double-stranded DNA and oligonucleotide probes have all been used for *in situ* hybridization experiments and can be generated with relative ease. However, the greatest sensitivity is generally achieved when using single-stranded RNA and DNA probes. The complementary strands of double-stranded DNA probes (de-natured prior to use) re-anneal at the temperatures used for hybridization and this reduces the concentration of probe available to hybridize to the target mRNA. Oligonucleotide probes carry fewer labels per molecule of probe and, because of their short length (20–40 bases), tend to cross-hybridize with genes containing sequence similar to the target. This issue is particularly relevant in developmental genetics where classes of genes, such as those containing homeodomains, exhibit incredible conservation within and between taxa.

From a purely theoretical standpoint, longer probes should be more effective than short probes. They incorporate a greater number of labeled nucleotides, which increases the signal intensity, and they re-nature with their complements more rapidly than shorter probes at the same concentration of molecules. However, longer probes have the major disadvantage of being unable to diffuse through or penetrate dense cellular matrices and may be more easily retained by non-specific interactions. Although a number of treatments can be carried out to permeabilize tissues and increase the probability that longer probes will reach their target, shorter probes between 50–1000 bases are most frequently used for *in situ* hybridization. It should also be noted that the choice of label will also influence the size of the probe; with a fluorescent label, the best results are obtained using probes that range in size from 100–500 bases, with 300 being optimal. Larger probes result in more background fluorescence [11].

2.2 Probe label

An *in situ* probe can either be labeled directly or indirectly. Directly labeled probes have a detectable molecule (or reporter), such as fluorescein, bound directly to the nucleic acid probe. When employing this type of label, the probe-target hybrids can be visualized immediately using fluorescent microscopy. Indirectly labeled probes contain a reporter molecule, such as digoxigenin, that is bound directly to the nucleotide, but that cannot be directly visualized by microscopy. In this instance, the reporter molecule is detected by using a specific antibody that is coupled to either a fluorochrome or an enzyme. The reporter molecule/antibody/fluorochrome/enzyme complex can then be visualized either by fluorescent microscopy or by introducing a substrate for the enzyme in a color reaction that yields a colored precipitate. For both directly and indirectly labeled probes it is important that the detectable molecule does not interfere with the hybridization and that they can survive the rigorous washing and hybridization conditions. It is noteworthy that directly and indirectly labeled fluorescent probes are less sensitive than those visualized by enzymatic yield of an insoluble colored product. However, the greater diversity of fluorescent labels provides a powerful set of tools with which to localize multiple abundant sequences in a single experiment. In the future, the coupling of enzymatic reactions and insoluble fluorescent products will provide additional sensitivity and result in preparations that lend themselves to analysis by confocal microscopy.

Investigators also have the choice of using radioactive or non-radioactive (hapten) labels. *In situ* protocols were originally developed using radioactive labels [9, 10]. ^{35}S labeled probes, the most commonly used labeling isotope, are still considered to be marginally more sensitive than hapten-labeled probes. However, radioactively labeled probes are expensive, deteriorate rapidly, are difficult to work with, and can only be used to visualize targets in *in situ* hybridization experiments of sectioned material. In contrast, hapten-labeled probes are cheap to produce, stable, and comparatively safe to work with. More importantly, they can be used to elucidate three-dimensional patterns of gene expression in whole mount which can be further clarified by sectioning. Given these advantages, it is not surprising that most developmental biologists generally use hapten rather than radioactive probes. The most commonly used hapten labels are fluorescein, digoxygenin and biotin.

2.3 Hybridization conditions

There are a number of issues to consider when optimizing hybridization conditions to achieve stable and specific probe:target hybrids. These include the temperature and duration of the hybridization, and the concentration of the probe. Hybridizations are usually conducted 20–30°C below the temperature at which 50% of the target:probe hybrids dissociate (T_m), a value that is influenced by a number of features of the experimental design. Target:probe hybrids that are rich in G:C bonds have higher melting temperatures and greater stability than those that are A:T(U)-rich, due to the additional hydrogen bond formed between G:C base pairs. Longer probes are more stable than shorter probes and the nature of the hybrid confers greater or lesser stability (RNA:RNA > RNA:DNA > DNA:DNA). Obviously, highly specific probes will form more stable hybrids than divergent probes and target sequences.

In addition to the properties of the probe and target, the T_m and stability of the hybrid can be altered by the chemical composition of the hybridization solution. High concentrations of monovalent cations, such as sodium, increase the stability of hybrids, and inclusion of formamide reduces the T_m and allows stringent hybridizations to be conducted at lower temperatures.

A number of formulae are available that allow the investigator to calculate T_m based on the GC composition of a given probe, probe length, and the percentage of formamide and concentration of monovalent cations used in the hybridization solution (see [12] for details). These formulae apply to homologous hybridizations (DNA:DNA or RNA:RNA) conducted in solution and do not consider issues relating to probe access to target in fixed tissues. However, RNA:RNA *in situ* hybridization experiments are thought to have a T_m approximately 5°C below the T_m of the same hybrid in solution [13]. It should be noted that many investigators do not calculate an exact T_m for their probes but arbitrarily select a temperature that has previously been effective for *in situ* hybridization experiments on the same tissue or using a similar type of probe. For many probes a temperature of 55°C, 50% formamide and 4–5 × SSC works well and is a good starting point when troubleshooting a new system (see protocols for detail). If no signal can be detected, the temperature should be lowered. However, if the background is high under these conditions, increase the specificity of the hybridization by raising the temperature.

The concentration of the probe and duration of hybridization will be dictated by the length of the probe, the accessibility of the target, and any secondary structure in the target sequence. The probe concentrations cited in the literature are extremely variable, ranging from 0.02 µg/ml to 5 µg/ml. However, the majority fall somewhere between 0.05–0.25 µg/ml. It is possible to use an excess of probe in the hybridization, as the stringent post-hybridization washes should remove any unbound probe, although there is an upper limit in probe concentration that will generate irreversible background (for review of hybridization kinetics see [14]). Thus, if the background is high using excess probe concentrations, it might be advantageous to decrease the amount of probe in the hybridization.

The duration of the hybridization is inversely related to the stringency of the conditions. For example, using a moderate hybridization temperature of 25°C below the T_m, RNA:RNA and DNA: DNA hybridizations are complete after approximately 6 h [13, 15]. In cases where high-stringency hybridizations are preferred (temperatures closer to the hybrid T_m), the duration varies from 12 h to 1 week, to allow for the reduced rate of hybridization under these more rigorous conditions.

2.4 Visualization

Directly labeled hapten probes are visualized using fluorescent microscopy. Indirectly labeled hapten probes are detected using specific antibodies that are conjugated to either a fluorochrome or an enzyme and visualized by fluorescent microscopy or in a color reaction that provides a chromogenic substrate for the enzyme that yields a precipitate. The latter serves to amplify the label signal and is considered to be the most sensitive method of visualizing hapten labels. The enzyme alkaline phosphatase is generally used for this purpose with 4-nitroblue tetrazolium chloride (NBT)/5-bromo-4-chloro-3-indolyl-phosphate (BCIP) as a substrate (other substrates are available but are less sensitive than NBT/BCIP), although galactosidase, horse-radish peroxidase and a streptavidin conjugate have also been used successfully.

Radioactive probes (most frequently ^{35}S) are visualized using autoradiography. For high-resolution signals, slides holding the hybridized ^{35}S labeled sections are dipped in nuclear track emulsion, dried, exposed and developed. The tissues are counterstained with a nuclear stain and examined using bright field and dark field microscopy. High densities of silver grains appear as dark spots under bright field illumination and as white dots under dark field illumination.

2.5 Controls

To determine if the signal visualized truly reflects the location of a transcript of interest, there are a number of controls that can be run. A sense probe generated from the same clone as that used to generate the antisense probe is commonly used as a control for hybridizations to specific mRNA. However, some genes, such as *Distal-less,* are transcribed on both the sense and antisense strand [12, 16], and as such, the sense control will also give a positive result. Alternative controls include using a probe of approximately the same length, melting point and G:C content generated for the same or a different gene in another organism, or a probe generated against a gene whose expression pattern is well documented in the organism of choice.

3 Taxon-specific Issues

3.1 Relaxation

Invertebrates such as cnidarians are contractile and must be relaxed prior to dissection or fixation for *in situ* hybridization. This can be achieved by briefly incubating the animals in 2% urethane [7] or gradually introducing menthol or magnesium chloride (0.36 M) into the growth medium (sea water or hydra medium) until the animal no longer responds to mechanical stimulation.

3.2 Membranes, mucus, chitin and calcareous structures

There are a number of external barriers that interfere with the penetration of *in situ* hybridization probes. These include extra-embryonic membranes such as those found in vertebrates, leech and fly, and the mucus, chitin and calcareous structures encountered in a diversity of invertebrates. Removal of these elements can be achieved using a variety of approaches.

Extra-embryonic membranes can be mechanically removed by dissection, dissolved by chemical treatment, or enzymatically digested. The membranes of larger vertebrate and invertebrate embryos are generally dissected away [17, 18], although a recent protocol for *Xenopus* uses a mixture of Proteinase K, collagenase A and hyaluronidase (10 µg/ml, 2 mg/ml and 20 U/ml respectively) to remove the vitelline membrane [19]. *Drosophila* embryos are de-chorionated using commercial bleach (50% solution; [20]) and the vitelline membranes of leeches are permeabilized using the protease pronase E (0.5–1 mg/ml at room temperature for 20–60 min; [21, 22]).

Most aquatic animals produce mucus and some contain structural elements such as chitin and shell. Generally, relaxing the organism prior to fixation can minimize the production of mucus. However, in taxa where this approach is not employed, such as planaria, mucus can be chemically dissolved by a brief treatment with hydrochloric acid (2% for 5 min; [23, 24]). Interestingly, the most recent protocol for this taxon combines relaxation, mucus removal and fixation into a single step (1% nitric acid, 2.25% formalin, 0.05 M magnesium sulphate [25]). Chitin can also interfere with probe penetration and can be removed enzymatically using chitinase after fixation (4 U/ml in 0.1 M HEPES for 60 min; Sigma EC 3.2.1.14). We have successfully used this technique to eliminate an ultrastructural layer of chitin in an investigation of *engrailed* expression in chiton embryos [4].

Likewise, the calcareous structure surrounding embryonic molluscs must also be removed to allow effective penetration of the probe, a procedure that should be carried out after fixation. We have successfully employed two methods in our work on the bivalve molluscs *Transennella* and *Crassostrea*. Embryos can be de-calcified either by treatment with EDTA overnight (0.05 M in phosphate buffered saline 0.1% Tween-20-PBT), or with glacial acetic acid for 10 min (1% in PBT). The former treatment is gentler than the latter and yields embryos that are less easily damaged during the multiple steps of *in situ* hybridization. However, both methods have yielded excellent results in our studies of *engrailed* expression in bivalve molluscs [4].

3.3 Fixation

Fixation for *in situ* hybridization must preserve cellular morphology while allowing the probe to penetrate the tissues and find its target. Fixation of tissues or embryos for whole mount *in situ* hybridization is generally conducted with 4% paraformaldehyde, 4% formaldehyde or 1–2.5% gluteraldehyde [the fixation of the planarian *Dugesia japonica* in Carnoys solution (60% ethanol, 30% trichloromethane, 10% acetic acid, 3 h at room temperature [23, 24]) is a rare departure from the more standard fixation methodologies] and the duration varied depending on the size of the specimen. Although the rate of fixation is temperature-dependent, it is generally carried out at 4°C to inactivate endogenous ribonucleases. To maintain the cellular integrity of the tissues being fixed it is very important to consider the osmolarity of the fixative relative to the composition of the organism. For vertebrates and some terrestrial invertebrates the fixative is usually diluted in phosphate-buffered saline [17, 26]. For aquatic and marine invertebrates the fixative is diluted in sea water (e. g., [27]), or in an isotonic culture medium (e. g., [7]), or in a buffer that is adjusted for osmotic conformity with salt (e. g., [28]). Fixation protocols may require extensive optimization depending on the species and developmental stage under investigation.

A greater variety of fixatives are used for tissues that are perfusion fixed-prior to sectioning, or cryo-sectioned and then used for *in situ* hybridization. These include the three aldehydes listed above, Carnoys solution (e. g., [29]), Bouins fixative (e. g., [30, 31]) and a formalin/picric acid/gluteraldehyde mix (e. g., [24]). The increased diversity of fixatives used for these experiments reflects a diminished concern regarding cross-linking and the associated problems of probe penetration, as sectioning the tissue decreases the diffusion distance of the probe to target.

3.4 Masking coloration

The localization of gene expression can be masked by the innate coloration of the taxon being investigated. A good example of this is the dark coloration of planarian flatworms. To overcome this problem, a post-fixation bleaching step is incorporated into the protocol. This involves incubating the samples for 10 hours in 10% hydrogen peroxide in methanol at room temperature under light [24].

3.5 Permeabilization

As already mentioned, one of the keys to successful *in situ* hybridization experiments is the effective penetration of the probe into the tissues. To increase the probability that this will occur, embryos or tissues are generally subjected to two pre-treatments that increase permeability. These treatments are both carried out post-fixation, and the first simply involves the removal of lipids by incubating tissues in organic solvents such as ethanol, methanol, heptane and xylene. The second treatment either involves the disruption or partial digestion of cellular proteins using a sonicator, heat followed by cold, detergents and/or proteases. Table 1 summarizes this treatment for selected vertebrate and invertebrate taxa. Although the vertebrate treatment is fairly standard, there is considerable variation in the type and concentration of the protease used, and in the timing of this permeabilization step in invertebrate protocols. Although these data can be used as a guideline when designing a protocol for a new taxon, the details of this treatment should be determined for the specific tissue being investigated. Note that fixation and permeabilization protocols should be co-optimized, as a stronger fixation will decrease permeability.

Table 1.

Organism	Stage	Prep	Permeablizing agent	Concentration	T°C	Duration	Reference
Vertebrates							
Mouse	7.5 d	WM	Proteinase K	10 µg/ml	RT	7 min	[17]
Mouse	8.5 d	WM	Proteinase K	10 µg/ml	RT	10 min	[17]
Mouse	9.5 d	WM	Proteinase K	10 µg/ml	RT	15 min	[17]
Mouse	10.5 d	WM	Proteinase K	10 µg/ml	RT	20 min	[17]
Mouse	11.5 d	WM	Proteinase K	10 µg/ml	RT	25 min	[17]
Chick	Stages 3–6	WM	Proteinase K	10 µg/ml	RT	5 min	[17]
Chick	Stages 6–12	WM	Proteinase K	10 µg/ml	RT	10 min	[17]
Chick	Stages 12–25	WM	Proteinase K	10 µg/ml	RT	20 min	[17]
Xenopus	Stages 1–14	WM	Proteinase K	10 µg/ml	RT	10 min	[17]
Xenopus	Stages >15	WM	Proteinase K	10 µg/ml	RT	15 min	[17]
Zebrafish	<20 somites	WM	Proteinase K	10 µg/ml	RT	5 min	[17]
Zebrafish	>20 somites	WM	Proteinase K	10 µg/ml	RT	10 min	[17]
Vertebrate		TS	Proteinase K	10 µg/ml	RT	5–10 min	[17]
Invertebrates							
Drosophila	embryos	WM	Proteinase K	50 µg/ml	RT	3–5 min	[20]
Brine shrimp	20–50h	WM	sonication			5–7 s	[32]
Triops	embryos	WM	sonication			2–4 s pulses	[33]
Sea urchin	embryos	WM	Proteinase K	5–40 µg/ml	RT	5 min	[34]
Sea urchin	embryos	WM	Proteinase K	5 µg/ml	RT	5 min	[35]
Sea urchin	embryos	WM	Proteinase K	2 µg/ml	RT	30 min	[36]
Leech – B	embryos	WM	Pronase	0.5 mg/ml	RT	20–25 min	[21]
Leech – B	embryos	WM	Pronase E	1.0 mg/ml	RT	40–60 min	[22]
Leech – A	embryos	WM	Pronase	0.5 mg/ml	RT	0–10 min	[37]
Leech – A	embryos	WM	Triton X-100 + 100 rpm	1.2%	RT	12 h	[18]
Earthworm	heads	TS	Proteinase K	10 µg/ml	RT	10 min	[38]
Clam	embryos	WM	Proteinase K	50 µg/ml	RT	5 min	[4]
Chiton	embryos	WM	Proteinase K	10 µg/ml	RT	5 min	[4]
Tapeworm	adult	TS	Pepsin	0.1% in 0.2 M HCL	37°C	20 min	[30]
Nematode	adult	WM	Proteinase K	100 µg/ml	37°C	20 min	[39]
Jellyfish	polyps	WM	Proteinase K	20 µg/ml	RT	15 min	[40]
Jellyfish	polyps	WM	Proteinase K	20 µg/ml	RT	7 min	[41]
Jellyfish	medusae	WM	Proteinase K	20 µg/ml	RT	5 min	[40]
Jellyfish	medusae	WM	Proteinase K	10 µg/ml	RT	3–7 min	[41]
Hydra	polyps	WM	Proteinase K	10 µg /ml	RT	10 min	[7]
Hydractinia	tissues	WM	Heat and ice		95°C	5 min	[8]

Note: The vitelline membrane of the two leech protocols designated A were removed by dissection prior to permeablization. These membranes were not removed in the protocol designated B.

3.6   Background issues

When conducting *in situ* hybridization experiments, the presence (and abundance) of both positively charged amino acids and endogenous phosphatases are two taxon-related issues to consider with regard to background. Nonspecific electrostatic binding of the probe to positively charged amino acid residues can dramatically increase the signal background and make the interpretation of *in situ* hybridization results difficult. This problem can be overcome by acetylating the positive residues using acetic anhydride. In our review of the literature it appears that this treatment is applied rather randomly. For example, within the cnidarians, it is a feature of most hydra protocols (Steele and Bode, pers. comm.; see detailed protocol below) but not other cnidarian protocols [8, 40, 41], and the same is true for crustaceans [32, 33] and flatworms [23, 25]. When employed, this step is carried out after the permeabilization but before the re-fixation that precedes hybridization.

The presence of endogenous phosphatases in the tissues under investigation is another potential source of background for protocols that use labeled probes detected using an antibody conjugated to alkaline phosphatase. As mentioned earlier, the activity, hence location, of the alkaline phosphatase conjugates are visualized using a substrate that yields an insoluble colored product. Endogenous phosphatases convert substrates similar to those of alkaline phosphatases, a property that may confound the interpretation of gene expression patterns localized using this approach. Endogenous phosphatases are often heat-inactivated after the re-fixation that follows permeabilization, but prior to prehybridization. For example, some crustacean and hydra protocols incorporate 30–40 min incubations at 75–80°C [7, 33]; see detailed protocol below). Additionally, the hybridized tissues are thoroughly washed with the endogenous phosphatase inhibitor levamisole prior to the addition of the enzyme substrate. Alternative methods are incubation in hydrogen peroxide (6%, see detailed mouse protocol below), acid treatment post-hybridization (0.1M glycine-HCL, pH 2.2, 0.1% Tween-20), or iodine treatment (Arenas-Mena, unpublished data. This treatment should be minimized as iodine also reacts with nucleic acids).

Another taxon-related issue that can increase the signal background is the trapping of reagents in heavily ciliated tissues like the ciliated bands of trochophore larvae, and in partially or completely enclosed cavities like the heart and neural tube in vertebrates and the gut of cnidarians. This problem can be avoided by dissecting the cavity open, cutting the organism in half ([17]; Bridge and Stover, pers. comm.) and/or by prolonging the post-hybridization washes.

4   Materials

Solutions for protocol 1: Generating labeled single stranded RNA probes
All solutions should be made up with molecular grade H_2O certified RNase-free, or with H_2O treated with the RNase inhibitor (and potent carcinogen) diethyl-pyrocarbonate (DEPC-treated H_2O, 0.5% DEPC in H_2O, stir vigorously in a fume hood for an hour and let sit for 2 h prior to autoclaving). All solutions except those containing Tris should be autoclaved.

- 10 × transcription buffer (supplied with RNA polymerase): 400 mM Tris-HCl pH 8.0 at 20°C, 100 mM Dithiothreitol (DTT), 60 mM $MgCl_2$, 20 mM spermidine.
- DIG RNA Labeling Mix (Boehringer Mannheim): 10 mM ATP, 10 mM CTP, 10 mM GTP, 6.5 mM UTP and 3.5 mM DIG-11-UTP (pH 7.5 at 20°C).
- 4M Lithium chloride: 169 g/ liter.

Solutions for protocols 2 and 3: Whole mount procedure for mouse
Note: All solutions should be made with molecular grade H_2O certified RNAse-free, or with H_2O treated with the RNase inhibitor (and potent carcinogen) diethylpyrocarbonate (DEPC-treated H_2O, 0.5% DEPC in H_2O, stir vigorously in a fume hood for an hour and let sit for 2 h prior to autoclaving). All solutions except those containing Tris should be autoclaved.

- PBS (Phosphate-buffered saline): 135 mM NaCl, 2.7 mM KCl, 10 mM Na_2HPO_4, 1.75 mM KH_2PO_4. Make a 10 × PBS stock solution: Dissolve 80 g NaCl, 2 g KCl, 14.4 g Na_2HPO_4, and 2.4 g KH_2PO_4 in less than 1 liter of RNAse-free H_2O; adjust to pH 7.4 and make up to 1 liter, autoclave.
- 4% paraformaldehyde: Dissolve 10 g paraformaldehyde in 200 ml of DEPC-treated H_2O at 65°C in a fume hood and cool on ice. Adjust pH to 7.5 with 5–10 µl 10 NaOH. Add 25 ml 10 × PBS and make volume up to 250 ml with DEPC-treated H_2O. Aliquots can be stored for several months at –20°C.
- PBT (Phosphate-buffered saline, Triton X-100): PBS 0.1% Tween-20.
- Methanol/PBT series: 25 ml methanol: 75 ml PBT; 50 ml methanol: 50 ml PBT; 75 ml methanol:25 ml PBT.
- 6% H_2O_2/PBT: 6 ml H_2O_2 in 94 ml PBT.
- Proteinase K: 10µg/ml in PBT, use an appropriate volume of a 10 mg/ml stock solution (store at –20°C).
- Glycine: 2mg/ml in PBT (store at –20°C).
- Glutaraldehyde (Sigma): Dilute an appropriate volume of a 25% stock solutiuon.
- 20 × SSC: 3 M NaCl, 0.3 M sodium citrate.
- Hybridization mix: 50% formamide, 5 × SSC (Use 20 × stock solution pH 4.5), 50 µg/ml yeast RNA, 1% SDS, 50 µg/ml heparin.
- Solution 1: 50% formamide, 5 x SSC (pH 4.5) 1% SDS.

- Solution 2: 0.5 M NaCl; 10 mM Tris HCl (pH 7.5) 0.1% Tween-20.
- Soultion 3: 50% formamide; 2 x SSC (pH 4.5).
- RNase A: Use a 10 mg/ml stock in RNase-free H_2O, dilute to 100 µg/ml in solution 2.
- TBST: 135 mM NaCl, 2.7 mM KCl, 25 mM Tris HCl (pH 7.5), 0.1% Tween-20, 2 mM levamisole (add on day of use). Make fresh or dilute from a 10 × stock solution.
- Sheep serum: Denature by heating to 70°C for 30 min.
- Embryo Powder: Homogenize day 12.5–14.5 mouse embryos in a minimum volume of PBS. Add 4 volumes of ice-cold acetone, mix and incubate on ice for 30 min. Centrifuge at 10.000 rpm for 10 min and remove the supernatant. Wash pellet with ice-cold acetone and centrifuge again. Grind the pellet into a fine powder and allow it to air dry on a sheet of filter paper. Store in an air-tight tube at 4°C.
- NTMT: 100 mM NaCl, 100 mM Tris HCl (pH 9.5), 50 mM $MgCl_2$, 0.1% Tween-20. 2 mM levamisole (add on day of use). Make fresh or dilute from a 10 × stock solution.
- NBT (4-nitroblue tetrazolium chloride, Boehringer Mannheim): Dissolve at 75mg/ml in 70% dimethyl formamide. Store aliquots at –20°C.
- BCIP (5-bromo-4-chloro-3indolyl-phosphate, Boehringer-Mannheim): Dissolve at 50 mg/ml in dimethyl formamide. Store aliquots at –20°C.

Solutions for protocol 4: Whole mount procedure for sea urchin
Note: All solutions should be made with molecular grade H_2O certified RNAse-free, or with H_2O treated with the RNase inhibitor (and potent carcinogen) diethylpyrocarbonate (DEPC-treated H_2O, 0.5% DEPC in H_2O, stir vigorously in a fume hood for an hour and let sit for 2 h prior to autoclaving. All solution except those containing Tris should be autoclaved.

- MOPS Buffer: 0.1 M MOPS (pH 7), 0.5 M NaCl and 0.1% Tween-20.
- Hybridization Buffer: 70% de-ionized formamide, 0.5 M NaCl, 0.1 M MOPS, 1 mg/ml BSA (pre-dissolved in H_2O), 0.1% Tween-20.
- Alkaline Phosphatase Buffer: 0.1 M Tris (pH 9.0), 0.05 $MgCl_2$, 0.1 M NaCl, 0.1% Tween-20.
- Staining Buffer: 0.1 M Tris (pH 9.0), 0.05 M $MgCl_2$, 0.1 M NaCl, 0.1% Tween-20, 10% dimethyl formamide, 0.337 mg/ml NBT (4-nitroblue tetrazolium chloride) and 0.175 mg/ml BCIP (5-bromo-4-chloro-3indolyl-phosphate), make fresh.
- NBT (4-nitroblue tetrazolium chloride, Boehringer Mannheim) stock solution: Dissolve at 75mg/ml in 70% dimethyl formamide. Store aliquots at –20°C.
- BCIP (5-bromo-4-chloro-3indolyl-phosphate, Boehringer-Mannheim) stock solution: Dissolve at 50 mg/ml in dimethyl formamide. Store aliquots at –20°C.

Solutions for protocols 5 and 6: Whole mount procedure for Drosphila
Note: All solutions should be made with molecular grade H_2O certified RNAse-free, or with H_2O treated with the RNase inhibitor (and potent carcinogen) diethylpyrocarbonate (DEPC-treated H_2O, 0.5% DEPC in H_2O, stir vigorously in a fume hood for an hour and let sit for 2 h prior to autoclaving. All solutions except those containing Tris should be autoclaved.

- Fixation Buffer: 0.1 M Hepes (pH 6.9), 2 mM $MgSO_4$, 1 mM EGTA (from a 0.5 M stock adjusted to pH 8.0 with NaOH).
- ME: 90% ethanol, 10% 0.5 M EGTA
- PP 4% paraformaldehyde in PBS: 4% paraformaldehyde: Dissolve 10 g paraformaldehyde in 200 ml of DEPC-treated H_2O at 65°C in a fume hood and cool on ice. Adjust pH to 7.5 with 5–10 μl 10 NaOH. Add 25 ml 10 × PBS and make volume up to 250 ml with DEPC-treated H_2O. Aliquots can be stored for several months at –20°C.
- PBS: 135 mM NaCl, 2.7 mM KCl, 10 mM Na_2HPO_4, 1.75 mM KH_2PO_4. Make a 10 × PBS stock solution: Dissolve 80 g NaCl, 2 g KCl, 14.4 g Na_2HPO_4, and 2.4 g KH_2PO_4 in less than 1 liter of RNAse-free H_2O; adjust to pH 7.4 and make up to 1 liter, autoclave.
- PBT (Phosphate buffered saline, Triton X-100): PBS 0.1% Tween-20.
- Proteinase K: 50μg/ml in PBT, use an appropriate volume of a 10 mg/ml stock solution (store at –20°C).
- Glycine: 2mg/ml in PBT, use an appropriate volume of a 100 mg/ml stock (store at –20°C).
- 4% paraformaldehyde: Dissolve 10 g paraformaldehyde in 200 ml of DEPC-treated H_2O at 65°C in a fume hood and cool on ice. Adjust pH to 7.5 with 5–10 μl 10 NaOH. Add 25 ml 10× PBS and make volume up to 250 ml with DEPC-treated H_2O. Aliquots can be stored for several months at –20°C.
- Hybridization Solution: 50% formamide, 5 × SSC, 50 μg/ml heparin, 0.1% Tween-20, 100 μg/ml sonicated and de-natured salmon sperm.
- 20 × SSC: 3 M NaCl, 0.3 M sodium citrate (1 × SSC 150 mM NaCl, 15 mM sodium citrate).
- Color Buffer: 100 mM NaCl, 50 mM $MgCl_2$, 100 mM Tris (pH 9.5), 1 mM levamisole, 0.1% Tween-20.
- NBT (4-nitroblue tetrazolium chloride, Boehringer Mannheim): Dissolve at 75mg/ml in 70% dimethyl formamide. Store aliquots at –20°C.
- BCIP (5-bromo-4-chloro-3indolyl-phosphate, Boehringer-Mannheim): Dissolve at 50 mg/ml in dimethyl formamide. Store aliquots at –20°C.

Solutions for protocols 7 and 8: Whole mount procedure for Hydra
Note: All solutions should be made with molecular grade H_2O certified RNAse-free, or with H_2O treated with the RNase inhibitor (and potent carcinogen) diethylpyrocarbonate (DEPC-treated H_2O, 0.5% DEPC in H_2O, stir vigorously in a

fume hood for an hour and let sit for 2 h prior to autoclaving. All solutions except those containing Tris should be autoclaved.

- HM: 1 mM $CaCl_2$, 1.5 mM $NaHCO_3$, 0.1 mM $MgCl_2$, 0.08 mM $MgSO_4$ and 0.03 mM KNO_3 in Arrowhead spring H_2O.
- 4% paraformaldehyde/HM: Dissolve 10 g paraformaldehyde in 250 ml HM at 65°C in a fume hood and cool on ice. Aliquots can be stored for several months at −20°C.
- PBS: 135 mM NaCl, 2.7 mM KCl, 10 mM Na_2HPO_4, 1.75 mM KH_2PO_4. Make a 10 × PBS stock solution: Dissolve 80 g NaCl, 2 g KCl, 14.4 g Na_2HPO_4, and 2.4 g KH_2PO_4 in less than 1 liter of RNAse-free H_2O; adjust to pH 7.4 and make up to 1 liter, autoclave.
- PBT (Phosphate-buffered saline, Triton X-100): PBS 0.1% Tween-20.
- 0.25% acetic anhydride, 0.1 M triethanolamine: make fresh by adding 2.5 µl of acetic anhydride per ml of 0.1 M triethanolamine.
- Hybridization Solution: 50% de-ionized formamide, 5 × SSC, 0.02% (w/v) Ficoll, bovine serum albumin (BSA, Fraction V) and polyvinylpyrolidone, 200 mg/ml yeast tRNA, 100 mg/ml heparin, 0.1% Tween-20, 0.1% Chaps (3-[(3-Cholamidopropyl) dimethylammonio]-1-propane-sulfonate).
- 20 × SSC: 3 M NaCl, 0.3 M sodium citrate (1 × SSC 150 mM NaCl, 15 mM sodium citrate).
- MAB: 100 mM maleic acid, 150 mM NaCl, pH 7.5.
- Sheep serum: Denature by heating to 70°C for 30 min.
- NTMT: 100 mM NaCl, 100 mM Tris HCl (pH 9.5), 50 mM $MgCl_2$, 0.1% Tween-20.
- Levamisole: make up a fresh stock of 1 M levamisole stock fresh by dissolving 60 mg in 250 µl of RNase-free H_2O.
- BM purple (Boehringer Mannheim). Proprietary mix of NBT/BCIP.

Solutions for protocols 9 to 12: In situ *hybridization on tissue sections*
Note: All solutions should be made with molecular grade H_2O certified RNAse-free, or with H_2O treated with the RNase inhibitor (and potent carcinogen) diethylpyrocarbonate (DEPC-treated H_2O, 0.5% DEPC in H_2O, stir vigorously in a fume hood for an hour and let sit for 2 h prior to autoclaving. All solutions except those containing Tris should be autoclaved.

- 4% paraformaldehyde: Dissolve 10 g paraformaldehyde in 200 ml of DEPC-treated H_2O at 65°C in a fume hood and cool on ice. Adjust pH to 7.5 with 5–10 µl 10 NaOH. Add 25 ml 10 × PBS and make volume up to 250 ml with DEPC-treated H_2O. Aliquots can be stored for several months at −20°C.
- PBS: 135 mM NaCl, 2.7 mM KCl, 10 mM Na_2HPO_4, 1.75 mM KH_2PO_4. Make a 10 × PBS stock solution: Dissolve 80 g NaCl, 2 g KCl, 14.4 g Na_2HPO_4, and 2.4 g KH_2PO_4 in less than 1 liter of RNAse-free H_2O; adjust to pH 7.4 and make up to 1 liter, autoclave.
- PK Buffer: 10 ml 1M Tris, 2 ml 0.5 M EDTA, adjust volume to 200 ml with RNase-free H_2O.

- Hybridization Buffer (15 ml): 7.5 ml deionized formamide (50%), 1.125 ml 4 M NaCl (0.3 M), 300 µl 1 M Tris-HCl pH 7.4 (20 mM), 150 µl 0.5 M EDTA, 150 µl 1 M NaH_2PO_4 pH 8.0 (10mM), 3 ml 50% Dextransulfate (10%), 300 µl 50× Denhardts, 0.75 ml 10mg/ml Yeast tRNA (5mg/ml) and 3.45 ml dH_2O.
- 20 × SSC: 3 M NaCl, 0.3 M sodium citrate. (1X SSC 150 mM NaCl, 15 mM sodium citrate).
- 5 × SSC, 10mM DTT: 63 ml 20×SSC, 0.4 g Dithiothreitol, add to 252 ml of RNase free H_2O. Make fresh as DTT is volatile.
- 50% formamide/2 × SSC/0.1 M DTT: 120 ml de-ionized formamide, 24 ml 20 × SSC, 0.4 g Dithiothreitol, add to 240 ml RNase-free H_2O. Make fresh as DTT is volatile.
- Washing solution: 0.4 M NaCl, 10 mM Tris HCl pH 7.5, 5 mM EDTA. Make a 10 × solution as follows: 233.8 g NaCl , 100 ml 1M Tris pH 7.5, 100 ml 0.5 M, dH_2O to 1liter.
- The developing solution: 40 g Kodak D-19 in 250 ml H_2O dissolved at around 65°C on a magnetic stirrer and then cooled on ice). Use at 16°C.

Microscopy for whole mount procedure
Clear the stained embryos in glycerol (70% in PBT) before photographing. Use a binocular microscope equipped with dark field optics for microscopic observation (black or blue background gives good contrast). For more detailed analysis, embryos can be embedded in gelatin-albumin, sectioned with a vibratome to a thickness of 30–50 µm and mounted on slides under nail varnish sealed coverslips in 70% glycerol/PBT.

Microscopy for tissue sections
Hybridization signals are visible as bright points in dark field microscopy or as black grains in bright field (only strong signals can be visualized in bright field). Expression patterns can be photographed using a suitable film (ILFORD PAN F, 50 ASA; Kodak Tmax100).

5   Sample Protocols

In this section we present protocols that have been employed to investigate the developmental biology of both vertebrate and invertebrate systems.

5.1   Generating labeled single stranded RNA probes

In order to generate single-stranded RNA probes the investigator must have a clone of a fragment of (>100 bases) of the gene of interest. The circular plasmid DNA containing this fragment is linearized on either side of the insert using a unique restriction enzyme in the polylinker of the cloning vector (use an enzyme that leaves a 5' overhang). Run-off sense (control) and antisense (positive) probes are generated using T7, SP6 or T3 RNA polymerases and the appropriate promoter in the vector, in the presence of a label (detailed protocols for generating double-stranded DNA and oligonucleotide hapten probes are available in Boehringer Mannheim [42]. For details on the synthesis of radioactively labelled probes, see Melton et al., [43] and Simeone [44]). Protocol 1 is modified from Boehringer Mannheim [42] and uses DIG-UTP as a label (alternatively, large amounts of template can be generated by PCR amplification (e.g., [45]) using one primer containing the T7 polymerase promoter and gene-specific sequence, and a second primer that contains only gene-specific sequence. Amplification using these primers and an appropriate template generates a linear fragment with the promoter at one end. It is necessary to include 5 bases 3' of the +1 end (3' end) of the promotor sequence for the promoter to function).

Protocol 1   Labeling single-stranded RNA probes

1. Grow up your clone containing your gene of interest and isolate (miniprep) the plasmid DNA from the bacterial cells.
2. Set up two restriction digests to linearize 20 µg of plasmid DNA on either side of the insert using vector-appropriate enzymes. Set up the digest under standard conditions with an excess of enzyme and digest at 37°C overnight.
3. Remove the components of the digest by phenol/chloroform and chloroform extraction and ethanol precipitate the DNA prior to resuspension in 40 µl of RNase and DNase-free H_2O. Check the efficacy of the digest by analyzing the linearized products on a 1.4% agarose checking gel.
4. To transcribe the RNA probes, prepare the following mix on ice: 2 µl 10X transcription buffer; 1 µg (2 µl) linearized plasmid DNA, 2 µl of 10X DIG RNA labeling Mix, 50 U RNase Inhibitor and 10 U RNA polymerase (SP6, T7 or T3. You must know the orientation of the insert DNA in the vector to determine the combination of restriction site and RNA polymerase that will generate the sense and antisense probe). Use RNase-free H_2O to achieve a final reaction volume of 20 µl. Incubate the reaction mixture for 2–3 h at 37°C.
5. To remove the DNA template, digest the reaction mix with 20 U of DNase I (2 µl). The DNase must be RNase-free.
6. Remove components of the transcription mix and the DNase I treatment by adding 2.5 µl of 4 M Lithium chloride and 75 µl of ice-cold RNase-free ethanol. Precipitate the RNA at –80°C for 30–60 min (-20°C for 2–3 h), pellet the RNA by centrifugation (13,000 g for 15 min at 4°C), remove the residual salt by washing the pellet with ice-cold 70% ethanol, centrifuge for 5 min (13,000 g for 15 min at 4°C), dry the pellet under vacuum and re-suspend the probes in 100 µl of RNase-free H_2O or 10 mM Tris (pH 7.0) by incubating them at 37°C for 30 min.

5.2 Whole mount *in situ* hybridization

In this section we present four methods for whole mount *in situ* hybridization. The first is a vertebrate method modified from Wilkinson [46] and Xu and Wilkinson [17]. This procedure, detailed in protocols 2 and 3, has been used successfully for expression studies in mice embryos up to 12.5 d of embryonic development [46, 47], xenopus [46], chicken and zebrafish [17, 48]. Here we describe the method for mouse embryos, but the procedure is essentially the same for all vertebrate embryos and isolated organs/tissues. The second method, detailed in protocol 4, has been developed for sea urchin larvae and juvenile rudiments by Arenas-Mena. This method is modified from Holland et al. [49] and incorporated both a long hybridization step to compensate for variation in hybridization rates in fixed tissues, and extended post hybridization washes to reduce non-specific retention of the probes in cavities. The third method, detailed in protocols 5 and 6, is the classic *Drosophila* protocol developed by Tautz and Pfeiffle [20]. Here we present two methods of fixation. The second fixation is more laborious than the first, but is thought to preserve morphology better and give a lower background signal. The final method, detailed in protocols 7 and 8, has been developed for *Hydra* and is modified from Grens et al. [7]. This protocol incorporates a number of unique pre-hybridization manipulations to reduce non-specific hybridization of the probes.

Protocol 2   Whole mount procedure in mouse – pre-treatments

Pretreatment of embryos
   1. Dissect the embryos out of the uteri in phosphate-buffered saline (PBS) and remove the extra-embryonic membranes by dissection.
   2. Fix the embryos in a 4% paraformaldehyde in PBS solution either overnight at 4°C or 3–4 h at room temperature. After washing twice for 5 min in PBT, dehydrate the embryos in a series of methanol/PBT (25%, 50%, 75%, 2 × 100% – 5 min each). The embryos can be stored in 100% methanol at –20°C for several months but must be in methanol for a minimum of 2 h to permeabilize the tissues (by removing the lipid membranes) prior to proceeding with the rest of this protocol.
   3. Re-hydrate the fixed embryos in methanol/PBT (75%, 50%, 25% – 5 min each) and wash 4 times in PBT for 5 min each.
   4. Incubate the embryos in a 6% H_2O_2/PBT to inactivate endogenous phosphatases and lower background staining.
   5. Treat the embryos for 15 min with proteinase K (10 μg/ml in PBT). This step is important to enhance the penetration of the probe into the tissue. The length of treatment may have to be optimized according to stage and size of embryos or tissues (see Table 1 for general guidelines). Stop the proteinase K digest by washing the embryos for 10 min in a freshly prepared glycine solution (2 mg/ml in PBT) followed by PBT (2 × 5 min).

6. Re-fix the embryos in a 0.2% glutaraldehyde/4% paraformaldehyde in PBT for 20 min. Wash twice in PBT (2 × 5 min) prior to hybridization.

Pre-hybridization and hybridization

The pre-hybridization and hybridization reactions can be carried out in a water bath or incubator.

1. Transfer the embryos in 6- or 24-well plates depending on the stage, the amount of embryos/tissue and number of different probes to be hybridized.
2. Replace the PBT with the hybridization mix and cover the plates with Saran wrap to prevent the embryos from drying out. Pre-hybridize the embryos for 1 hour at 70°C. Longer pre-hybridizations may help to reduce background staining.
3. Replace the pre-hybridization solution with hybridization mix pre-warmed to 70°C and containing1 µg/ml digoxygenin labeled RNA-probe. Incubate overnight at 70°C.

Protocol 3   Whole mount procedure in mouse – staining reactions

Washes and antibody reaction

1. Wash the embryos twice for 30 min at 70°C in pre-warmed solution 1. Wash the embryos once for 10 min at 70°C in a 1:1 mix of pre-warmed solutions 1 and 2. Wash the embryos three times for 5 min each in room temperature solution 2.
2. To remove un-hybridized single-stranded RNA, incubate in 100 µg/ml RNase A in solution 2 for 30 min at 37°C. Wash once with solution 2 for 5 min at room temperature, then twice with solution 3, each for 15 min at 65°C. Wash three times in TBST, 5 min each at room temperature.
3. Pre-block the embryos with 10% sheep serum in TBST for 1–3 h. The sheep serum must be denatured by heating to 70°C for 30 min prior to use.
4. Pre-absorb the antibody using mouse embryo powder (see [17] for details on powder preparation). Add 0.5 ml 1% sheep serum TBST (heat treated) to 3 mg of embryo powder and add 1 µl anti-digoxygenin Fab fragments (Boehringer-Mannheim). Shake the mixture gently for 1–2 h at 4°C to pre-absorb the antibody.
5. Centrifuge the mixture for 10 min to pellet the embryo powder, remove the supernatant, make it up to 2 ml with 1% sheep serum in TBST corresponding to an antibody dilution of 1:2000.
6. Replace the blocking solution with the pre-absorbed antibody and incubate the embryos overnight at 4°C. The diluted antibody is stable under these conditions and can be re-used three times when stored at 4°C.

Post-antibody washes and staining reaction

All the following washes and the staining reaction are carried out at room temperature.

1. Perform three 5 min washes and at least five 1 h washes in TBST (add 2 mM levamisole on the day of use).
2. Wash three times for 10 min each with NTMT. The embryos can be stored in NTMT at 4°C overnight prior to staining if desired.
3. Perform the staining reaction by incubating the embryos in NTMT solution including 4.5 μl NBT and 3.5 μl BCIP (50 mg/ml; Boehringer-Mannheim) per ml solution. Keep the staining reaction in the dark and monitor the staining intensity intermittently using a binocular microscope.
4. When the desired staining intensity has been achieved, stop the reaction by washing at least twice in PBT, 5 min each.

Protocol 4   Whole mount procedure in sea urchins

Formaldehyde Fixation

1. Fix late stages (with several adult spines), in 10 volumes of 4% formaldehyde in 0.1 M MOPS (pH 7), 0.5 M NaCl. Incubate for 1 hour at room temperature. Earlier stages with pentameral rudiments should be fixed in 2% paraformaldehyde and 2% formaldehyde under the same conditions. This improves morphological preservation despite some loss of hybridization signal.
2. Wash 5 times with 10 volumes of MOPS buffer. Transfer directly into 70% ethanol and store at −20°C until needed. The samples can be stored indefinitely.

Pre-hybridization and hybridization

1. Rehydrate the embryos by washing them 3 times for 15 min each in 10 volumes of MOPS buffer.
2. Pre-hybridize the embryos by rinsing them in hybridization buffer and washing them twice for 3 h in fresh hybridization buffer at 50°C.
3. Hybridize for 1 week with 0.1 ng/ml of digoxigenin labeled riboprobe in hybridization buffer at 50°C.

Post-hybridization washes, antibody binding and color reaction

1. Wash the embryos 5 times in MOPS buffer at room temperature to remove the probe.
2. Incubate the embryos in hybridization buffer for 3 h at 50°C.
3. Wash the embryos 3 times with MOPS buffer.
4. Block the embryos first with 10 mg/ml BSA in MOPS for 20 min at room temperature and then with 10% goat serum, 1mg/ml BSA in MOPS buffer at 37°C for 30 min.
5. Incubate with a 1/1500 dilution of alkaline phosphatase conjugated Fab fragments (Boehringer-Mannheim) overnight at room temperature.
6. Remove unbound antibody by washing 5 times with MOPS buffer over a period of 12 h at room temperature.

7. Wash the embryos in several changes of alkaline phosphatase buffer (AP) over a period of 1 h.
8. Replace the AP buffer with the staining buffer and place in the dark to develop the color. Check the progress of the reaction regularly. The color should develop during the next 12–24 h. Note that inclusion of dimethyl formamide in the AP buffer enhances the staining reaction.
9. Stop the color reaction by repeatedly washing with MOPS buffer.

Protocol 5   Whole mount procedure in *Drosophila* – fixation

Formaldehyde fixation

1. Transfer embryos into a vial containing 4 ml of fixation buffer, 0.5 ml of 37% formaldehyde and 5 ml of heptane. Shake the vial for 15–20 min.
2. Remove the lower phase and add 10 ml of methanol and the embryos will sink to the bottom. The fixed embryos can be stored in methanol at 4°C for several weeks. Note: if the embryos do not sink to the bottom of the vial after the addition of methanol, remove the heptane upper phase and add more methanol.

Paraformaldehyde fixation

1. Transfer embryos into a vial containing 1.4 ml of fixation buffer, add 0.4 ml of liquified 20% paraformaldehyde and 8 ml of heptane. Shake the vial for 15–20 min.
2. Remove the lower phase and add 10 ml of methanol and the embryo will sink to the bottom.
3. Transfer embryos to microfuge tubes containing ME. Re-fix and de-hydrate the embryos by washing them for 5 min each in 70, 50, and 30% methanol:PP, and for 20 min in PP.
4. Wash for 10 min in PBS. (Note: If the embryos are not going to be used immediately, dehydrate for 5 min each in 30, 50 70% ethanol and store at – 20°C).

Protocol 6   Whole mount procedure in *Drosophila* – post fixation

Post-fixation treatments

1. Wash embryos 3 times in PBT, 5 min each, at room temperature.
2. Digest embryos for 3–5 min with 50 µg/ml Proteinase K at 37°C. Remove the Proteinase K solution and stop the digestion by washing twice with 2 mg/ml glycine.
3. Wash twice with PBT, 5 min each and re-fix the embryos for 20 min with 4% paraformaldehyde. Remove excess fix by washing 3 times in PBS for 5 min each.

Pre-hybridization and hybridization

1. Wash embryos for 20 min in 1:1 hybridization solution (HS):PBT. Replace the solution with HS pre-warmed to 45°C and pre-hybridize for 20–60 min at 45°C.
2. Remove the pre-hybridization solution from the embryos and dilute the heat-denatured probe to 0.5 µg /ml in HS. Add the probe, mix and hybridize overnight at 45°C.

Post-hybridization washes, antibody binding and color reaction

1. Wash the embryos for 20 min each in 4:1 HS:PBT, 3:2 HS:PBT, 2:3 HS:PBT; 1:4 HS:PBT, and twice for 20 min each in PBT.
2. Incubate the embryos on a rotating wheel for 1 h at room temperature with 500 µl of alkaline phosphatase conjugated anti-DIG antibody (1:2000–1:5000 in PBT).
3. Wash the embryos in PBT four times for 20 min each followed by three 5 min washes in the color buffer.
4. Add 4.5 µl NBT and 3.5 µl BCIP and mix thoroughly. Monitor the color reaction over the next 10–60 min and stop the reaction by adding PBT when the background starts to develop. This reaction should proceed in the dark when the color reaction is not being monitored.
5. De-hydrate the embryos in an ethanol series and mount in a medium of your choice.

Protocol 7   Whole mount procedure in *Hydra* – fixation and post-fixation

Fixation

1. Put 24 h starved polyps into 6-well microtiter dishes and place them on light box to make animals extend to maximum length. Relax the animals by replacing the hydra medium (HM) with 2% urethane/HM for 1–2 min.
2. Fix overnight with fresh 4% paraformaldehyde/HM at 4°C.

Post-fixation treatments

1. To prevent trapping and background signal in the gut, cut the animals in half prior to transferring them to mesh-bottomed baskets (Netwell Carriers, available from Costar Scientific) in a 12-well microtiter plate.
2. Re-hydrate by passage through an ethanol series, 5 min each in 100, 100, 75, 50, 25% ethanol/PBT followed by three 5 min washes in PBT
3. Permeabilize the fixed animals for 10 min at room temperature with 10 µg/ml Proteinase K in PBT. Stop the digestion by replacing the proteinase K solution with 4 mg/ml glycine in PBT and incubating for 10 min at room temperature. Wash twice in PBT for 5 min each.
4. To acetylate positively charged amino acids wash twice for 5 min in 0.1 M triethanolamine, pH 7.8 and twice for 5 min in 0.25% acetic anhydride, 0.1 M triethanolamine.

5. Wash twice for 5 min in PBT and re-fix the animals for 20 min in 4% paraf-ormaldehyde/HM. Remove the fixative by washing five times for 5 min in PBT.
6. Transfer the fixed animals to a 1.5 ml microfuge tube and heat-inactivate endogenous phosphatases by incubation at 80°C for 40' in PBT, 5 mM EDTA.

Protocol 8   Whole mount procedure in *Hydra* – staining and color reaction

Pre-hybridization and hybridization

1. Wash animals for 10 min at room temperature in a 1:1 mixture of PBT and hybridization solution, followed by 10 min in hybridization solution pre-warmed to 55°C. Replace with fresh hybridization solution and pre-hybri-dize for a minimum of 2 h at 55°C (an overnight pre-hybridization may help to reduce background).
2. Dilute probe as necessary in RNase-free H_2O. Heat the probe at 65°C for 5 min and add to hybridization solution preheated to 55°C. Hybridize 2.5–3 days at 55°C. Note that the temperature can by lowered one or two degrees if signal is faint.

Post-hybridization washes and antibody binding

Solutions need not be RNase-free from this point on.
1. Wash at 55°C for 5–10 min each in 100% hybridization solution; 75% hybridization solution: 25% 2× SSC; 50% hybridization solution: 50% 2× SSC; and, 25% hybridization solution: 75% 2× SSC.
2. Wash twice in 2× SSC, 0.1% CHAPS for 30 min at 55°C.
3. Transfer polyps from microfuge tube to a 24-well microtiter plate and wash twice for 10 min in MAB at room temperature. Pre-block for 1 h in MAB, 1% BSA, 0.1% azide at room temperature.
4. Block for 2 h (or more) at room temperature in 80% MAB, 1% BSA, 0.1% azide: 20% sheep serum which has been heat-inactivated for 30 min at 55°C and had 0.1% azide added prior to use.
5. Incubate animals with pre-absorbed antibody overnight at 4°C on a nutator, or for 1 h at room temperature.

Staining reaction, destaining and mounting

1. Transfer animals to the mesh-bottomed baskets in a 2-well microtiter plate. Wash at least 8 times for at least 30 min each in MAB at room temperature.
2. Wash twice in NTMT and once in NTMT, 1 mM levamisole each for 5 min at room temperature. Replace the last wash with 500 µl BM purple (Boehrin-ger Mannheim), wrap the microtiter plate with foil and incubate at 37°C for 30 min or longer. Stop the staining reaction by washing with 100% ethanol for 30 min.
3. Wash twice in 100% ethanol for 5 min each. Place polyps on a slide and mount in a drop of Euparal (Asco Laboratories). Cover polyps with a weighted coverslip and dry overnight.

5.3   Two-color whole mount *in situ* hybridization

It is possible to detect two different RNA transcripts by two color whole mount *in situ* hybridization [50–52] using any of the protocols described above. For this application one RNA probe is labeled using fluorescein-12-UTP and the other using digoxigenin-11-UTP (according to the protocol described above). Both probes are hybridized together and detected sequentially in two rounds of antibody and staining reactions.

It is essential to inactivate the first Fab-AP conjugate prior to the application of the second. This is achieved either by heating or by incubating the embryos for 10 min at room temperature in 0.1 M glycine-HCl, 0.1% Tween-20 at pH 2.2. This will prevent the second substrate from precipitating over the first probe in addition to localizing the second probe. The commonest alkaline phosphatase substrates for two color *in situ* hybridization are NBT/BCIP and Fast Red. Fast Red is less sensitive than NBT/BCIP and thus should always be detected first and used to label the stronger probe.

5.4   *In situ* hybridization on tissue sections

Protocols 9–12 are modified from Wilkinson et al. [53] and have been used successfully for *in situ* hybridization experiments on paraffin sections of embryonic, postnatal and adult mouse tissues [54–56] and can also be used for cryosectioned material. When using the latter, the hybridization begins after the re-hydration step described for paraffin sections (step 5).

Protocol 9   Procedure for tissue sections – preparation of sections

Preparing paraffin sections
  1. Fix the embryos (or tissues) overnight in 4% paraformaldehyde/PBS at 4°C. Wash with PBS for 30 min, followed by 0.85% NaCl for 30 min.
  2. Dehydrate the embryos by sequentially washing in 1:1 mix of 0.85% NaCl/ 100% EtOH for 15 min, twice in 70% EtOH for 15 min, 85% EtOH for 30 min, 95% EtOH for 30 min and two final washes in 100% EtOH for 30 min each. Note that the larger the tissue the longer the duration of the washes required to completely dehydrate the tissue. Well-dehydrated mouse tissue is nearly white.
  3. Clear the embryos by washing twice in 100% xylene for 30 min each (the tissue should be transparent following this step).

4. Incubate the embryos in a 1:1 mix of xylene/paraffin for 45 min at 60°C. Preheat the xylene before mixing it with the paraffin, because paraffin solidifies just below 60°C (use xylene-resistant containers). Transfer the embryos through three changes of 100% paraffin at 60°C, ensuring that the volume used is 10 times as great as the volume of the tissue being infiltrated. The first paraffin change should be made after 30 min, the second after an overnight incubation and the last after 8 h. These long washes are thought to reduce background.

5. Embed the embryos in the correct orientation and store the paraffin block for 12 h at room temperature and 24 h at 4°C before sectioning.

6. Cut sections of an appropriate width (10 µm) using a microtome and collect them on slides that have been coated with an adhesive such as 3-Amino-propyltriethoxy-silane (TESPA) according to manufacturer's protocol.

7. Dry the sections by incubating overnight at 37°C. Sections for *in situ* hybridization experiments can be stored for at least one year at room temperature.

Preparing cryosections

1. Fix tissues in 4% paraformaldehyde/PBS overnight at 4°C.

2. Freeze the tissues directly in liquid nitrogen and using OCT medium to glue the tissue to the chilled block, cut sections using a cryostat and collect them on TESPA-coated slides.

3. Dry the sections at room temperature and store desiccated at –80°C. Prior to use warm to room temperature.

Protocol 10   Procedure for tissue sections – pre-hybridization

1. De-paraffinize the slides by washing twice in xylene, 10 min each.

2. Rehydrate the sections by immersing the slides in 100% EtOH for 2 min; then sequentially dipping the slides for 30 s in 100, 95, 85, 70, 50 and 30% EtOH.

3. Wash once in 0.85% NaCl for 5 min, followed by PBS for 5 min.

4. Re-fix the tissue (paraffin and cryostat sections) in fresh 4% paraformalde-hyde for 20 min and wash twice in PBS for 5 min prior to the proteinase K treatment.

5. The proteinase K solution is applied on each slide individually and the extent of the digestion varied by increasing or decreasing the time in the enzyme. Overlay each slide with 1 ml of 10 µg/ml in PK buffer. Incubate at room temperature for 5–10 min. Stop the digestion by immersing the slides in PBS.

6. Wash the sections in PBS for 5 min and re-fix in 4% paraformaldehyde for 5 min.

7. Dip the slides in dH$_2$O and transfer them into 250 ml of fresh 0.1 M dithiotheitol. While stirring, introduce 625 µl of acetic anhydride and continue to stir for 10 min. This acetylation step can be skipped, but is thought to reduce background particularly when the slides have been coated with poly-D-lysine.
8. Wash the sections in PBS for 5 min and 0.85% NaCl for 5min. Dehydrate the sections by dipping them in 30, 50, 70, 85, 95, 100, 100 and 100% ethanol.
9. The sections have to be air dried for at least 30 min before hybridization and should be used the same day. If necessary they can be stored at room temperature for 1–2 days or desiccated at –20°C for several days.

Protocol 11   Procedure for tissue sections – hybridization and washes

1. Dilute probes to 3–5 × 10^4 c.p.m/µl in hybridization buffer.
2. Heat the diluted probe to 80°C for 2 min in a heating block and cool on ice for 1 min.
3. Apply 60–70 µl of probe per slide and spread over the sections with a wick of parafilm. Cover the sections with a dust-free coverslip and place the slides onto a paper towel soaked in a solution of 50% formamide/4 × SSC solution in a plastic hybridization chamber. Seal the hybridization chamber and incubate at 50–52°C for about 16 h. The soaked paper towel maintains the humidity during hybridization.

Post-hybridization washes

1. Remove the slides from the hybridization chamber and place them into a slide rack. Immerse the rack in pre-warmed 5 × SSC/10 mM DTT and wash for 30 min at 50°C. The coverslips will fall off during this step.
2. Wash for 20 min in pre-warmed 50% formamide/2 × SSC/0.1M DTT at 65°C and then in 1 × Washing Solution (2 × 10 min, 37°C).
3. Remove un-hybridized single-strand RNA from the sections by incubating in RNase A solution for 30 min at 37°C.
4. Immerse the slide rack in pre-warmed washing solution for 5 min at 37°C, twice in 2 × SSC for 15 min at 37°C, and once in 0.1 × SSC for 15 min at 37°C.
5. De-hydrate the sections by passing them rapidly through an ethanol series 30, 60, 80, 95, 100, and 100%). Air-dry the sections and remove any lint.

Protocol 12   Procedure for tissue sections – autoradiography and staining

The following steps must be conducted in a dark room with an appropriate safe-light.

1. Pre-heat the autoradiographic emulsion (for example Kodak NTB-2, stored carefully sealed from light at 4°C) in a 45°C water-bath in the dark room. When liquid, pass a clean blank slide through the emulsion to remove bubbles or allow bubbles to clear for 20–30 min.

2. Dip slides with sections into the emulsion. Drain, wipe the back of the slides and dry in a vertical position for 1–3 h.

3. Place the slides in a light-tight box with a desiccant, seal the box tightly and wrap it in two layers of aluminum foil. Store at 4°C for 7 days.

4. Prior to developing and staining the slides, remove the box from the refrigerator and allow it to assume room temperature. Place slides gently into a slide rack and immerse them in Kodak D-19 developing solution for 3.5 min, in dH_2O briefly, in Kodak Rapid Fix solution for 2 min (following the fixation it is no longer necessary to work in the dark room), and then through multiple changes of cold dH_2O.

5. Lightly stain the sections in 0.2% toluidine blue/dH_2O for 30 s and rinse several times in dH_2O. If necessary, 0.01 N HCl can be used to destain. A light stain allows the investigator to visualize faint signals.

6. Dehydrate the sections by passing them through an ethanol series, 50, 70, 85, 95, 100, 100% (last step must be absolute ethanol) and remove the ethanol by immersing twice in xylene. Mount slides individually using a mounting media compatible with autoradiography, such as Entellan or Polymount. Dry the mounted sections overnight.

References

1 Wray CG, Langer MR, DeSalle RL, Lee JJ. et al. (1995) Origin of the Foraminifera. *Proc. Nat. Acad. Sci. USA* 92: 141–145

2 Keys DN, Lewis DL, Selegue JE, Pearson BJ et al. (1999) Recruitment of hedgehog regulatory circuit in butterfly eyespot evolution. *Science* 283(5401): 532–534

3 Irvine SQ, Martindale MQ (1997) Novel early patterns and regional restrictions in the expression of Hox genes in the polychaete annelid *Chaetopterus*. *Dev. Biol.* 210: 187

4 Jacobs DK, Wray CG, Wedeen C J at al. (2000) Molluscan *engrailed* expression, serial organization, and shell evolution. *Evolution and Development 2(6): 340–347*

5 Lowe CJ, Wray GA (1997) Radical alterations in the roles of homeobox genes during echinoderm evolution. *Nature* 389: 718–721

6 Arenas-Mena C, Martinez P, Cameron RA, Davidson EH (1998) Expression of the Hox gene complex in the indirect development of a sea urchin. *Proc. Natl. Acad. Sci. USA* 95: 13062–13067

7 Grens A, Gee L, Fisher DA, Bode HR (1996) *CnNK-2*, an NK-2 homeobox gene, has a role in patterning the basal end of the axis in Hydra. *Dev. Biol.* 180: 473–488

8 Mokady O, Dick MH, Lackschewitz D, Schierwater B et al. (1998) Over one-half billion years of head conservation? Expression of an ems class gene in *Hydractinia symbiolongicarpus* (Cnidaria: Hydrozoa). *Proc. Natl. Acad. Sci. USA* 95(7): 3673–3678

9 Pardue ML, Gall JG (1969) Molecular hybridization of radioactive DNA to the DNA of cytological preparations. *Proc. Nat. Acad. Sci. USA* 64(2): 600–604

10 John HA, Birnstiel ML, Jones KW (1969) RNA-DNA hybrids at the cytological level. *Nature* 223(206): 582–7

11 Kearney L (1998) Detection of genomic sequences by fluorescence *in situ* hybridization to chromosomes. In: DG Wilkinson (ed): *In Situ Hybridization: A Prac-*

tical Approach. Oxford University Press, Oxford, 165–180

12 Wilkinson DG (1998) The theory and practice of in situ hybridization. In: DG Wilkinson (ed): *In Situ Hybridization: A Practical Approach.* Oxford University Press, Oxford, 1–21

13 Cox KH, DeLeon DV, Angerer LM, Angerer RC (1984) Detection of mRNAs in sea urchin embryos by *in situ* hybridization using asymmetric RNA probes. *Dev. Biol.* 101(2): 485–502

14 Davidson EH (1986) Gene activity in early development (3rd ed). Academic Press, Orlando

15 Britten RJ, Graham DE, Neufeld BR (1974) Analysis of repeating DNA sequences by reassociation. In: L Grossman, K. Moldave (eds): *Nucleic Acids and Protein Synthesis.* Academic Press, New York

16 Liu JK, Ghattas I, Liu S, Chen S et al. (1997) Dlx genes encode DNA-binding proteins that are expressed in an overlapping and sequential pattern during basal ganglia differentiation. *Dev. Dyn.* 210: 498–512

17 Xu Q , Wilkinson DG (1998)) In situ hybridization of mRNA with hapten labeled probes. In: DG Wilkinson (ed): *In Situ Hybridization: A Practical Approach.* Oxford University Press, Oxford, 87–106

18 Holten B, Wedeen CJ, Astrow SH, Weisblat DA (1994) Localization of polyadenylated RNAs during telophase formation and cleavage in leech embryos. Roux's Arch. *Dev. Biol.* 204: 46–53

19 Islam N, Moss T (1996) Enzymatic removal of vitelline membrane and other protocol modifications for whole mount *in situ* hybridization of *Xenopus* embryos. *Trends in Genetics* 12(11): 459

20 Tautz D, Pfeiffle C (1989) A nonradioactive *in situ* hybridization method for the localization of specific RNAs in Drosophila embryos reveals a translational control of the segmentation gene hunchback. Chromosoma (Berl.) 98: 81–85

21 Nardelli-Haefliger D, Shankland M (1992) *Lox2*, a putative leech segment identity gene, is expressed in the same segmental domain in different stem cell lineages. *Development* 116: 697–710

22 Master VA, Kourakis MJ, Martindale MQ (1996) Isolation, characterization, and expression of *Le-msx*, a maternally expressed member of the *msx* gene family from glossiphoniid leech, *Helobdella. Dev. Dynamics* 207: 404–419

23 Umesono Y, Watanabe K, Agata K (1997) A planarian *orthopedia* homolog is specifically expressed in the branch region of both the mature and regenerating brain. *Dev. Growth Differ.* 39: 723–727

24 Agata K, Soejima Y, Kato K, Kobayashi C et al. (1998) Structure of the planarian central nervous system (CNS) revealed by neuronal cell markers. *Zool. Sci.* 15: 443–440

25 Kato K, Orii H, Watanabe K, Agata K (1999) The role of dorsoventral interaction in the onset of planarian regeneration. *Development* 126: 1031–1040

26 Imase A, Kumagai T, Ohmae H, Irie Y et al. (1999) Localization of mouse type 2 *Alu* sequences in schistosomes. *Parasitology* 119(3): 315–321

27 Yanze N, Groger H, Muller P, Schmid V (1999) Reversible inactivation of cell-type-specific regulatory and structural genes in migrating isolated striated muscle cells of Jellyfish. *Dev. Biol.* 213: 194–201

28 Peterson KJ, Harada Y, Cameron RA, Davidson EH (1999) Expression pattern of *Brachyury* and *Not* in the sea urchin: Comparative implications for the origins of mesoderm in the basal deuterostomes. *Dev. Biol.* 207: 419–431

29 Friedrich AB, Merkert H, Fendert T, Hacker J et al. (1999) Microbial diversity in the marine sponge *Aplysina cavernicola* (formerly *Verongia cavernicola*) analyzed by fluorescence *in situ* hybridization (FISH). *Mar. Biol.* 134: 461–470

30 Wahlberg MH (1997) Three main patterns in the expression of six actin genes in the plerceroid and adult *Diphyllobothrium dendriticum* tapeworm (Cestoda). *Mol. Biochem. Parisotol.* 86: 199–209

31 de Lange RPJ, van Minnen J (1998) Localization of the neuropeptide APGWa-

mide in gastropod molluscs by *in situ* hybridization and immunocytochemistry. *Gen. Comp. Endocrinol.* 109: 166–174

32 Manzanaras M, Marco R, Garesse R (1993) Genomic organization and developmental pattern of expression of the *engrailed* gene from the brine shrimp *Artemia*. *Development* 118: 1209–1219

33 Nulsen C, Nagy LM (1999) The role of *wingless* in the development of multibranched crustacean limbs. *Dev. Genes Evol.* 209: 340–348

34 Harkey MA, Whiteley HR, Whiteley AR (1992) Differential expression of the msp130 gene among skeletal lineage cells in the sea urchin embryo: a three dimensional *in situ* hybridization analysis. *Mech. Devel.* 37: 173–184

35 Mitsunaga-Nakatsubo K, Akasaka K, Sakamoto N, Takata K (1998) Differential expression of sea urchin *Otx* isoforms ($HpOtx_E$ and $HpOtx_L$) mRNA during early development. *Int. J. Dev. Biol.* 42: 645–651

36 Onodera H, Kobari K, Sakuma M, Sato et al. (1999) Expression of a *src*-type protein tyrosine kinase gene, *AcSrc1*, in the sea urchin embryo. *Dev. Growth Differ.* 41: 19–28

37 Bruce AEE, Shankland M (1998) Expression of the head gene *Lox22-Otx* in the leech *Helobdella* and the origin of the bilaterian body plan. *Dev. Biol.* 201: 101–112

38 Satake H, Takuwa K, Minakata H, Matsushima O (1999) Evidence for conservation of the vasopressin/oxytocin superfamily in Annelida. *J. Biol. Chem.* 274: 5605–5611.

39 Gomez-Saladin E, Wilson DL, Dickerson IM (1994) Isolation and *in situ* localization of a cDNA encoding a Kex2-like prohormone convertase in the nematode *Caenorhabditis elegans*. *Cell. Mol. Neur.* 14 (1): 9–24

40 Pan TL, Groger H, Schmid V (1998) A toxin homology domain in an astacin-like metalloproteinase of the jellyfish *Podocoryne carnea* with a dual role in digestion and development. *Dev. Genes Evol.* 208: 259–266

41 Baader CD, Heiermann R, Schuchert P, Schmid V et al. (1995) Temporally and spatially restricted expression of a gland cell during regeneration and *in vitro* transdifferentiation in the hydrozoan *Podocoryne carnea*. Roux's Arch. *Dev. Biol.* 204: 164–171

42 Boehringer Mannheim Biochemicals (1996) Nonradioactive in situ hybridization application manual. http://biochem.boehringer-mannheim.com/prod_inf/manuals/InSitu/InSi_toc.htm

43 Melton DA, Krieg PA, Rebagliati MR, Maniatis T et al. (1984) Efficient *in vitro* synthesis of biologically active RNA and RNA hybridization probes from plasmids containing a bacteriophage *SP6* promoter. *Nucleic Acids Res* 12(18): 7035–56

44 Simeone A (1998) Detection of mRNA in tissue sections with radiolabelled probes. In: DG Wilkinson (ed): *In Situ Hybridization: A Practical Approach*. Oxford University Press, Oxford, 69–86

45 Technau U, Bode HR (1999) *HyBra1*, a *Brachyury* homologue, acts during head formation in *Hydra*. *Development*. 126: 999–1010

46 Wilkinson DG (1992) Whole mount *in situ* hybridization of vertebrate embryos. In: DG Wilkinson (ed): *In Situ Hybridization: A Practical Approach*. Oxford University Press, Oxford, 75–84

47 Hadrys T, Braun T, Rinkwitz-Brandt S, Arnold HH et al. (1998) *Nkx5-1* controls semicircular canal formation in the mouse inner ear. *Development* 125(1): 33–9

48 Herbrand H, Guthrie S, Hadrys T, Hoffmann S et al. (1998) Two regulatory genes, *cNkx5-1* and *cPax2*, show different responses to local signals during otic placode and vesicle formation in the chick embryo. *Development* 125(4): 645–54

49 Holland LZ, Holland PWH, Holland ND (1996) Revealing homologies between body parts of distantly related animals by *in situ* hybridization to developmental genes: Amphioxux versus vertebrates. In: JD Ferris, SR Palumbi (eds) *Molecular zoology, advances, strategies and*

protocols. Wiley-Liss, Inc, New York, 267–282

50 Hauptmann G and Gerster T (1994) Two-color whole-mount *in situ* hybridization to vertebrate and *Drosophila* embryos. *Trends Genet* 10(8): 266

51 Hauptmann G, Gerster T (1996) Multi-colour whole-mount *in situ* hybridization to *Drosophila* embryos. *Development Genes and Evolution* 206(4): 292–295

52 Jowett T (1998) Two color *in situ* hybridization. In: DG Wilkinson (ed): *In Situ Hybridization: A Practical Approach*. Oxford University Press, Oxford, 107–126

53 Wilkinson DG, Bailes JA, Champion JE, McMahon AP (1987) A molecular analysis of mouse development from 8–10 days post coitum detects changes only in embryonic globin expression. *Development* 99: 493–500

54 Bober E, Baum C, Braun T, Arnold HH (1994) A novel *NK*-related mouse homeobox gene: expression in central and peripheral nervous structures during embryonic development. *Dev. Biol.* 162: 288–303

55 Rinkwitz-Brandt S, Justus M, Oldenette ll, Arnold HH et al. (1995) Distinct temporal expression of mouse *Nkx-5.1* and *Nkx-5.2* homeobox genes during brain and ear development. *Mech. Dev.* 52(2–3): 371–81

56 Rinkwitz-Brandt S, Arnold HH, Bober E (1996) Regionalized expression of *Nkx5–1*, *Nkx5–2*, *Pax2* and *sek* genes during mouse inner ear development. *Hear. Res.* 99(1–2): 129–38

Guide to Solutions

Guide to Protocols

Troubleshooting Guide

Index

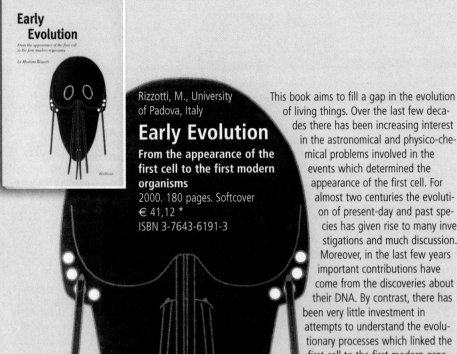